Cultivation and Processing of Medicinal Plants

Cultivation and Processing of Medicinal Plants

edited by

L. HORNOK
University of Horticultural Sciences, Budapest

John Wiley & Sons
Chichester · New York · Brisbane · Toronto · Singapore

This is the revised version of the Hungarian
Gyógynövények termesztése és feldolgozása
published by Mezőgazdasági Könyvkiadó, Budapest

Translated by

K. Raffalszky

Published in co-edition with Akadémiai Kiadó, Budapest
by John Wiley & Sons Ltd, Baffins Lane, Chichester, UK

Library of Congress Cataloging-in-Publication Data

Hornok, László.
[Gyógynövények termesztése és feldolgozása. English]
Cultivation and processing of medicinal plants / László Hornok: [translated by Katalin Raffalszky]
p. cm. — (A Wiley medical publication)
Translation of: Gyógynövények termesztése és feldolgozása.
Bibliography: p.
Includes index.
ISBN 0–471–92383–4
1. Medicinal plants. 2. Medicinal plants—Processing. 3. Medicinal plants—Hungary. 4. Medicinal plants—Hungary—Processing. I. Title. II. Series.
SB293. H6513 1989
633.8'8—dc20

British Library Cataloguing in Publication Data
Cultivation and processing of medicinal plants.
1. Medicinal plants
I. Hornok, László II. Gyógynövények termesztése és feldolgozása. English
581.6'34

ISBN 0–471–92383–4

Printed in Hungary by Akadémiai Kiadó és Nyomda Vállalat, Budapest

Contents

List of Contributors. IX

Foreword *(A. Somos)*. XI

PART 1 GENERAL

General Aspects of Medicinal Plants *(L. Hornok)*. 3
 Definition of medicinal plant cultivation 3
 The role and importance of medicinal plants 5
 Medicinal plant production in Hungary 7
References

Biological Aspects. 10
 Heredity and variability of medicinal plants *(P. Tétényi)* 10
References
 Active substances of medicinal plants *(D. Vágújfalvi)*. 19
References
 Taxonomy of medicinal plants *(A. Terpó)*. 42
 Environmental factors *(J. Bernáth and L. Hornok)*. 56
References
 Plant breeding and variety control of medicinal plants *(P. Tétényi)*. . . . 69

Technical and Technological Conditions for Medicinal Plant Cultivation . . . 80
 Possibilities and conditions for mechanization *(D. Földesi and L. Hornok)* 80
 Possibilities and conditions for introducing chemicals *(D. Földesi and L. Hornok)*. 82

Primary Processing of Medicinal Plants 86
 Drying *(L. Hornok and A. Gulyás)*. 86
 Extraction processes for essential oils *(R. Kolta and L. Hornok)* 93
References
 Storage, packing and certification *(L. Hornok, R. Kolta and A. Gulyás)*. . 105

PART 2 THE CULTIVATION OF MEDICINAL PLANTS

Ergot *(L. Hornok and O. Lenchés)* 111
Poppy *(D. Földesi)* 119
Rue *(L. Hornok)* 128
Coriander *(L. Hornok)* 131
Dill *(L. Hornok)* 137
Anise *(L. Hornok)* 143
Angelica *(P. Szalai)* 147
Lovage *(P. Szalai and L. Hornok)* 151
Caraway *(J. Sváb)* 154
Cumin *(L. Hornok)*. 160
Fennel *(D. Földesi and L. Hornok)* 162
Periwinkle *(L. Hornok)* 166
Valerian *(J. Sváb and L. Hornok)* 170
Lavender *(J. Szalai and L. Hornok)* 176
Lemon balm *(L. Hornok and O. Lenchés)* 183
Peppermint *(L. Hornok)* 187
Hyssop *(J. Sváb)* 196
Garden thyme *(D. Földesi and L. Hornok)*. 200
Clary sage *(J. Sváb)* 206
Sage *(J. Sváb)* 210
Marjoram *(O. Lenchés and L. Hornok)* 213
Summer savory *(J. Sváb and L. Hornok)* 218
Sweet basil *(L. Hornok and O. Lenchés)*. 220
Henbane *(O. Lenchés and L. Hornok)*. 224
Grecian foxglove *(L. Hornok)* 228
Common foxglove *(O. Lenchés)* 231
Mustard *(D. Földesi)* 232
Safflower *(L. Hornok)* 235
Blessed thistle *(L. Hornok)* 238
Pot marigold *(O. Lenchés)*. 241
Variegated thistle *(D. Földesi)* 243
German camomile *(J. Sváb)* 246
Tarragon *(P. Szalai)* 254
Pyrethrum *(P. Szalai)*. 258
Roman camomile *(P. Szalai and L. Hornok)*. 262
References

PART 3 MISCELLANEOUS

Belladonna *(J. Sváb)* 271
Buckthorn *(L. Hornok and O. Lenchés)* 273
Cardamom *(L. Hornok)*. 275

Catnip *(L. Hornok)* 278
Cinnamon and cassia *(L. Hornok)* 280
Common juniper *(J. Sváb)* 283
Dog rose *(L. Hornok and D. Földesi)* 285
European elder *(D. Földesi and L. Hornok)* 287
Fenugreek *(L. Hornok)* 289
Ginger *(L. Hornok)* 290
Ginseng *(L. Hornok)* 292
Hawthorn *(J. Sváb)* 295
Hollyhock *(L. Hornok and O. Lenchés)* 296
Horehound *(L. Hornok)* 298
Liquorice *(L. Hornok and J. Sváb)* 299
Linden tree *(O. Lenchés)* 301
Marshmallow *(D. Földesi)* 303
Mullein *(L. Hornok)* 305
Oregano *(L. Hornok)* 307
Plantain *(L. Hornok)* 308
Rosemary *(L. Hornok)* 310
Wormwood *(J. Sváb)* 311
Yarrow *(L. Hornok and J. Sváb)* 313
References

Tables (Table 17–20) 316

Subject index 331

List of Contributors

BERNÁTH, Jenő D.Sc.	*Research Institute for Medicinal Plants Budakalász. Deputy-director*
FÖLDESI, Dezső	*Research Institute for Medicinal Plants Budakalász. Head of Agronomy Department*
GULYÁS, András	*Agricultural Cooperation, Kerepestarcsa. Head of Medicinal Plant Processing Department*
HORNOK, László Ph.D.	*University of Horticulture and Food Industry, Budapest. Head of Medicinal Plant Production Department*
KOLTA, Rezső	*CAOLA Company for Chemicals, Cosmetics and Household Products Budapest. Head of Technical Development Department*
LENCHÉS, Ottó	*HERBÁRIA Cooperation for Medicinal Plant Production, Processing and Trade, Budapest. Head of Agricultural Development Department*
SVÁB, János (Mrs) Ph.D.	*Agricultual Cooperation, Kerepestarcsa. Head of Agricultural Development Department*
SZALAY, Pál	*Agricultural State Farm of Herceghalom. Head of Medicinal Plant Production and Processing Department*
TERPÓ, András D.Sc.	*University of Horticulture and Food Industry, Budapest. Head of Botany Department*
TÉTÉNYI, Péter D.Sc.	*Research Institute for Medicinal Plants Budakalász. Director*
VÁGÚJFALVY, Dezső, Ph.D.	*Eötvös Loránd University of Arts and Sciences, Budapest. Head of Biochemistry Department*

Foreword

Although Hungarian medicinal plant cultivation has a significant past and traditions, we can talk about a new field of production wrote the late Dr. József Kerekes, university professor and outstanding personality of Hungarian medicinal plant cultivation and research, in 1969, in the foreword of his book. In his book the large-scale production of medicinal plants was described for the first time in Hungary. In the meantime tremendous advances have also been made in this field of horticulture. The cultivation and processing methods for medicinal plants have been modified significantly, and the first producing systems have been established for several species (ergot, poppy, hop).

The base materials of plant origin are again extensively used in the pharmacological industry. The publication of this book has been long overdue, given the widening possibilities of using medicinal plants, the increasing demand for medicinal plants, and also the accumulation of new information about the improvement of the present methods of cultivation.

In addition to the general description of the cultivation of medicinal plants, the book deals with the controlling possibilities of biological, economical and technical parameters which influence the efficiency of cultivation. The special biological requirements, technology and equipment of the primary processing of medicinal plants (in the cultivating organizations) are discussed in a separate chapter (Primary Processing of Medicinal Plants).

This book serves different requirements. It can be used as a guide by the users of medicinal plants and by people who are interested in natural sciences, as well as by the cultivators of medicinal plants.

The intention of the authors, who were the students and colleagues of Dr. József Kerekes, was to publish the most recent advances in the cultivation and processing of medicinal plants on the basis founded by Dr. József Kerekes. The book can be used either as a university textbook or as a manual. It gives me great pleasure to introduce this book, which is the work of a team involving mostly young authors who have enriched the literature with a high-level, up-to-date publication including much new information.

Dr. András Somos

PART 1 GENERAL

General Aspects of Medicinal Plants

Definition of Medicinal Plant Cultivation

Healing with plants probably dates back to the evolution of *Homo sapiens*. Consequently, medicinal plants are among those that have been cultivated and utilized for the longest period of time. In spite of this fact, the cultivation of medicinal plants is the most recent branch of agriculture and of horticulture. One of the reasons for this phenomenon is that the gathering, preparation and application of medicinal plants were associated with religious cults and magic for a very long time.

It was only the rapid development of chemistry, phytochemistry and pharmacology in the nineteenth century which provided possibilities for the scientifically supported application of plants. Owing to the widening knowledge on plants, metabolism of plants, products of metabolic processes and the physiological effects of these, the field of application has been ever increasing. Plants hardly known previously became significant, while other plants were put aside as their ineffectiveness was proved. After establishing more accurate requirements with widened criteria, the cultivation of wild-growing species could start, and also species cultivated previously over small areas are now cultivated on large fields, with industrial methods, so the specific branch of medicinal plant cultivation has evolved. This process in continuing.

According to Somos (1975) the independent branches of plant cultivation can be defined with suitable accuracy only if several factors are taken into consideration. What are the common features of plant species belonging to the group of medicinal plants?

(1) One of the main features of these plants is that they are cultivated and used because they contain accumulated, biologically active materials, the active ingredients of which have various effects. These active ingredients, representing the value in use, are produced by biological syntheses in the plant and are accumulated in the body of the plant in very small concentrations, sometimes less than 1 per cent of the dry material content of the plant.

(2) Only those portions of the plant containing the active ingredient are used (e.g. leaf, fruit, root) and not the whole plant.

(3) These plants are generally not suitable for being consumed in a fresh state. That part of the plant which contains the active ingredient is processed (by drying, extracting, extraction of essential oil, etc.) prior to utilization. Processing con-

sists of several stages, and the first stage, the preparation of the drugs, is the responsibility of the producer himself.

(4) Medicinal plants or the active ingredients extracted from medicinal plants are consumed to a limited extent compared to other horticultural or agricultural grown plants, and they are consumed in the forms of extracts and as ingredients of different products.

The area of application of plants belonging to this branch is very wide. For this reason, these plants are classified into three main artificially defined groups, according to their general utilization, as follows:

Medicinal plants. These are feral plants and cultivated plants which can be used for medical purposes directly or indirectly, owing to their active ingredient contents.

Spice plants. These are plants used for seasoning, spicing, flavouring and colouring foods, drinks and different products of the food-processing industry, i.e. making a product more enjoyable.

Essential oil plants. Plants which accumulate essential oil in some of their parts can be used for the production of essential oil.

The practical classification and separation of the plants according to the above categories is impossible because a great number of plants and plant portions are used equally for the three quoted purposes. For instance, the peppermint is widely used today for medical purposes (medicinal plant), a high amount of essential oil is processed from peppermint (essential oil plant) and its applications as a seasoning and spicing material is also significant (spice plant). The case is similar for many other plants, like thyme, anise, dill, lavender, and coriander. It is therefore advisable to give these plants the collective name *medicinal plants* because of their general common features.

Drug is the name of the yield of a plant, in terms of technology, which contains the active ingredient; it is generally preserved by drying. Essential oil is also a drug. Therefore, *when determining the cultivation of medicinal plants, activities aimed at producing drugs should be meant.*

According to their botanical and physiological properties, medicinal plants form a more heterogeneous group than, for example, the vegetable plants or fruit-bearing plants. Similarly to ornamental plants, medicinal plants also include one-, two- or several-year-old species with weedy or wood-like stalks, shrub and half-shrub species, and species with or without fruit.

Another feature of the cultivation of medicinal plants is that the number and composition of the cultivated plants are still changing. This is due to widening knowledge and to other factors (labour shortage, environmental problems, etc.) so the insignificant wild-grown plants may become important cultivated medicinal plants. Only 5 per cent of the whole universal flora has been investigated regarding active ingredients and chemical and pharmacological aspects (Silva, 1986). Out of 1600 medicinal species from the European flora known in folk medicine, approximately two-thirds have been up to now selected qualitatively and 320–350 have currently been evaluated by European industry. This also shows that a tremendous amount of natural sources is available and further species are expected to be utilized in medicine. The number of species used for medical purposes in Hungary is around 300. Out of these, 50–70 species are imported (in the form of drugs), 160–180 grow wild and 50–55 are cultivated in Hungary.

Yet another feature of the cultivation of medicinal plants is that certain operations are complicated (e.g. picking flowers and gathering leaves). For this reason the mechanization of these operations is difficult. The majority of medicinal plants can only be cultivated with a high amount of labour. An additional feature of the cultivation of medicinal plants is that the last operational phase of the cultivation technology is in each case the primary processing (on site).

The Role and Importance of Medicinal Plants

When defining the present state and significance of medicinal plant cultivation, several factors (area, yield, profit, etc.) should be taken into consideration. Therefore it is reasonable to analyse in detail the value in use of the species belonging to the group of medicinal plants.

At the time when chemical sciences started to develop dynamically and when the application of synthetic materials was greatly extended, many people thought that the significance and utilization of natural materials, including medicinal plants, would diminish gradually. These forecasts proved to be wrong. The medicinal plant consumption in Europe has not decreased during recent years. According to statistical data from all over the world, nearly 50 per cent of medicines on the market are made of natural basic materials (Rácz, 1982). An increasing role is played by medicinal plants in the improvement of health care.

According to Farnsworth and Bingel (1977) the population of the USA paid three billion dollars during one year (1973) for medicines containing active ingredients derived from plants. Available information shows that the 1532 billion prescriptions issued included 25.2 per cent of plant material-based, 13.3 per cent of microbe-based and 2.7 per cent of animal-based medicines.

Although these data refer to the USA, it is undeniable that the significance of plant-based materials has been increasing all over the world.

Italy, for instance, contains in its list of drugs over 2400 species on the basis of vegetal raw materials, which represent 28 per cent of all the drugs in use. Three hundred and fifty species are used as sources of phytotherapeutics and/or phytopharmaceutics. France, West Germany and Switzerland present a similar picture. Their proportions vary from year to year and from country to country between 25 and 37 per cent.

The natural remedies, officially indexed as species or in therapeutic formulations in the USSR, Poland, Bulgaria, Czechoslovakia, Hungary and Romania, represent a proportion of 32–42 per cent. Out of the total amount of species in use, in the investigated material the share of the drugs originating from plants is—on the basis of pharmacological groups—29–45 per cent for the alimentary system, 21–26 per cent for the cardiovascular system, 8–14 per cent for the genitourinary system, 2–5 per cent for the ophthalmic preparations, and 2–8 per cent for the dermatological preparations (Silva, 1986).

The green wave in the utilization of medicinal plants resulted in higher consumption. From a database compiled by the United Nations Statistical Office and ITC, it appears that the value of total imports of medicinal plants increased from dollar 355 million in 1976 to dollar 551 million in 1980. Over 400 botanicals are used commercially in Western Europe and the total European imports were 80.738 tons in 1980. The overall USA internal trade in medicinals and botanicals reached dollar 3.912 million in 1981. According to Tyler (1987), herbs are big business in the USA where in 1985 the sales of all forms of herbs exceeded dollar 190 million. Japan is a major importer of medicinal plants among the Asian countries. Japanese imports rose from 21 000 tons in 1979 to 22 640 tons (valued at 10.8 billion yen) in 1980. A renewed interest in traditional medicine in Asia and in health foods in Europe and North America has provided new outlets for a number of botanical products.

During the past 20 years there has been a resurgence of interest in medicines prepared from plants, but modern scientific knowledge about plants and their constituents makes it essential that more effective herbal products should be made available for sale to the general public (Shellard, 1987).

The trade in botanicals has increased following their greater use in the food and cosmetics industries. Natural essential oils, because of their complex and unique structures, which add richness to their aromas, continue to play an important role in perfumery. The demand for essential oils and fragrances worldwide has increased very significantly during the last 20 years (Pisano, 1986).

The reasons for the increasing demand for medicinal plants and their extracts can be summarized as follows:

(1) The medicinal plants are sources of many biologically active ingredients which cannot be ignored by advanced medical practice, owing to their healing effects. Because of their unknown features and complexity, these active ingredients cannot be prepared synthetically yet, or the synthetic production is excessively expensive (heart glycosides, ginseng, catharanthus, ergot alkaloids, etc.).

(2) There are certain plants (e.g. *Solanum*) which contain some special materials that cannot be directly applied for healing but are used as basic materials for semisynthetic processes in the pharmacological industry (e.g. corticosteroids).

(3) As opposed to the simple uses of some synthetic materials, the use of drugs is more sophisticated, including the application of essential oils for healing, flavouring or seasoning purposes as a part of biopharmacology, which has become more and more significant (galeniss, teas, biocosmetics, etc.).

(4) In certain areas of the chemical industry (cosmetics, household chemicals) and in the food industry, the essential oils extracted from plants cannot be replaced and sometimes these are the initial basic materials for producing a new scent or flavour material (coriander, linalool).

(5) The food industry is also a widening application area. In addition to the liquor

industry, which traditionally uses a lot of drugs, other branches, like the canning industry, meat industry and sugar-processing industry, also require more and more plant-based materials for the purpose of flavouring, seasoning and colouring. An advanced diet requires a variety of spices instead of limited spicing using salt, paprika and pepper; more sophisticated flavour compositions are required. More and more people realize the advantageous effects of natural spices (marjoram, tarragon, coriander, etc.) on the digestive system.

(6) According to Minker and Szendrei (1986) the 'green wave' means in the developing countries the use of the healing potential of plants for essentially medical purposes, because in these countries most true drugs are not available to the poor, for financial reasons. In the developed countries medicinal plants are used for maintaining health and stamina, or for widening the limits of stamina.

The significance of medicinal plants nowadays is also shown by the increasing number of established research centres for analysing the active ingredients of plants and for discovering new, biologically active natural materials.

Medicinal Plant Production in Hungary

The cultivation and utilization of medicinal plants have a significant background and historical tradition in Hungary. This is based on the rich flora of the country. Nevertheless, industrial cultivation and the development of this special branch started to develop only after World War II, following the reorganization of agriculture. After a dynamic development in the 1960s, the rate of development somewhat decreased during the 1970s, and since then slight development has been continuous. Compared to the total horticultural production in Hungary, the share of medicinal plant production, amounting to 7.2 per cent in area and to 7.3 per cent in gross production value, seems less significant. The importance of the medicinal plants branch is represented by its role in health care rather than by its financial benefits. An additional advantageous feature of some of the medicinal plants (camomile, lavender, etc.) is that they facilitate the economical utilization of low-quality soils (eroded, stony hillsides, alkaline soil, sand). The extension of exports is in the interests of the whole country. The Hungarian medicinal plants branch is mostly oriented towards foreign trade. The balance of foreign exchange with Western countries is favourably influenced by the annual 20–25 million dollars income earned by sales of medicinal plant products on Western markets.

The area utilized for the cultivation of medicinal plants had been stabilized by the 1980s at 42 000 ha (see Table 1). The data in Table 1 illustrate the changes in the composition and show that during recent years more than 75 per cent of the total cultivation area of medicinal plants has been occupied by six main species (mustard, caraway, poppy, ergot, dill and fennel).

The amount of drugs produced in Hungary has increased from 9 500–10 000 t (yearly average between 1970 and 1975) to 35 000–40 000 t (yearly avarage between 1980 and 1985). The produced essential oil volume is 80–100 t/year. Seventy-five per

Table 1 Sown area of medicinal plants in Hungary 1975–1985

Species	Sown area (ha)		
	1975	1980	1985
Ergot	2 630	4 400	2 100
Poppy	2 400	4 500	6 250
Mustard	10 900	18 500	14 600
Dill	1 430	2 100	1 600
Lavender	610	300	250
Peppermint	150	300	350
Caraway	1 310	1 600	6 750
Fennel	590	900	1 200
Marjoram	160	400	870
German camomile	260	350	380
Clary sage	560	400	810
Hop	640	690	620
Other species	1 100	3 100	6 500
Total	22 740	37 540	42 280

cent of the volume is obtained from cultivated plants; the percentage collected from spontaneously grown plants has decreased to 25 per cent from the previous 34–35 per cent. A large volume of the produced drugs is sold on foreing markets: 20 000–30 000 t/year. The annual exports of essential oils amount to 45–50 t. The Hungarian pharmaceutical industry processes 8 000–10 000 t/year of drugs, and the rest is processed by the food industry and distributed by the home trade.

The average value of products made directly by the medicinal plants branch during the last three years amounted to 1 200 million forints per year. There was an increase of 76 per cent between 1975 and 1985.

The most valuable drug exports have increased by 70 per cent between 1970 and 1980 and by an additional 30 per cent between 1980 and 1985. It is worth mentioning that 80 per cent is exported to Western markets. Hungary's principal trade partners are: Germany, Austria, Italy, Switzerland and France. The Hungarian medicinal plants are marketed not only in their natural states (drugs, essential oils) but also as basic materials and products following the primary processing (extraction), and as medicines, food or scents. This multiplies the value of the medicinal plants.

Considering other branches which are users of medicinal plants (like the meat and canning industries, the liquor-processing industry, etc.) the production value of the medicinal plants branch was 2.2 billion forints as a yearly average between 1980 and 1985.

The medicinal plant products made in Hungary (drugs, seeds, essential oils) are obtained mostly from cultivated plants (75 per cent). The larger portion (50 per cent) of the cultivation land belongs to the agricultural co-operatives. State farms own 6 per cent of the land and small-scale, private cultivation is performed on 22 per cent of the area. Small-scale farms cultivate mainly those plants which require a lot of labour which is difficult to mechanize, and for which it is complicated to introduce chemicals. Production is performed in all three sectors according to local cultivation traditions, under specific circumstances and usually on the basis of contracts signed with trading or user companies.

The whole range of activities concerning medicinal plants is accompanied by the certification of drugs and essential oils. For this branch the certifying system and its institutions are the same as for other branches of agriculture.

Marketed drugs and essential oils are certified by the Medicinal Plant Certification Department of the Research Institute for Medicinal Plants as an authorized organization. The certification, which is performed in accordance with the Hungarian Pharmacopoeia (Ph. Hg) and with the relevant national standards, is carried out prior to the marketing and distribution. In the case of any debates, the Quality Control Department of the National Institute for Pharmacology also takes part in approving the home distribution.

The obligatory residue tests for directly consumed agricultural products are also in effect for the drugs and are carried out by the Research Institute for Medicinal Plants.

References

Farnsworth, N. R., and Bingel, A. S. (1977). 'Problems and prospects of discovering new drugs from higher plants by pharmacological screening'. In H. Wagner and P. Wolff (eds) *New Natural Products and Plant Drugs with Pharmacological or Therapeutical Activity,* Vol. 7, Springer-Verlag, New York, 32–47.

Minker, E. and Szendrei, K. (1986). 'About the "Green Wave"' *Gyógyszerészet,* **30** (9) 323–5.

Pisano, R. C. (1986). 'The future of natural essential oils in the flavour and fragrances industry', *Parfumer and Flavorist,* **11** (5) 35–42.

Rácz, G. (1982). *Modern Use of Medicinal Plants,* Napoca, Kolozsvár (In Hungarian).

Shellard, E. J. (1987). 'Medicines from plants with special reference to herbal products in Great Britain', *Planta Medica,* **53** (2) 121–3.

Silva, F. (1986). 'The place of natural remedies in the medicine of the future', *Herba Hungarica,* **25** (1), 151–62.

Somos, A. (1975). *Cultivation of Vegetables,* Mezőgazdasági Kiadó, Budapest (In Hungarian).

Tyler, V. E. (1987). 'Herbal medicine in America', *Planta Medica,* **53** (1) 1–4.

UNCTAD/GATT (1982). *Markets for Selected Medicinal Plants and their Derivatives,* International Trade Centre, UNCTAD/GATT, Geneva.

Biological Aspects

Heredity and Variability of Medicinal Plants

Biological phenomena are closely related to the chemical and physical characteristics of macromolecules occurring in living organisms. Their dynamic equilibrium may be attributed to the alterations which continuously occur in the living organisms, to the specified enzymes which control these alterations, and to the different natures, ratios and activities of the enzymes, depending on well-defined conditions. Enzymes are not the only source of inherited characters, because the sum of chemical qualities, the so-called chemism, is actually the full system of biochemical processes which is characteristic of the organism. This includes DNA and RNA being the bearers of genetic information and having the ability to transcribe and translate it (Fig. 1), proteins (enzyme stock), as well as the universal steps of metabolism (photosyntheses, respiration) and special steps of metabolism (biosynthesis, accumulation and decomposition of alkaloids, flavonoids, terpenes, etc.).

The (intrinsic) reaction of the living in an organism at a certain stage of phylogeny is directly connected with its chemism. It is a quite correct idea that the differentiation of any taxon is fundamentally a chemical process. This statement has been supported by molecular biology, too: the so-called enzyme stock can be traced back to the composition and structure of the proteins of cells, which are in fact the fundamentals of all inherited properties. The basis of the differentiation of cells can be found in chemodifferentiation at a molecular level. The fundamental differences appearing in the course of ontogeny manifest themselves mainly in the differences between proteins present in the organisms and, in connection with that, in differences between the enzyme systems. All other differences (morphological, anatomical) are only the consequences of these.

In investigating the rise of chemical taxa from the numerous aspects of molecular biology, the biochemical process of metabolism and its genetic control are of particular interest. According to Mothes (1957), real genetics of metabolism began only when the genetical analysis of a larger collection of races chemically differing from each other became possible. Though genetic research was primarily directed to the synthesis of proteins and to the role of DNA and RNA in this process, there exist some data on the genetically determined synthesis, accumulation and control of the specialized metabolites.

Thus, it was experimentally shown in the 1950s that it is impossible to produce viable alkaloid-free individuals of any alkaloid-bearing species by irradiation. The

Fig. 1 General scheme of protein biosynthesis (DNA→RNA→protein) (Spirin, 1987)

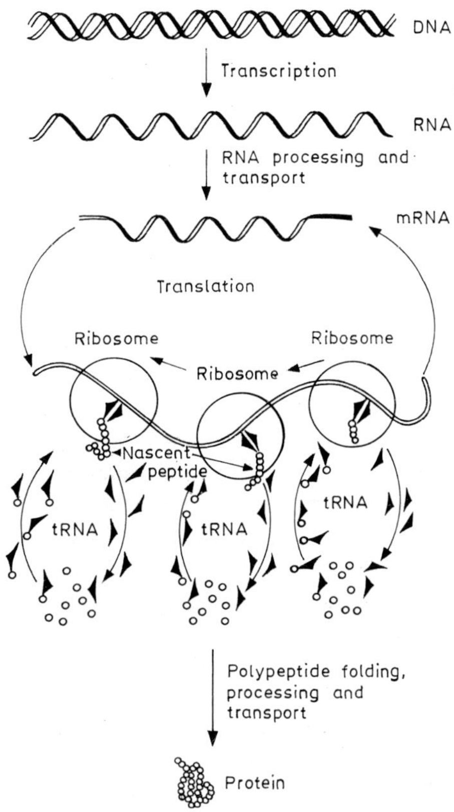

production and accumulation of alkaloids to a limited extent is an indispensable part of plant metabolism because mutants unable to form alkaloids sooner or later die out.

Owing to the large number of processes of chemism and to the variability of interconnections between these processes, the study of any system of chemism is still performed through intermediate and/or end products. Investigation of amino acid sequences can be regarded as a first step towards the full understanding of the chemism, and this can also be said about the study of isozymes. An example of this type can be seen in Fig. 2.

End products of metabolic processes, if accumulated, can be directly observed (colour, taste, odour) and can be revealed indirectly by the methods of chemical analysis. However, these chemical characters, because of their basic nature, are mostly hidden—whereas flower-colour variations are very convenient for so-called 'eyeball taxonomy'—and are therefore a source of difficulty, because they need to be examined by suitable technical methods. Characters of this type are closely connected with physiological phenomena; they are at the root of the structural features. It is recognized that hidden biochemical characters are accompanied by morphological divergence; for example, in the case of *Trifolium repens,* the correlation of cyanogenic glycoside presence or absence with the white spot on the leaflet.

Chemism is very important for other cultivated plants, besides medicinal plants and those producing aromatics and spices. Plants used as sources of foods, beverages, textiles, timber, etc., all have their special chemical features which must be taken into account when dealing with their evaluation for theoretical and practical purposes. Differences between these products are attributed to the dissimilarities of metabolism occurring in plants. The complexity of chemism can be characterized by its elaborated constituents. These are results of the genetic background, and can be determined like fingerprints. For this reason the term *chemosyndrome* was accepted to indicate this phenomenon.

Identical materials, accumulated in different plants, however, do not in themselves imply a relationship in chemism. The alkaloid of the toads, bufotenine, can also be found, for example, in the flowering plant *Piptadenia falcata* Benth. Plants containing anthraquinones are not connected in terms of heredity just, because of the common substance; for example, fungi form their anthraquinones in a quite different way from that of higher plants. Different plant species accumulate anabasine and nicotine through different metabolic biosyntheses. The same constituent can be produced by different routes; therefore, synthesis pathways are characteristic, not the compound itself (Fig. 3).

Neither is the frequency of occurrence of a certain substance a determining factor as regards relationship, because rare substances may also have homologous biosyntheses (such as, for instance, the formation of thebaine in certain *Papaver* species), as may ubiquitous structures, e.g. chlorophylls.

The inherited chemism cannot be categorized according to the micro- and macrostructure of a substance since the life functions are biochemically complex and cannot

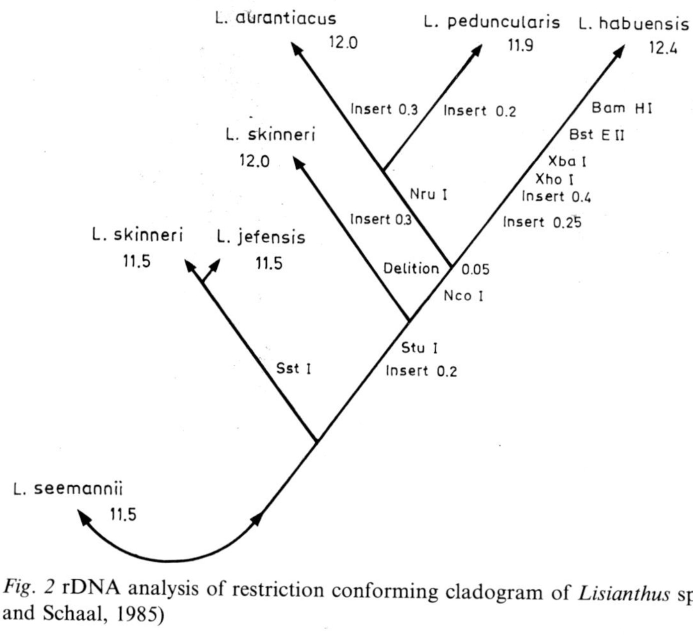

Fig. 2 rDNA analysis of restriction conforming cladogram of *Lisianthus* species (Gentianaceae) (Sytsma and Schaal, 1985)

Fig. 3 Two fundamentally different pathways of anthraquinone biosynthesis (Zenk, 1967)

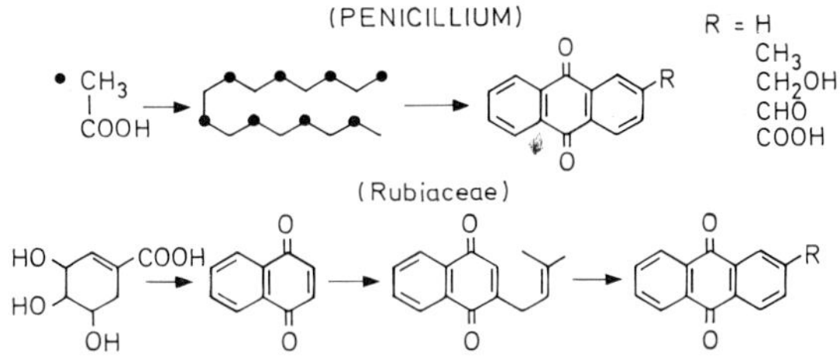

be fully explained by reference to simple compounds. It cannot be stated that amino acids are simpler or more complex than terpenes, in spite of the fact that the essential oil complex of terpenes and the protein complex of amino acids differ significantly with regard to their structures and to their evolution.

The concept that in the case of medicinal plants only the active substances accumulated in them are important, and while these are 'packed' differently in various plants, this 'packing' is unimportant, cannot be accepted. The accumulated substance is never independent of the plant; it is always the result of its metabolic processes. For this reason the evolutionary evidence for development and differentiation is the homology of the biosyntheses of the substances rather than that of the produced molecules.

The inherited metabolism may be similar to that of the direct ancestors, and in this case the reaction norm is the same. However, the plants are able to adapt themselves to changing conditions, and can even change under the influence of sudden or prolonged effects. According to Soó (1953) all species have an intrinsic contradiction, namely they have constant inherited characters, and at the same time characters are alterable (producing new taxa). This law has been in action during evolution since the appearance of life, and the inheritance as well as the differentiation of chemism can be recognized through it.

Chemisms of living organisms at different levels of evolution are different. The differences are small within species, but are important at the level of species or family. Conclusions can be drawn about evolution from the times and sequences of biosynthesis, as the biochemical process reflects the past.

According to the *aromorphosis* defined by Severcov—the chemism is also changing ontogenetically from the initial simple processes to the more complex but favourable processes with regard to their energy levels, up to the enzymic actions. This ontogenetic reflection of the change of the chemism through the ages, and the inherited change in the biosynthetic pathways during the phases of ontogenesis, is the proof of the evolution, and at the same time it is the basis for the further variability of morphological and chemical features.

Ivanov (1915) was the first to recognize the effect of external conditions on the chemical composition of plants. He succeeded in showing that the examined plant species were chemically variable within an area; in the centre of the area they showed an average content, while moving away from the centre, the contents of active substances became more and more different. The experimental evidence for this concept was later many times confirmed. The connection between active substances and external conditions has been substantiated by Soviet botanists through experiments carried out on the Pamir plateau.

The correlation between the geographical distribution and metabolism of plants has been thoroughly studied. It was shown that the geographical variability of plants was based on variations in their metabolism which, in turn, are due to alterations in the external environmental conditions. It has been pointed out, for example, that linalool, the main component of the essential oil of coriander, is present in the plant in varying amounts depending on the climatic and soil conditions.

According to the theory of the 'ecologically directed adaptation mechanism' evolved by Hegnauer (1959), under various ecological and geographical conditions, different local plant populations and ecotypes are developed, due to the adaptability of plants. Ecotypes, according to Hegnauer, are the raw material of evolution, and if, owing to certain causes, hybridization is inhibited in some directions or becomes entirely impossible (e.g. owing to geographical limits, to differences of floral biology and periods of flowering, and to intersterility), ecotypes may develop into species – ecospecies or into series of species – species aggregates. This dynamic Darwinian interpretation of the 'little species' theory of Jordan and of the 'ecotype' theory of Turesson is undoubtedly a positive feature of the concepts of Hegnauer.

Ecotypes have been formed through the effect of plant habitat, and are characterized by specific inheritable properties, e.g. in respect of the vegetation period and stages of development. Just for this reason we cannot accept the attempt of Hegnauer to assign certain ecological categories in an obligatory way to systematic categories, even if such a procedure is completely justified in certain cases. Another feature of Hegnauer's concept which can be criticized is that he considers the changes taking place because of ecological effects as being those which automatically gradually lead to new species. It is more likely that these changes occur after a preparatory period of accumulation, more or less simultaneously, and this is why a *hiatus* is reached. This sudden change based on the accumulation of effects is defined as mutation by classical genetics. Subsequently, the adequately adapted plant individuals and taxa, respectively, emerge from these mutants through the effect of natural selection and gene drift.

A number of researchers suppose that the changes induced by different climates, soils or other ecological and geographical conditions, are only temporary. This is true, at least in some cases. It is assumed, however, that the prolonged environmental influence might cause heritable modifications as well. As an example it is pointed out by Hegnauer that among the Solanaceae species producing tropane alkaloids, the tropical ones (which are presumed to be older in terms of evolution) produce scopolamine, while extratropicals (in colder regions) accumulate hyoscyamine. In spite of this, the first alkaloid in the initial period of ontogenesis of extratropicals is also scopolamine! According to the basics of biogenetics, this can be regarded as evidence for ecological adaptation during evolution and also for biochemical variability.

There is also experimental evidence for the effect of climatic and edaphic conditions on alterations in the composition of plants. In the polar botanical gardens of the USSR, alterations of plant colour with environment were observed over a period of 10–15 years. Thus, the originally golden-yellow flowers of *Erigeron aurantiacus* Rgl. in the second generation appeared in a diversity of colours, from white through yellow to dark brown and violet. Colour changes were also observed in the flowers of second generation *Rhodiola linearifolia* Boriss and *Phyrethrum carneum* Bieb.

It has been pointed out that *Pycnanthemum lanceolatum* Pursh, a plant indigenous to the USA and containing carvacrol in its essential oil, did not produce any carvacrol after being introduced to and after having multiplied in the USSR. The principal components of the essential oil of the variety grown in the USSR, were menthone and pulegone. It was reported that the plant *Cocculus laurifolius* DC., growing in Japan and containing the alkaloid coclaurine, was completely free of coclaurine when grown in the botanical gardens of Batumi. This latter variety, in turn, contained cocculine and cocculidine, two alkaloids of entirely different structure. *Hyoscyamus muticus* L. grown along the upper course of the Nile contains 1.8 per cent alkaloid, while the

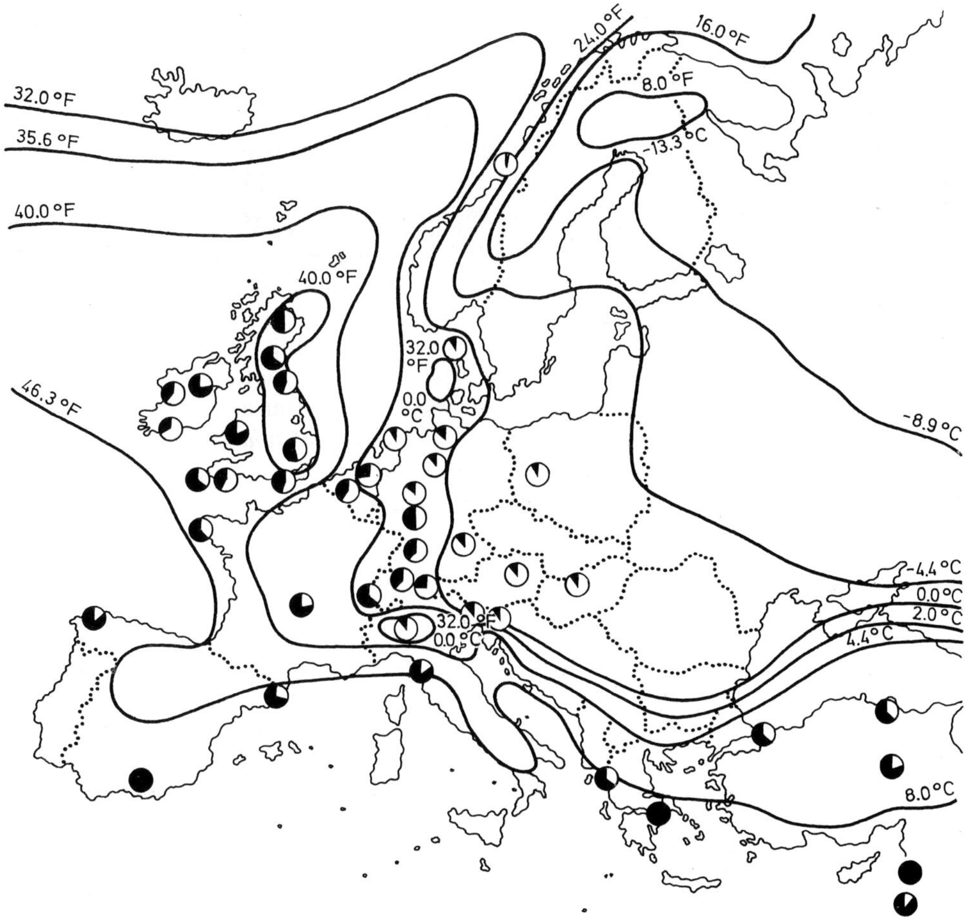

Fig. 4 Distribution of cyanogenetic glycoside-bearing and free *Trifolium repens* L. populations; glycosides present; ◦ glycosides absent; – mean temperature isotherms in January (Daday, 1954)

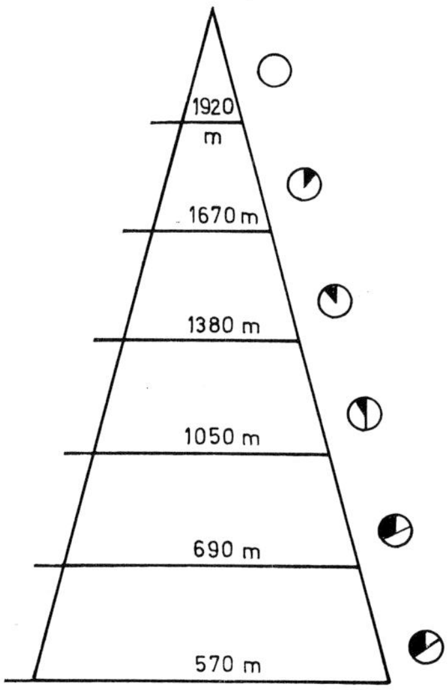

Fig. 5 Distribution conforming to height above sea level

alkaloid content of the plants grown in the Nile Delta is only around 0.1 per cent. The species *Ammi visnaga* L. in the Mediterranean basin contains much coumarin and cromone. In North America (Arizona), in turn, these compounds are present in the plant in just detectable traces. A Catalan research group has recently traced the differences in the essential oils of *Thymus* species back to the different conditions in the Mediterranean region. In the case of *Trifolium repens,* some populations contain cyanogenetic glycoside while at other sites this molecule is entirely absent. This chemical character can be correlated with the 0° median temperature isotherm of January or with the altitude in the Alps (Figs 4 and 5). Plants of this species contain cyanogenetic glycoside when grown in warmer areas, while those grown in cool areas free of glycoside.

As can be seen from the above discussion, the chemical changes and the infraspecific chemical modifications are created by the ecological and geographical conditions. These chemical differences are called *polychemism* by Gurvich. This denomination seems to be arbitrary, as a plant is being investigated only with regard to an economically valuable final product at the time of harvesting or gathering. Actually, the polychemism is the materialized result of all the life processes of the plant, and it is the realization of differences between the chemism of taxonomic units, namely *chemotaxa,* which have been established during the chemical differentiation. The hereditary feature of polychemism and the fact that it is related to the ontogenesis is evidence that polychemism is of evolutionary origin.

Polychemism may appear in mutually exclusive differences such as the appearance of thymol or citral in the essential oils of different chemotaxa of thyme. It can also occur as a continuous transition, like the citronellal component (5–90 per cent) in the essential oil chemovarieties of *Eucalyptus citriodora* Hook (Fig. 6). If the polychemism is the same for more species of a genus for instance,—the main component of the essential oils of several infraspecific taxa of thyme is citral instead of thymol or linalool—then the identically different chemotaxa form a chemically homologous series. It is thus justifiable to extend to these chemotaxa the hypothesis of Vavilov, conforming to the morphological homologous series. If the chemism is interpreted in this way, then we are actually searching for the evolutionary homologous series of biosyntheses, and thus we make chemical characters the basis of plant taxonomy.

Polychemism is frequent in the case of autogamous species, as the sudden chemical change can be easily inherited, and it often occurs also in the case of cross-fertilized species, if they are isolated, and in the case of woody species, where the modified composition can be inherited without interference for years.

The chemical characters forming the differences of chemism are thus developed just like other features of any organism. While investigating the heredity of these features, researchers have found a number of types of inheritance. Most of the time the addition of chemical characters has been observed, as in the case of the terpene components of *Pinus* species, or in the case of the glycoside spectrum of *Brassica juncea*. In the case of cross-breeding of *Papaver,* the modification of alkaloid composition has been observed, which is an indication of the dominance of the chemical characters of one parent. In the case of *Mentha* hybrids, the menthol content of the mother plant is dominant, while in the case of *Digitalis* species, the glycoside content of the wild parent dominates. In spite of these facts, theories stating that the mother's active ingredient is dominant, or that dominants are chemical characteristics of the parent, cannot be accepted as laws because opposite examples can also be found. One

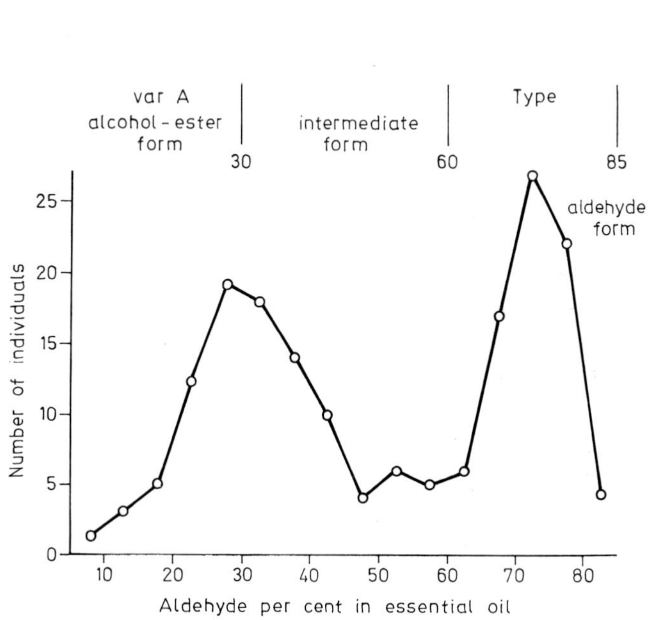

Fig. 6 Variation curve of aldehyde content in the F_1 generation individuals of *Eucalyptus citriodora* (Penfold, McKern and Willis, 1953)

cannot generalize from the observation for *Ocimum* species that the more complex chemical constituent is dominant, as this very same phenomenon was excluded by research into *Digitalis* species. The heterosis effect could have been observed also in the heredity of chemical properties; increases in quinine *(Cinchona)* and essential oil (*Lavandula, Pelargonium,* etc.) have been demonstrated by researchers. In a number of cases this kind of quantitative increase is accompanied by a decline in quality (*Mentha, Chenopodium,* etc.), but this indicates the incompleteness of our knowledge and the need for further research rather than having the status of a kind of hereditary law.

It is evident from these phenomena that chemical characters are as important as other features, which have been disclosed earlier and much more easily. Their inheritance and variability are governed by the same laws. The differentiation of chemism is the general feature of living organisms.

Chemical characters and their differences are especially important in the case of medicinal plants (gathered or cultivated for their active substances) because these serve as the real biological base of the breeding, as opposed to the dogmatic opinion that species are chemically uniform.

The efficiency of medicinal plant cultivation depends fundamentally on the productivity of plant material within which our compounds are accumulated. Medicinal plants available for cultivation always represent only a small part of the whole diversity existing worldwide. Thus, the research work being carried out with different collections of medicinal and aromatic plants is of special importance and great significance in the conservation of nature. Its main task is to preserve the treasure of today's medicinal plants throughout the world and the germplasm for posterity. Among the several efforts, I underline the gene-bank of 220 species in the Philippines, the special germplasm-bank for medicinal and aromatic plants in India, and, because of its peculiar interest, Paulrott's study in Germany dealing with the medicinal plants cultivated in medieval gardens. It is well known that such collections as these ensure the base for defining the infraspecific differences of medicinal and aromatic plants, and furthermore provide a starting point for conscious breeding work.

References

Daday, H. (1954). 'Gene frequencies in wild populations of Trifolium'. Part I. *Heredity,* **8,** 61–78; Part II. *Heredity,* **8,** 376–384.

Hegnauer, R. (1959). 'Rassenbildung in der Natur und die Bedeutung für die Pharmakognosie'. *Pharmazeutische Zeitung,* **104,** 382.

Ivanov, Sh. L. (1915). Pravilnost v raspredelenie zapasnovo masla v rastitelnom carstve. Published by the Plants' growing Bureau of Russia. St. Petersburg.

Mothes, K. (1957). 'Einführung in die Diskussion über chemische Rassen'. *Pharmazeutisch Weekblad,* **92,** 818–820.

Penfold, A. R., McKern, H. H. G. and Willis, J. L. (1953). 'Studies on the physiological forms of *Eucalyptus citriodora*'. *Research on Essential Oils of the Australian Flora,* **3,** 15–20.

Soó, R. (1953). *Fejlődéstörténeti Növényrendszertan* (Phylogenetical Taxonomy), Tankönyvkiadó, Budapest.

Spirin, A. S. (1987). 'Protein-synthezising structures of prokaryotic and eukaryotic cells'. *Biology International,* No. 14, 16–23.

Sytsma K. J. and Schaal, B. A. (1985). 'Phylogenetics of the *Lisianthius skinneri* (Gentianaceae) species complex'. *Evolution,* **39,** 286–303.

Zenk, M. H. (1967). 'Biochemie und Physiology sekundärer Pflanzenstoffe', *Berichte der Deutschen Botanischen Gesellschaft.* **80,** 573–591.

Active Substances of Medicinal Plants

Medicinal plants contain various materials possessing a number of activities. These materials are referred to as active substances. The discovery of naturally occurring chemical compounds was initiated by the chemical analyses of medicinal plants performed in the nineteenth century. It has been revealed that not only do medicinal plants contain compounds with special structures and pronounced properties (colour, odour, taste, physiological effect), but almost any plant contains some kind of special material. The materials occurring in living organisms (amounting to many thousands of compounds) are called natural substances. The active substances are plant substances grouped according to certain practical aspects. The boundaries of the categories, of course, are not defined strictly: new natural substances might be used as active substances after demonstrating new pharmaceutical effects; new materials with better properties might be discovered, or the interest in some substances might decline for other reasons. In order to avoid limiting the study to the present situation, the plant substances are introduced here in a general context, while keeping the active substances of medicinal plants, of course, in the centre of attention.

The natural substances are classified into two categories, namely primary substances and secondary substances. The primary substances are considered necessary for living organisms, and are formed primarily during metabolic processes. The other category includes the special materials not absolutely necessary to sustain life (e.g. fats, sterols, alkaloids). According to another definition, the primary substances are the functional, operating components of all living organisms, while the secondary substances do not occur in all living organisms.

These classifications, however, will not be valid for all situations. How do we classify chlorophyll, for instance? It is not present in all living organisms, only in plants; thus it is a secondary substance according to the second classification. The chlorophyll actually takes part in photosynthesis, which may be regarded as the most primary metabolic process. Metabolism cannot be divided categorically, as it involves processes universal or special. At various levels there are processes which are common in all living organisms whereas others occur in all plants or in large groups of plants, and certain processes take place only in a given species (Bell and Charlwood, 1980; Conn, 1981; Luckner, 1984; Manitto, 1981; Mann, 1978; Schütte, 1982; Vickery and Vickery, 1981).

For the expression of this gradualness I prefer to use the terms universal and special processes and universal and special substances, accompanied by the taxonomical level within which the substance or process is regarded as being universal or special. The universal processes and materials of a certain taxon are the metabolic functions and their products, respectively, which occur in all individual plants belonging to the group of plants. The special processes and substances of a taxon are metabolic functions and their products which occur only in smaller units within the group of plants. According to this definition we discuss the universal substances of living organisms instead of 'primary substances'. The 'secondary materials' might be the special substances of plants or animals, but in smaller taxons these may be again universal or special. For example, among the special substances of plants the alkaloids are universal in family Amaryllidaceae, while they are special in family Liliaceae. This is also true in the opposite sense, i.e. the universal substances of a taxon are the special substances of a higher taxon; for example, chlorophyll is a special substance of living organisms and at the same time it is universal in plants. The universal and special features may be observed not only in the production of the substances, but also in their accumulation, as the production of substances is more general in most of the substance groups than accumulation.

The classification of active substances

Attempts have been made to classify natural substances into further classes within the frame of the two main groups ever since their discovery. A number of classification aspects have even been mixed within a system (i.e. chemical, biological or physical aspects) (Haslam, 1979; Robinson, 1983; Thomson, 1985).

The active ingredients of medicinal plants are still often classified into the four traditional groups; alkaloids, glycosides, essential oils and other active substances.

The term alkaloid was initially used by Meissner for plant substances which contain nitrogen, are basic, form salts with acids, and have strong and characteristic physiological effects on the human organism. Owing to their extremely variable composition, the classification of alkaloids in precise groups is an evasive goal. The former definitions have been rendered partly invalid by the results of investigations. The alkaloids have been divided into three groups by Hegnauer according to chemical and biochemical considerations; i.e. 'true' alkaloids (N-heterocyclic compounds formed from amino acids), protoalkaloids (compounds formed from amino acids, which contain nitrogen in aliphatic chains), and pseudoalkaloids (plant compounds containing nitrogen, but not formed from amino acids).

The glycosides are compounds with various chemical structures and physiological effects which are formed in a number of extremely divergent metabolic processes. The glycosides today are not regarded as a uniform group of active ingredients. Their only common feature is that one or more identical sugar molecules are bound to a non-sugar-type compound (aglycone). Aglycones may be many different types of compounds (Table 2).

The essential oils are not simple compounds, but a mixture of various compounds (mainly terpenes and terpene derivatives). The term 'essential oil' corresponds only to the practical technological features of these compound mixtures. Those plant active substances are reffered to as essential oils which can generally be distilled with water steam, do not dissolve in water or dissolve only very poorly, evaporate at room

Table 2 The traditional active ingredient groups in the biogenetic system of substances

Biogenetic class	Traditional active ingredient groups			
	Essential oil	Alkaloid	Glycoside	Other active ingredient
Saccharides				
Monosaccharides and their derivates				Carbohydrates
Polysaccharides				Reserve, structural and mucilaginous substances
Phenoloids				
Aromatic intermediary substances				Amino acids, aromatic acids
Cinnamoids	Essential oil components		Glycosides of cinnamic acids and coumarins	Lignin: structural substances
Degradative products			Phenol glycosides	Tannins, melanin, plant pigments
Polyketide cinnamoids			Flavonoid glycosides and anthocyanins	
Quinones			Anthraquinone glycosides	Vitamins (K) bitter substances (humulone)
Polyketides				Fatty acids, lipids, antibiotics (tetracyclines)
Terpenoids				
Monoterpenoids	Essential oil components			
Sesquiterpenoids				Bitter substances,
Diterpenoids		Diterpene alkaloids		Bitter substances, growth substances (gibberelic acid)
Triterpenoids		Steroidal pseudoalkaloids	Triterpene and steroidal saponins, cardenolides	Sterols, lipids
Tetraterpenoids				Carotenoids, plant pigments
Polyterpenoids				Rubber
Azotoids				
Universal azotoids				Amino acids, proteins, nucleic acids
Simple special azotoids		Purine protoalkaloids		Special amino acids

cont. of Table 2

Biogenetic class	Traditional active ingredient groups			
	Essential oil	Alkaloid	Glycoside	Other active ingredient
Cyanogenic compounds			Cyanogenic glycosides	
Isorhodanidogenic compounds	Mustard oils		Mustard glycosides	
Alkaloids		True alkaloids		

temperature without residues, often have characteristic and strong odours and tastes and whose specific weights are mostly lower than that of water.

Plant active substances which cannot be classified into the above three groups are put into the group of 'other active ingredients'. These substances may have either simpler or more complicated compositions and effects. Carbohydrates, mucilaginous substances, aromatic acids, tannins, plant pigments, vitamins, bitter substances, sterols and rubber are such substances.

It can be concluded from the above description that the traditional categories of active ingredients are based on a combination of various diagnostic aspects. The actual genealogical connections between these substances are indicated only by the recently established biogenetic groups. The main frame of such a biogenetic system is discussed in the following section. The connection between the traditional active ingredient groups and the biogenetic system is shown in Table 2. It should be noted that the biogenetic substance groups involve other compound types as well as those given in Table 2.

Biogenetic system of natural substances

In the biogenetic system the natural substances are classified according to main pathways of the universal metabolism and to the connecting special metabolic pathways. The five main metabolic pathways and the corresponding five plant substance groups are shown in Fig. 7: (1) saccharides; (2) phenoloids; (3) polyketides; (4) terpenoids; (5) azotoids. In the schemes of biosynthesis, the connections of special metabolic pathways to the universal metabolic processes of plants, to the reaction series of photosynthesis, to respiration, and to protein and nucleic acid biosynthesis, can be traced. The biosynthesis of amino acids and the formation of nucleotide bases, purines and pyrimidines, is also connected to the two main cycles, the pentose phosphate cycle and tricarboxylic acid cycle, as well as to the pyruvic acid connecting the two main cycles. The bottom zone indicates the introduction of basic structural elements (C, N, S) into the metabolic process from inorganic materials.

The initial, universal part of the biosynthesis is followed by special stages which are no longer general in the whole plant kingdom but which in many taxons (orders,

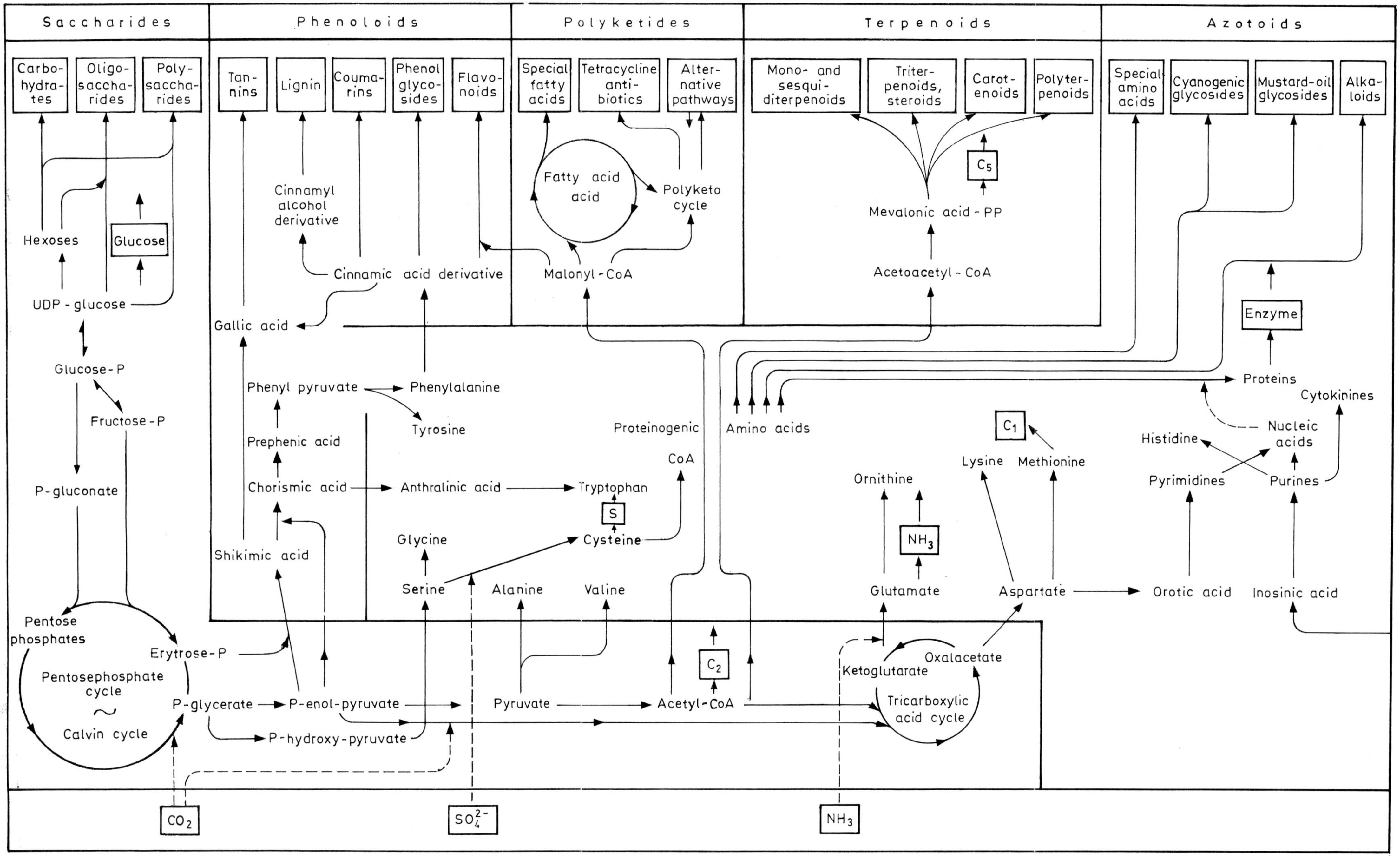

Fig. 7 Scheme of main pathways of special plant metabolism

families) take place in identical ways. The most important groups within the five classes of substances are formed during these processes, e.g. the tannins, lignins, coumarins, phenolglycosides and flavonoids within the group of phenoloids. At this stage of the discussion we can disclose yet another unclarified problem of classification of natural substances; owing to the undefined taxonomic levels of substances, the alkaloids are arbitrarily referred to as 'alkaloid class, family or group'. In order to eliminate this problem, a substance taxonomy is outlined here as follows. According to the five main metabolic pathways, five classes of substances are contemplated (e.g. azotoids); and within these classes the orders of substances (e.g. alkaloids), then families of substances (e.g. alkaloid derived from tyrosine), then genera of substances (e.g. benzylisoquinolines), and finally species of substances, are defined, the last-mentioned being chemically uniform compounds (e.g. morphine). These substance taxons include the substances which are formed in the same pathway; the substances of smaller and smaller taxons are the products of more and more specialized metabolic pathways.

One of the features of metabolism is the possibility of alternative pathways of transformation. A typical example of these 'branching points' in Fig. 7 is acetyl-CoA, this compound being the starting point of the polyketide pathway and the terpenoid pathway and at the same time the connecting point to the tricarboxylic acid cycle. Additionally, it can take part in a number of other processes, and its activity as a donor of C_2 is indicated with the symbol C_2 in Fig. 7. Similarly general C_1, NH_3, S and glucose donors are also involved in Fig. 7.

A typical mode of specialization of biosynthetic processes is the joining of metabolic pathways. This phenomenon occurs when an important intermediate of the polyketide pathway, malonyl-CoA, enters the phenoloid pathway, where flavonoids are formed. Other examples will be introduced in the detailed description of biosynthetic processes. The substances formed in mixed pathways are classified according to the biosynthetic pathway of the more important structural part (e.g. the flavonoids are put into the class of phenoloids, but these are distinguished from the cinnamoids of purely phenoloid origin, and are put into the order of 'polyketide cinnamoids'). It can be seen from the above considerations that no strict boundary exists between the 'primary' and 'secondary' processes; the universal stages are followed by increasingly specialized processes; in this way the metabolism also reflects in a philosophical sense 'the unity and antagonism of the general and the special'.

In the following sections the five main pathways of natural plant material formation are introduced, together with the five classes of substances, and their most important orders and families.

Saccharides

Sugars and derivatives form this group. Saccharides are primarily formed by photosynthetic carbon assimilation, and secondarily by the dissimilation of already formed sugars and the decomposition of reserve carbohydrates (Preiss, 1980). Further classification is made according to the polymerization and special transformations as shown Fig. 8.

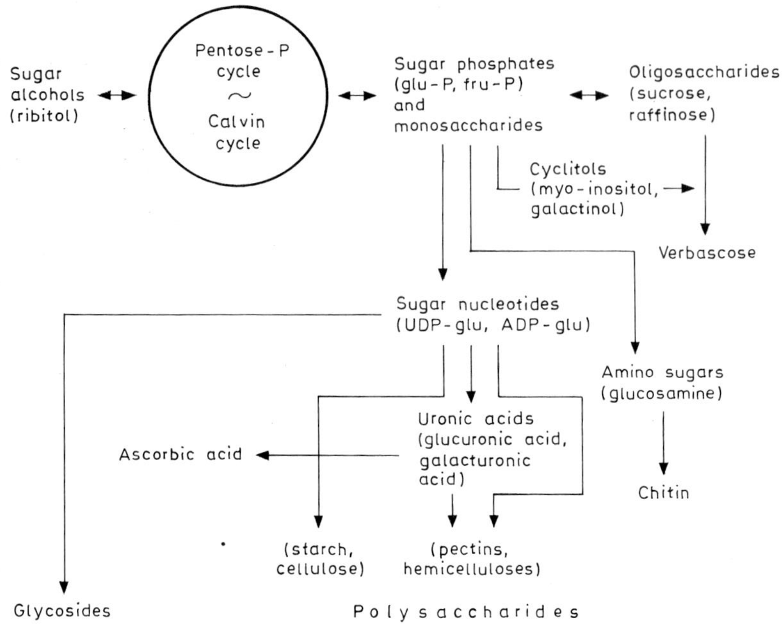

Fig. 8 Biosynthesis of saccharides

The sugar phosphates (substance order No. 1) are compounds having 3–6 carbon atoms, and are formed in two basic processes; in the Calvin cycle of photosynthesis and in the glycolysis and pentose phosphate cycles. These are ubiquitous compounds of plants as are their derivatives, the monosaccharide sugars (glucose, fructose), which can be put into the group of monosaccharides (order No. 2). The monosaccharide sugars, however, can also be formed by the hydrolysis of already formed oligo- and polysaccharides. Among the sugar alcohols, which are derivatives of monosaccharides, glycerol, containing three carbon atoms, can be generally found in fats, while the sugar alcohols containing more carbon atoms are special substances (e.g. in Rosaceae).

The cyclic sugar alcohols, the cyclitols, are all C_6 compounds; the most widely found one is myo-inositol. This is formed from glucose-6-phosphate by ring closing. Galactinol, which is myo-inositol bound to galactose, is able to transfer the galactose. The intermediary derivatives of sugar (order No. 3) have a biogenetic relationship with sugar phosphates and monosaccharides. The formation and further transformations of sugar nucleotides, which belong to this order, take place along the main, universal pathway of saccharide metabolism; these compounds are the best sugar donors for the formation of complex sugars. The most important ones are the UDP-sugars, although the GDP, ADP and TDP derivatives have also some limited donor role. By the oxidation of the sugar part the uronic acids, which are important elements of polysaccharides, are formed. From glucuronic acid the highly important ascorbic acid is formed. The amino sugars containing the NH_2 group are sugar derivatives, which are seldom found in plants (although they are very significant in the animal kingdom).

The bound forms of sugars are the glycosides (order No. 4). In the case of sugar–sugar bonding, the holosides are formed, which are oligosaccharides containing 2–7 (units of constituent) monosaccharide units. Among these compounds the most important is the disaccharide sucrose, owing to its central role in sugar metabolism and also because it can act as a sugar donor (similarly to the sugar nucleotides), thus enabling it to take part in starch synthesis as well. The other oligosaccharides can be found in plants in a great number of varieties. These are probably produced through sugar nucleotides, and cyclitol, mentioned above; galactinol takes part in the formation of oligosaccharides containing galactose (e.g. verbascose, gal-gal-gal-glu-fru, which can be found in *Verbascum* species). Heterosides are formed when mono- or oligosaccharide sugar is bound to a non-sugar compound, to the aglycone. The aglycones are various compounds, which can be classified into any of the four other substance classes. For this reason it is not justified to regard the heterosides as a uniform group of compounds, so these compounds are preferably dealt with at the location corresponding to the aglycone section.

The polysaccharides (order No. 5), although they exhibit a great variability in the various species, are ubiquitous and highly important compounds of plants, as they constitute the cell walls as well as having other roles. They are formed from sugar nucleotides, primarily from the corresponding UDP derivatives, and less frequently from the GDP or ADP analogues. Starch and cellulose are built up from glucose, a neutral sugar, while pectins and hemicelluloses consist of sugars of various types and uronic acids bound together in highly variable ways. It is to be noted that amino sugars also take part in the biosynthesis of chitin (which can be found in the cell walls of fungi, although it is a characteristic polysaccharide of animals). A number of plants contain mucilaginous substances, which consist mainly of polysaccharides characteristic of the respective species, and the composition of which is not usually known accurately.

The acidoids (order N. 6) are acids formed from the dissimilation of carbohydrates, such as universal pyruvic, acetic and tricarboxylic acids, and the more specific products like ascorbic acid and oxalic acid.

After reviewing the saccharides it can be stated that the characteristic active substances of medicinal plants occur only rarely among them (ascorbic acid, mucilaginous substances).

Phenoloids

The name of the second class of plant substances is an allusion to the primarily phenolic feature of the constituent compounds. (This feature may be secondarily modified, e.g. by methylation, or by glycosidation of the phenolic hydroxyl group.) The initial phase of the biosynthetic pathway is common for plants and micro-organisms, and its main products are aromatic amino acids (phenylalanine, tyrosine, tryptophan) (Van Sumere and Lea, 1985; Weiss and Edwards, 1980).

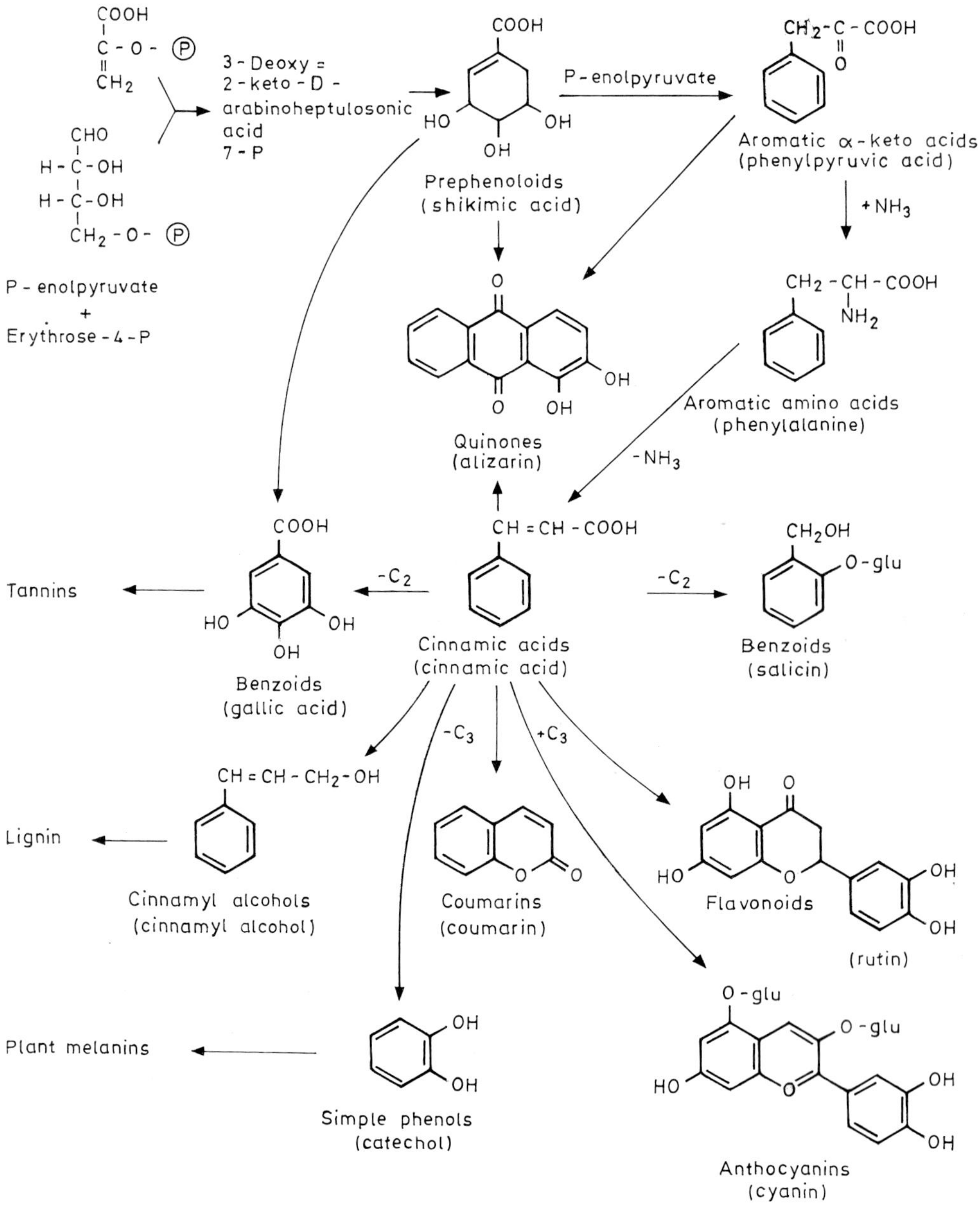

Fig. 9 Scheme of phenoloid metabolic pathways

The further reaction series leading to the special phenoloids are branched from the universal phase. A few intermediates and main products are illustrated in Fig. 9.

The aromatic biosynthetic pathway (shikimic acid pathway) starts with the production of a heptulosonic acid derivative, which is formed by the combination of phosphoenolpyruvate and erythrose-4-phosphate from the pentose phospate cycle or from glycolysis and the pentose phosphate cycle, respectively. The pathway then goes through shikimic acid to prephenic acid. The intermediates of this part of the pathway constitute the substance order No. 1, the prephenoloids. The essence of the shikimic acid pathway is that the mentioned straight-chained heptulosonic acid derivative is transformed into cyclic compounds containing six carbon atoms. Shikimic acid is one of these compounds. A further phosphoenolpyruvate molecule connects the ring, forming a C_3 side chain, and two double bonds are also established; prephenic acid has such a structure. The further transformation of this leads to the aromatic intermediates (order No. 2). Initially, aromatic α-keto acids are formed, by establishing the third double bond in the ring, which provides the aromatic feature (one of these compounds is phenylpyruvic acid in Fig. 9). The aromatic amino acids are formed from these compounds by transamination, and are the main products of the phenoloid biosynthetic pathway. The two universal proteinegenic aromatic amino acids, phenylalanine and tyrosine, are formed from the two primary α-keto acids. These compounds can be put into the substance class No. 5, together with the whole substance family, since these are important N-compounds. It is, however, justifiable to treat them here owing to their very important role as intermediates in the biosynthesis of phenoloids.

This overlapping of two metabolic pathways is a good indication of the fact that the classification of substances by any method is to some degree simplistic because of the complex relationships between metabolic processes, and cannot reflect the complexity of living organisms completely. The third universal aromatic protein amino acid, tryptophan, does not take part in the phenoloid metabolism, and for this reason it can be put into the group of azotoids as a whole. DOPA (3,4-dihydroxyphenylalanine), however, belongs to the aromatic intermediates; it is formed from tyrosine by oxidation (by the action of *o*-kiphenol oxidase, or polyphenol oxidase as it was called earlier). The main pathway of phenoloid metabolism starts from phenylalanine and tyrosine; cinnamic acid and *p*-coumaric acid are formed from them by deamination. These are the characteristic representatives of substance order No. 3, the cinnamoids, and can be put into the family of cinnamic acid derivatives. The unsaturated C_3 chain is characteristic of the family, and makes possible further transformation in a number of ways. The hydroxylation of the benzene ring, and the subsequent partial methylation, glycosidation or esterification of this group, leads to diversification of the compounds of the family. The most important compounds in terms of further biosynthesis are the coenzyme-A esters. Most—and probably all—of the pathways for the further transformations of cinnamic acids start from these compounds.

It is probable that they also take part in the formation of the No. 2 family of cinnamoids. The coumarins, which are included in this family are formed by closing the side chain of cinnamic acid to yield *O*-heterocyclic compounds. Family No. 3, the lignoids, are the products of cinnamic alcohol polymerization, and the cinnamic alcohols are also formed from coenzyme-A esters of cinnamic acids. Their most important genus, the lignins, consists of highly polymerized substances, consisting of a ring system with lattice-like bonds.

Order No. 4 of phenoloids includes the degradative transformation products of phenoloids. The decomposition of phenoloids takes place in the side chain, with decarboxylation and or by the cleavage of C_2 units, and compounds of types C_6C_2, C_6C and C_6 are formed. The most widely occurring benzoids appear mainly in the form of glycosides (phenolglycosides), e.g. thesalicin. An important member of the family is gallic acid, which can also be formed directly from shikimic acid, i.e. in short-circuited mode, additionally to the way described above. One group of tannins, the hydrolysable tannins, are formed from gallic acid. They are built up from sugar and two gallic acid molecule derivatives bound in several ways. The decomposition products of type C_6 constitute the family of simple phenols and their derivatives. It is highly probable that the melanins of plants are formed from catechol, a simple *O*-dihydric phenol, while the *O*-dihydric phenolic amino acid, DOPA, has been regarded as the general starting material of animal melanins (DOPA has been mentioned as being in order No. 2).

Another route for the biosynthesis of phenoloids is represented by the formation of the next order of substances. To the C_6C_3-type cinnamoid part, yet another C_6 part is connected; the latter part has been formed from three acetate units from the polyketide pathway; this is indicated by the polyketide cinnamoids denomination of order No. 5. The cinnamoids taking part in the bonding are the CoA esters of cinnamic acids, and the triacetate part transforms a second aromatic ring. In the chalcones thus produced one of the C_6 rings comes from cinnamic acid, while the other comes from the acetates. The C_3 chain joining the two rings also comes from cinnamic acid. In most plant species a third ring is also formed; the C_3 chain bonds to the first ring with an oxygen, and in this way the third ring becomes a six-membered *O*-heterocyclic ring.

The compounds thus formed are the flavonoids, which exhibit a high variability (Harborne and Mabry, 1982). The connection of the aromatic ring of acetate origin and the *O*-heterocyclic compound may take place at not only the second carbon atom, as in most cases (in flavone compounds), but also at the third one (in isoflavone compounds), and even at the fourth one (in neoflavonoids). This is a secondary bonding taking place by the migration of a phenyl group from the second carbon atom. There is a further possibility in the case of flavone compounds, i.e. the state of oxidation of the heterocyclic compound might change, and the various flavone types are named accordingly (flavonone, flavone, flavonol); the widely distributed anthocyanidins differ from the mentioned flavones also in this respect. These flavylium salts can be found free or as glycosides and esters. Their complex compounds involving metal ions (Al^{3+}, Fe^{3+}), other flavones, or even macromolecules, e.g. polysaccharides, take part in the establishment of colours of flowers. Iso- and neoflavones are very rare relative to the flavone compounds.

Quinones and one group of their derivatives are included in the last order of phenoloids; order No. 6. (Other derivatives are formed in the polyketide pathway.)

The quinones discussed here are not purely the products of the phenoloid pathway either, because the key substance of the terpenoid pathway, mevalonic acid, also takes part in their biosynthesis (by providing C_5 units). These compounds, nevertheless, are dealt with here, because in spite of their many dissimilarities they have a common biogenetic feature, namely the phenoloid pathway is present in the formation of any one of them. Their categorization according to the number of rings (benzo-, naphtho-, anthroquinones) cannot be applied to the classification according to biogenetic features. For the formation of the same chemical structure there are probably not so many alternative pathways for any compound type as exist for the quinones.

One group of quinones is formed from the aromatic intermediaries of the phenoloid pathway (e.g. from p-hydroxyphenylpyruvic acid). The ubiquinones, which are universal substances taking part in respiration, are formed in this way. In these compounds a side chain consisting of 5–12 C_5 units of mevalonic acid origin is bound to the benzoquinone ring. The units of mevalonic acid origin joined to the benzene part share also in the establishment of rings; in this way the second ring of a naphthoquinone-type compound, alkannin, originates from the mevalonic acid. Other quinones are formed directly from shikimic acid by a special connection. This shortened way leads for example, to the formation of vitamin K; in this case a 'C_3 part' originating from succinic acid is joined to the ring of shikimic acid, and takes part in the establishment of a naphthoquinone ring, while the mevalonic acid units develop the side chain. The third ring of the three-ringed anthraquinone-type compound, alizarine, is also formed from mevalonic acid. The quinonic central ring here also originates from C_3, and the other aromatic ring develops from shikimic acid.

In the diversified substance class of phenoloids a number of characteristic medicinal plant active substances can be found. The structures of some compounds indicated in Fig. 9 are introduced here, such as coumarine (coumarines), gallic acid (benzoids), salicin (benzoids, phenolglycosides), rutin (flavonoids, flavones), and cianin (flavonoids, anthocyanins).

Polyketides

The members of this class are well-recognized β-polyketo derivatives, thus they are built up from C_2 units having the structure

$$(\text{—}CH_2\text{—}\underset{\underset{O}{\|}}{C}\text{—})_n .$$

The original chain with a β-polyketo structure can then be modified in a number of ways by condensation, cyclization or addition. The formation of compounds of this substance class takes place in the living organisms in relation to fatty acids. (These are formed from acetyl-CoA through malonyl-CoA in the fatty acid cycle.)

As well as the most widely distributed seven major fatty acids, the minor fatty acids can also be found in small amounts in most of the plants. Special fatty acids, however, are formed only in smaller taxons, although their levels are frequently high. These compounds involve the saturated, unsaturated, acetylenic (triple bond) fatty acids

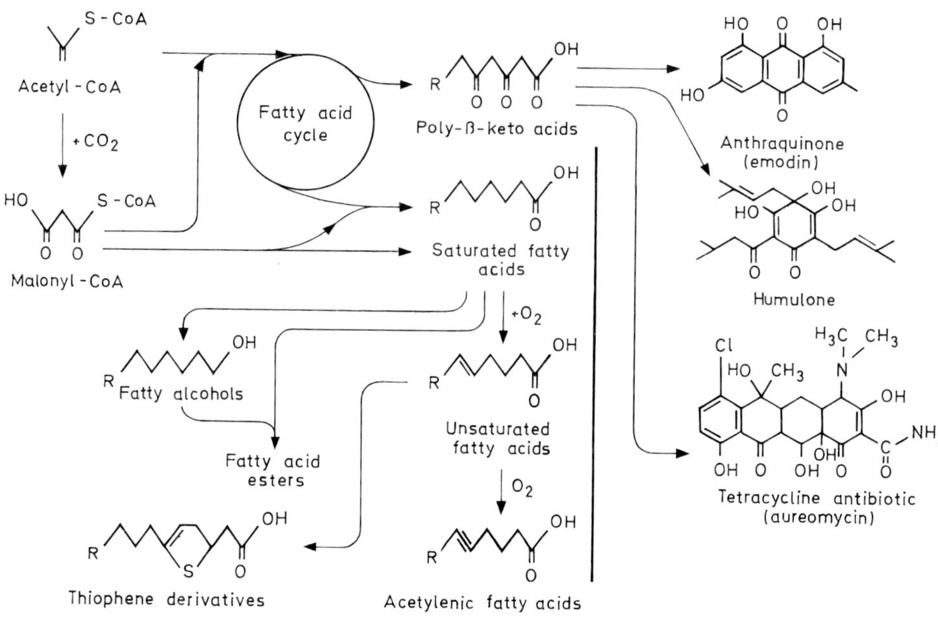

Fig. 10 Main pathways of formation of polyketides

with various chain lengths, and the hydroxy, keto, epoxy and thiophene derivatives. The main pathways of their formation are shown in Fig. 10. It can also be seen in the Figure that the fatty alcohols formed by the reduction of fatty acids take part in the establishment of esters (occurring in waxes).

It is beyond doubt that the special groups of this substance class, which are synthesized only in certain plant taxons (mainly in micro-organisms), are built up from acetate and malonate subunits C_2 by C_2, but the mode of this building up is not yet known in detail. It is possible that at the formation of the fatty acid chain the primary carbonyl (C=O) group is not reduced in these cases, although it is also possible that the reduction takes place and the keto groups are secondarily formed by the oxidation of saturated fatty acids. Yet another interesting problem is the survival of the formed β-keto acids, in view of the high reactivity of these compounds. They have to undergo some kind of stabilization, e.g. joined to protein, or as metal chelates of their enol form.

Owing to the many uncertainties, the classification of polyketides is not yet resolved, unlike that of the other substance classes. One of the reasons for the difficulties is that the main metabolic pathways differ from the those of the other four classes in many respects. The special polyketides are the most widely distributed and most diversified in the micro-organisms. As our main objective is the study of plant biosynthesis, just one example of an important microbe substance is shown in Fig. 10, namely a tetracycline antibiotic which is formed from nine malonyl-CoAs when the polyketo acid chain bends and joins in a four-ringed structure.

In plants of higher order, the special polyketide products are not primarily formed along this pathway; the importance of the polyketide pathway emerges from the fact that it acts as an alternative route for the biosynthesis of special substances, which otherwise are formed by other routes. The emodine of anthraquinone structure, for example, is synthesized in this way in *Rhamnus* and *Rumex,* from the keto acid which has been formed by the linear condensation of acetate units. The modified benzoquinone ring of humulone and lupulone probably also originates from a polyketide, and the side chains evidently originate from mevalonic acid. There are thus the alternative possibilities of quinone formation by the interconnection of phenoloid and terpenoid pathways. The polyketide pathway may provide a kind of spare route for the formation of many other special plant substances. An extreme case of this possibility occurs in the fungus *Trychophyton rubrum;* in this species the complete shikimic acid pathway is missing, and the proteinogenic aromatic amino acids are formed from acetate along the polyketide pathway. Another feature of the polyketide pathway is that it is frequently connected to other metabolic pathways, as was mentioned in the discussion of the biosynthesis of flavonoids.

Terpenoids

The compounds of class No. 4 of plant substances are formed along the same biosynthetic pathway as the terpenes, and are characteristic components of essential oils and resins. These compounds are constructed from C_5 units (Nes, Fuller and Tsai, 1984; Porter and Spurgeon, 1981; 1983). Their biosynthesis starts from acetyl-CoA; two of its molecules join to form acetoacetyl-CoA, then a third acetyl-CoA joins to form a branched-chain C_6 compound. Mevalonic acid, also having six carbon atoms, is formed from this, and is the key substance of this biosynthetic pathway. This is further transformed through its phosphate, then through its pyrophosphate, derivatives, to form a C_5 compound, Δ^3-isopentenyl pyrophosphate, by decarboxylation.

This latter compound is partly transformed into another C_5 compound by an equilibrium-controlled reaction (about 87 per cent), namely into dimethylallyl-PP. These two C_5 compounds represent the starting components for the further pathway of terpenoid biosynthesis. Depending on the number of such C_5 units linked, the terpenoids are categorized into eight substance orders; 1, 2, 3, $2\times2=4$, $2\times3=6$, $2\times4=8$, 6–15 or 4000–40 000 units are in the respective orders, which are called hemiterpenoids, monoterpenoids, diterpenoids, triterpenoids, tetraterpenoids, oligoterpenoids and polyterpenoids, respectively. In most cases the linking of the two starting C_5 compounds takes place in such a way that the head part of dimenthylallyl-PP joining the pyrophosphate group condenses with the tail part of a branch chain end of Δ^3-isopentenyl-PP ('head-to-tail' linkage). The C_{10} compound thus formed, geranyl-PP, is the general precursor of monoterpenoids on one hand, and on the other hand it may be transformed to a C_{15} compound, farnesyl-PP, if it attaches to a new molecula of Δ^3-isopentenyl-PP. Analogously, farnesyl-PP is a starting point for sesquiterpenoids, or it can be transformed via another condensation to the C_{20} compound geranylgeranyl-PP, which is the precursor of diterpenoids. It is highly probable that the oligoterpenoids and the polyterpenoids are formed in a similar

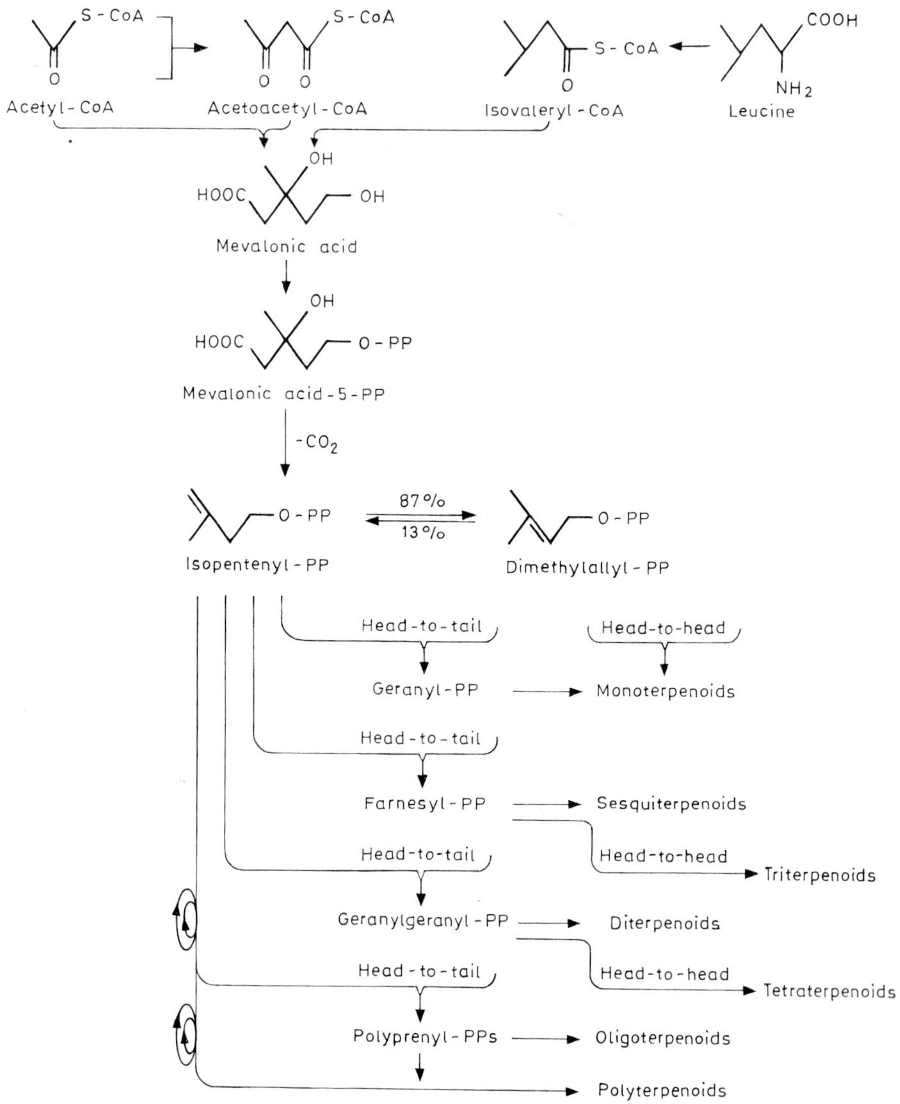

Fig. 11 Biosynthetic scheme of terpenoids

manner, i.e. by linking further isopentenyl-PP units. Besides the mentioned head-to-tail type of condensation, two farnesyl-PPs or two geranylgeranyl-PPs may be linked by head-to head condensation; the triterpenoids and tetraterpenoids are formed in this way. (In some of the literature the terms 'head' and 'tail' are used in opposite senses, so 'head-to-tail' and 'tail-to-tail' connections are mentioned.) These relationships are summarized in Fig. 11, and the corresponding significant medicinal plant active substances are shown in Fig. 12.

The already mentioned C_6 and C_5 compounds belong to the substance order No. 1 among the intermediaries and hemiterpenoids. Most of the monoterpenoids (order N. 2) are formed from geranyl-PP or linalool-PP; the latter is formed by the isomerization of geranyl-PP. This is also a way of forming the open-chained geraniol, linalool and citral. Geranyl-PP can close into a ring in several ways, the most widely occurring cycle being that which produces menthol, carvon, thymol and carvacrol. The ring of thujon and borneol is much more complicated. Some pyrethrine I, chrysanthemic acid, which is an irregular monoterpene, is probably formed by the head-to-head connection of two dimethylallyl-PPs, instead of the ordinary way (the origin of the ringed part is even more uncertain).

The sesquiterpenoids (order No. 3) also involve open-chain active substances (farnesol) and cyclic compounds (chamazulene, cnicine). From order No. 5 (diterpenoids), what are worth mentioning are the openchain phytol, which can be found in chlorophyll, and the phytohormone gibberellic acid, which has a more complicated ring structure. Among the compounds of order No. 6 (tetraterpenoids), the carotenoids have significance. There is no compound of much importance among oligoterpenoids (order No. 7), while the polyterpenoids (order No. 8) include the highly significant rubber.

The triterpenoids (order No. 4) are among the most widely distributed and most diversified plant substances. The starting compound of this order is formed by the head-to-head condensation of two farnesyl-PPs. This is a C_{30}-type compound, squalene, which is formed through the intermediary presqualyl-PP. The squalene then closes into a cyclic compound in a characteristic manner, also through an intermediary (2,3-epoxysqualene). One of the possible modes of attaining the cyclic feature is introduced here through the example of glycyrrhetinic acid, which is an active substance belonging to the triterpenes; the compound is formed by five linked C_6 rings.

There is a more general type, in which the squalene is closed to form three 6-membered and one 5-membered ring, and a side chain is also connected to the 5-membered ring. These four rings constitute the skeleton of plant steroids. Their most widely distributed representatives are the sterols, among which β-sitosterol and cholesterol are important owing to accumulation and further transformations, respectively. Cholesterol is the precursor for all the further plant steroids; its side chain makes possible the formation of rings of various types. In this way, cholesterol closes to form *O*-heterocyclic compounds in the family of steroid saponins (e.g. dioscin), while *N*-heterocyclic compounds are formed in the case of *N*-compounds of steroid origin, but even oxygen-holding rings may be in the molecule, as in the case of solasodine. In the case of oxidized steroidal derivatives, the formation of rings is not as characteristic as the presence of many *O*-function groups (e.g. ecdysone, which is well known as a hormone of insects but also forms in certain plants). In the case of pregnane derivatives, the side chain of cholesterol has shortened by six carbon atoms; in this way pregnelonone and progesterone, which are well known in human biochemistry are formed also in plants. From these compounds the extremely important cardenolides are formed (lanatoside-C) by the addition of a C_2 unit and subsequently by the formation of a 5-membered *O*-heterocyclic ring (γ-lactone).

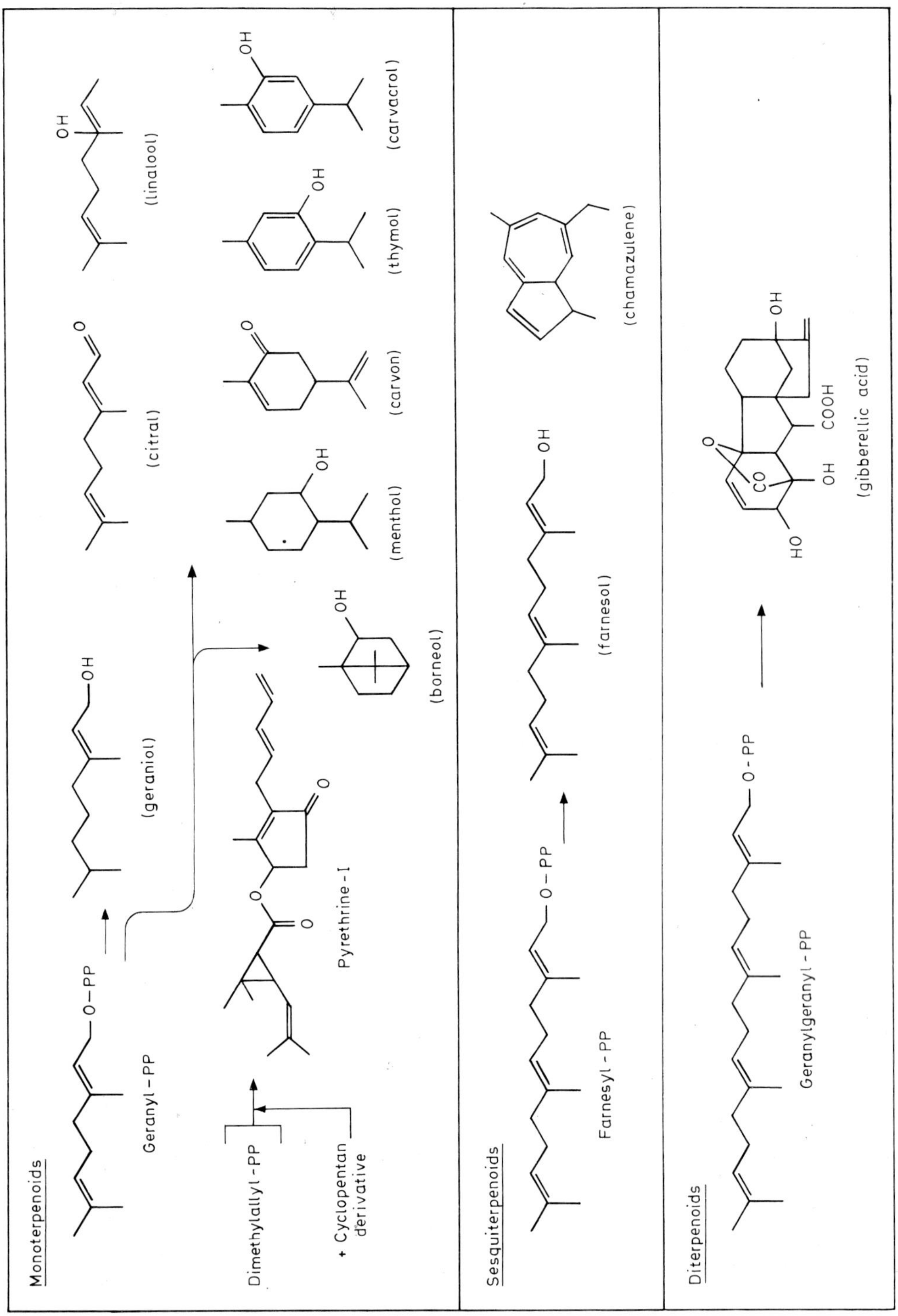
Monoterpenoids
Geranyl-PP
O-PP
(geraniol)
OH
(citral)
(linalool)
Dimethylallyl-PP
+ Cyclopentan derivative
Pyrethrine-I
(borneol)
(menthol)
(carvon)
(thymol)
(carvacrol)
Sesquiterpenoids
Farnesyl-PP
(farnesol)
(chamazulene)
Diterpenoids
Geranylgeranyl-PP
(gibberellic acid)
COOH
CO
HO

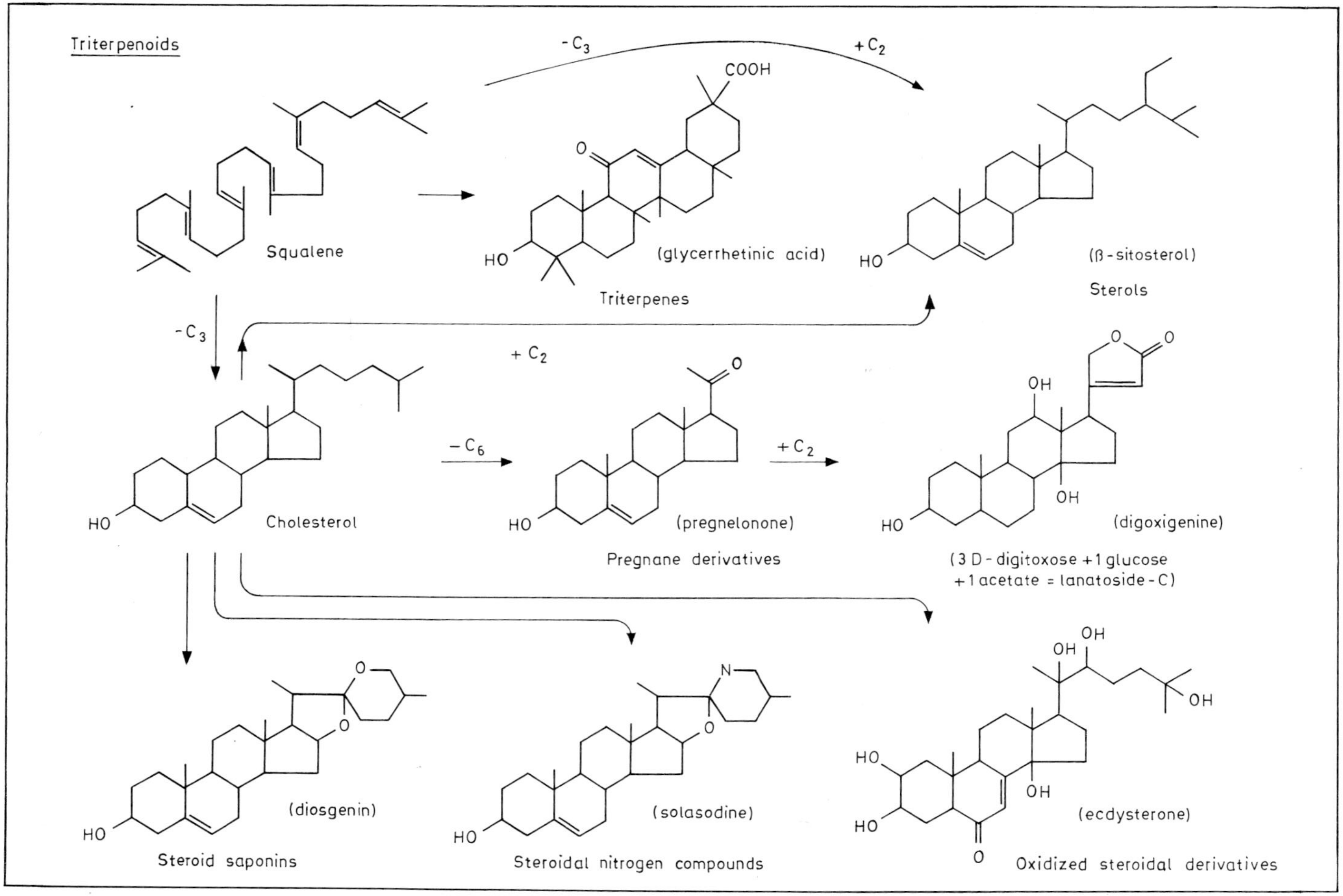

Fig. 12 Significant terpenoid active substances

The extremely rich class of terpenoids includes most of the active substances of medicinal plants, because most of the medicinal plants cultivated today contain essential oils, and a large portion of essential oil constituents are mono- and sesquiterpenoids. The active substances of steroid-containing medicinal plants are also included here (triterpenoids). If, however, the range of active substances is not limited to medicinal plants cultivated in Hungary, then it can be stated that most of the applied active substances in the world should be included in the azotoid substance class.

Azotoids

The name of the fifth substance class refers to the presence of nitrogen. Although the presence of nitrogen is a necessary condition to include compounds here, it is not sufficient of itself. What is determining in this respect is the fact that the compounds are formed and take part in nitrogen metabolism. (Nitrogen-containing substances have already been mentioned, i.e. amino sugars and steroidal nitrogen compounds. These compounds, however, join nitrogen metabolism only at one point, during transaminations when nitrogen is introduced. Prior to this process, as well as afterwards, these compounds take part in the processes which are characteristic of their own group of substances, namely in saccharide and steroid metabolism, respectively.)

The formation of the main types of azotoids, and their interrelationships, are summarized in Fig. 13. Their most diversified order, the alkaloids, are outlined in Fig. 14.

This is the most diversified class of plant substances. As has been pointed out, the universal phase of the former substance classes joins the metabolism only at one point (or two). All their groups are formed from a simple key substance (shikimic acid, malonyl-CoA, mevalonic acid). As opposed to this simplicity, the biosynthesis of the universal precursors of azotoids joins carbohydrate and acid metabolism at many points, e.g. at six and two points in the cases of amino acids and purine and pyrimidine bases, respectively. In this way even the universal azotoids (order No. 1) are extremely diversified. Proteinogenic amino acids and nucleotide bases—which are fundamental materials of living organism—have further universal derivatives, such as porphyrins, CoA and cytokinins.

The simple special azotoids are placed in order No. 2, these compound types being formed in one or a few steps from the universal azotoids.

The amines are formed from amino acids in a single step, with decarboxylation; e.g. ethanolamine from serine, and choline from ethanolamine by methylation. The formation of amides also takes place in one step, which may start from an amino acid, as with the two universal amides of amino acids, glutamine and asparagine, but can also start from an amine such as the pungent substances of paprika, the capsaicines. The most diverse family of the order comprises the special amino acids. About 300 compounds are included here. The compounds may be *N*- or *S*-methyl derivatives of universal proteinogenic amino acids, they may be longer or shorter than normal by one carbon and the position of the amino group may be β or γ instead of α. D-amino acids and chloro and phospho derivatives can also be found, and analogues even exist in which the sulphur is replaced by selenium. There are compounds among them with

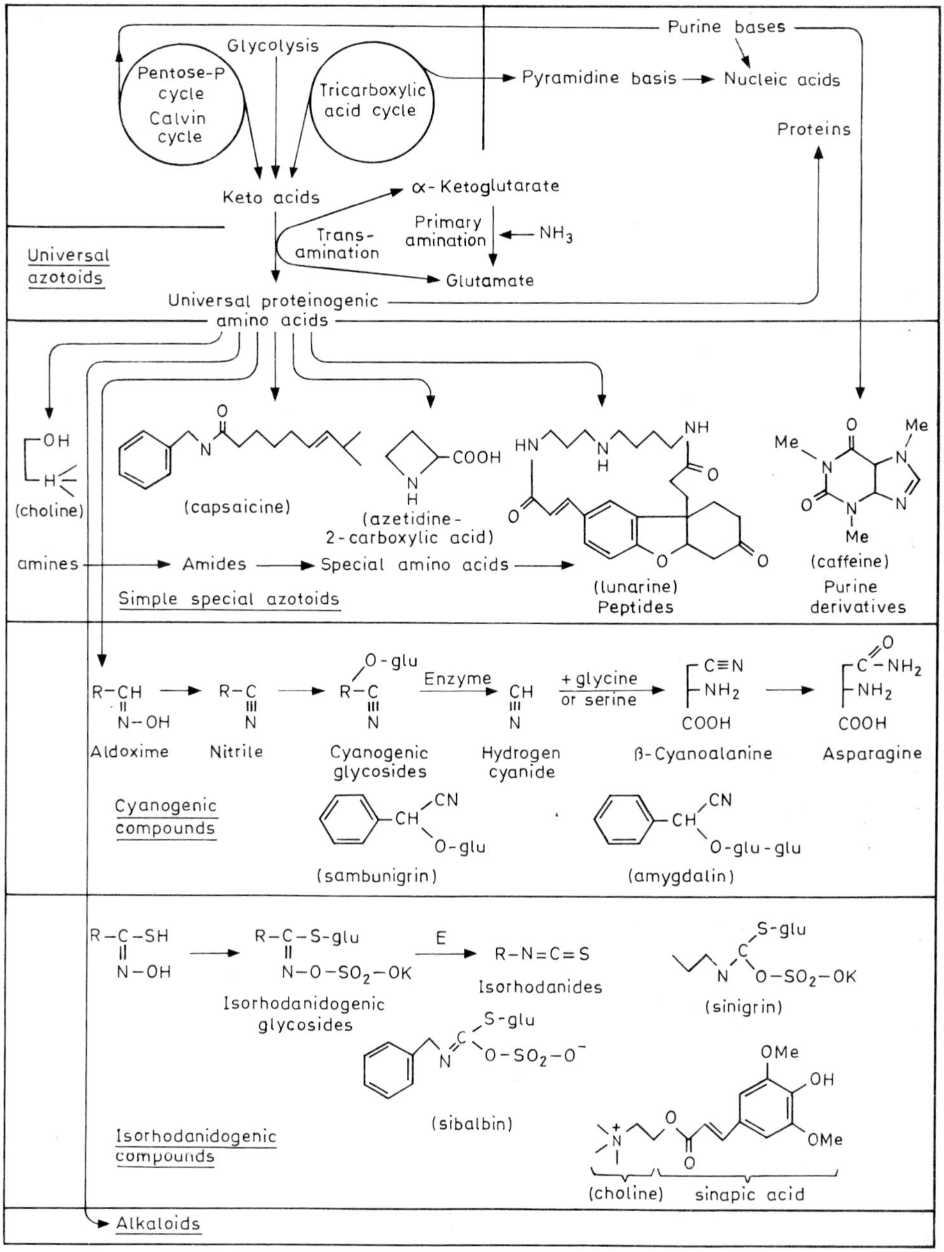

Fig. 13 Formation of main azotoid types

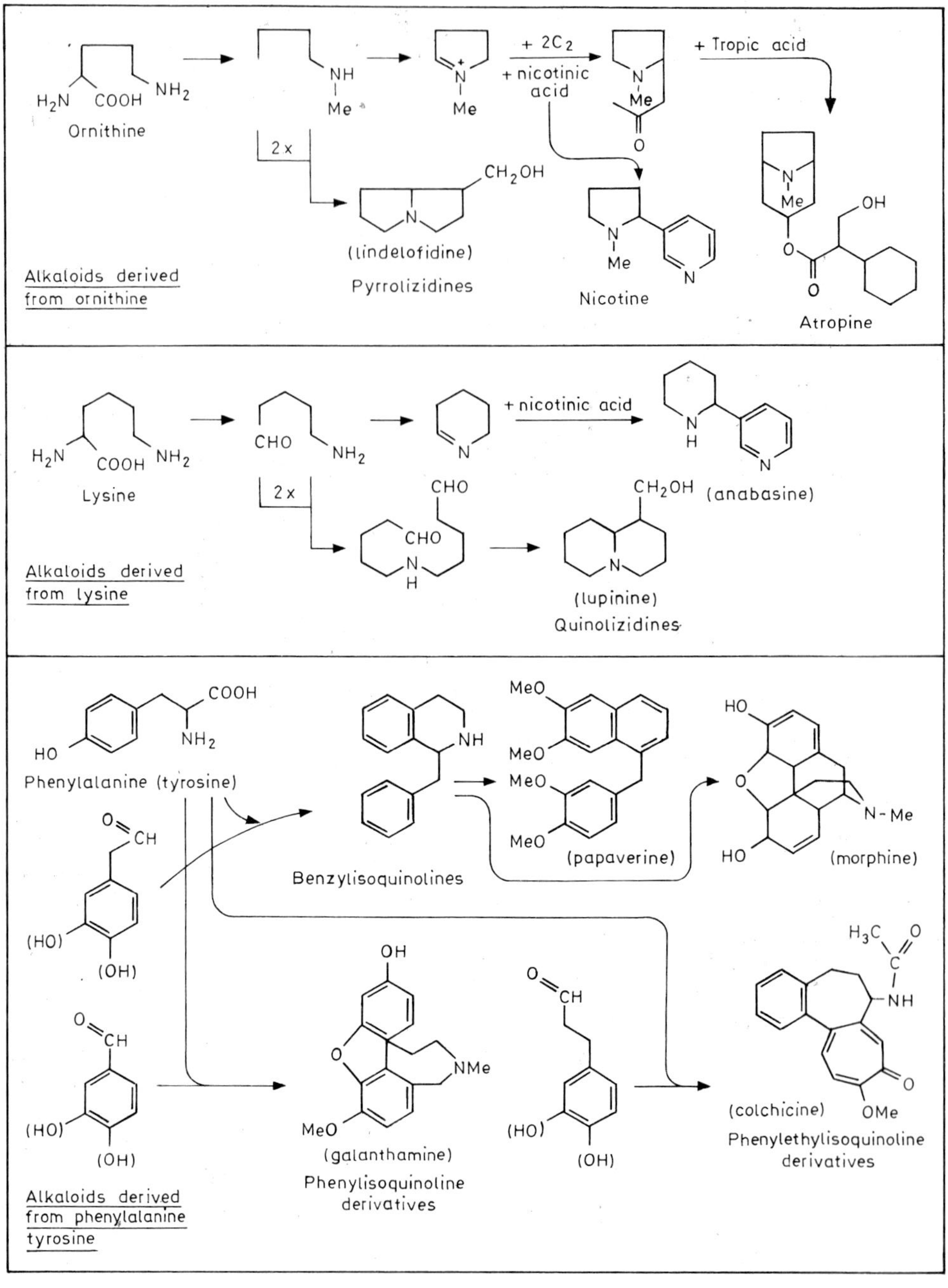

Fig. 14 Structures of significant alkaloids

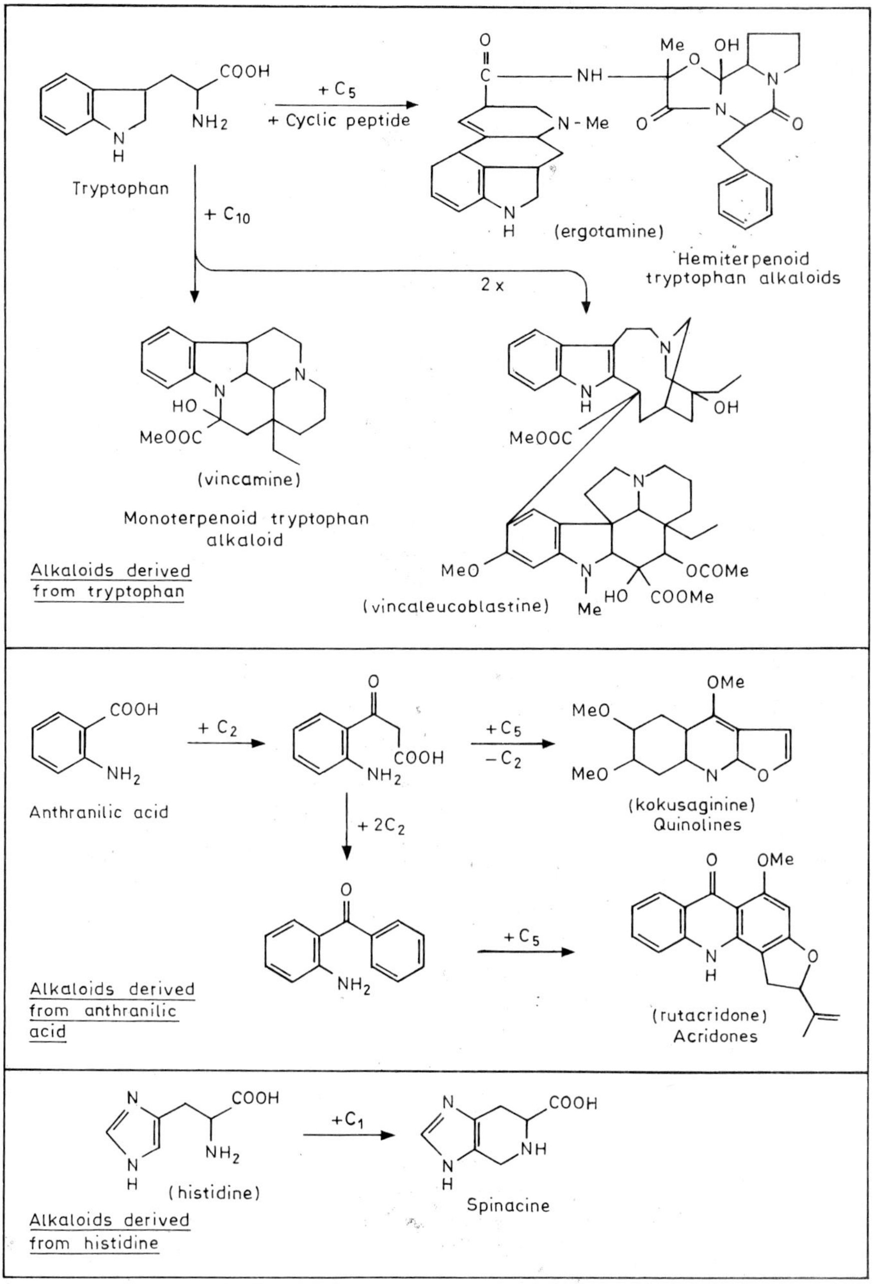
Tryptophan
+ C5
+ Cyclic peptide
(ergotamine)
Hemiterpenoid tryptophan alkaloids
+ C10
2 x
(vincamine)
Monoterpenoid tryptophan alkaloid
(vincaleucoblastine)
Alkaloids derived from tryptophan
Anthranilic acid
+ C2
+ C5
− C2
(kokusaginine)
Quinolines
+ 2C2
+ C5
(rutacridone)
Acridones
Alkaloids derived from anthranilic acid
(histidine)
+C1
Spinacine
Alkaloids derived from histidine

universal or at least general occurrence, while others are special substances of species, genera or families. Special compounds can be found among the purine derivatives as well, e.g. methyl derivatives such as caffeine. (These so-called purine alkaloids are not placed among the alkaloids, because no new *N*-heterocycle is formed when these are developed from the universal precursor compound.) Finally, the special peptides are included in this order of substances. Figure 13 shows lunarine, which is a cyclic peptide (generated by a flowering plant).

Cyanogenic compounds comprise order No. 3 of the azotoids. These are also synthesized from amino acids. The best known, amygladin, is formed from phenylalanine, while others are formed from tyrosine, valine and isoleucine. Following the decarboxylation, the amino group is transformed to nitrile, then to hydroxynitrile, and this is stabilized as *O*-glycoside; these are the cyanogenic glycosides. Enzymic decomposition cleaves the HCH from it; this process has long been known. It has been recently established that the HCN may go back into the metabolic process.

The —SH group of cysteine or the —OH group of serine is exchanged for nitrile, and then this may be transformed to asparagine. This can take place not only with HCN of endogenic origin, i.e. in plants containing cyanogenic glycosides, but also, in certain plants, with HCN of external origin.

For the production of isorhodanidogenic compounds (order No. 4), the reaction series is even longer than that for cyanogenic glycosides. The amino acid changes to an oxime, then an —SH group is attached to it; this is what bonds the sugar component in the form of *S*-glycoside, and a sulphate group is also joined to the molecule. The isorhodanides are formed from glycosides by the action of enzymes. (Owing to their characteristic occurrence these compounds are also called mustard oils, and the sugar derivatives are called mustard oil glycosides.) As well as the proteinogenic compounds (methonine, phenylalanine, tryptophan), the precursor amino acids also include compounds which have been formed by lenghtening the carbon chain by one, two or three carbon atoms (the homologues of methionine).

Order No. 5 of the azotoids comprises the alkaloids. According to the classification of substances on the basis of biogenetic features, this groups of substances seems to be more uniform than it does using the view of the term alkaloid found generally in the literature (Mothes, Schütte and Luckner, 1985; Cordell, 1981, Dalton, 1979). According to our definition the alkaloids are the products of one pathway of specialization of nitrogen metabolism, the main forming pathway of which starts from the primary *N*-compounds, and are primarily related to the formation of *N*-heterocycles. The first part of the definition places the alkaloids among the azotoids, while the other features jointly determine the specificity of alkaloids. These features are follows:

(1) The starting point of the main pathways of biosynthesis are the respective proteinogenic amino acids (lysine, phenylalanine, tyrosine, tryptophan, histidine, according to present knowledge), and ornithine and anthranilic acid, which have a direct biogenetic relationship with them. In theory the purine and pyrimidine bases may also have alkaloid derivatives, but these have not yet been discovered (the so-called purinalkaloids have already been mentioned).

(2) In theory, living organisms have the ability to form alkaloids—as far as the precursores are concerned—but the biosynthesis takes place mostly in plants of higher orders.
(3) The characteristic of the main pathways of alkaloid biosynthesis is their origin from the mentioned compounds. In the biosynthesis of more complex molecules the products of other processes may take part, such as acetate and mevalonic acid units, and phenoloids.
(4) Alkaloid biosynthesis is generally a longer process, consisting of several steps. The formation of the heterocycle is preceded by some kind of modification of precursor amino acid (e.g. decarboxylation and, in the case of diamino acids, oxidation).
(5) Further rings might be formed additionally to *N*-heterocycles, but secondarily the *N*-heterocycle might also decompose (e.g. in the case of *Colchicum* alkaloids).

The alkaloids are categorized into families according to the precursor amino acids of their biosynthesis. A high degree of homology exists between the alkaloids formed from the amino acids ornithine and lysine (families Nos 1 and 2, Fig. 7), because five-membered and six-membered *N*-heterocycles are formed from the diamino acid as the initial products of cyclization, and being homologous provide the possibility of connecting identical items.* Among the ornithine-derived alkaloids, the tropane alkaloids, such as atropine and scopolamine, should be mentioned. From two ornithines and lysines the pyrrolizidines and quinolizidines are formed, respectively. The skeleton of isoquinoline derivatives originated from phenylalanine and tyrosine (family No. 3) is the product of so-called Mannich condensation taking place between a carbonyl compound and a phenyl alkyl amine, the latter being formed by decarboxylation of amino acid.

In the case of the three most important substance genera, the carbonyl compound is an aromatic derivative of a C_6C_1, C_6C_2 or C_6C_3 skeleton and the formed products are phenylisoquinoline, benzylisoquinoline and phenylethylisoquinoline derivatives, respectively. Among the phenylisoquinolines, the so-called Amaryllidaceae alkaloids should be mentioned, and the colchicines are worth mentioning among the phenylethylisoquinolines. The benzylisoquinolines are much more diversified and at the same time much more important: e.g. the alkaloids of the poppy, papaverine and morphine. The most complex and numerous family incorporates the alkaloids derived from tryptophan (family No. 4). The variability of the structure is ensured by the large number of potential coupling compounds and by the formation of intermediaries which possess the ability to change in many directions. In the alkaloids of ergot a C_5 hemiterpenoid unit is coupled to the tryptamine. In the most widely distributed alkaloid genus a C_{10}-monoterpenoid unit is joined to the tryptamine (the joining itself may take place in a number of ways, and the possibilities of further changes are numerous); such compounds are, for example, vincamine and vincaleucoblastine. The precursor compound of class No. 5 is anthranilic acid. From this the quinoline alkaloids are formed by the introduction of acetate units, and after the introduction

* Such compounds are nicotine and anabasine, for which the connecting item is nicotinic acid.

of two additional acetate units the acridines are formed. The smallest family (No. 6) includes the alkaloids formed from histidine. In the case of spinacine, which is mentioned as an example, the new heterocycle is formed by the introduction of one C_1 unit.

In this section the plant special substances have been identified as chemical compounds, and the chemical and biochemical processes have been outlined which lead to the formation of these compounds in the plants.

References

Bell, E. A. and Charlwood, B. V. (eds) (1980). *Secondary Plant Products,* Springer Verlag, Berlin.

Conn, E. E. (ed.) (1981). *Secondary Plant Products,* Vol. 7, *The biochemistry of Plants* (P. K. Stumpf and E. Conn eds), Academic Press, New York.

Cordell, G. A., (1981). *Introduction to Alkaloids. A Biogenetic Approach,* J. Wiley, New York.

Dalton, D. R., (1979). *Alkaloids,* M. Dekker, New York.

Harborne, J. B. and Mabry, T. J. (eds) (1982). *The Flavonoids: Advances in Research,* Chapmann and Hall, London.

Haslam, E. (ed.) (1979). *Biological Compounds,* Vol. 5. *Comprehensive Organic Chemistry* (D. R. Barton, and W. D. Ollis, eds). Pergamon Press, Oxford.

Luckner, M. (1984). *Secondary Metabolism in Microorganisms, Plants and Animals,* Fischer Verlag, Jena.

Manitto, P. (1981). *Biosynthesis of Natural Products,* Ellis Horwood, Chichester.

Mann, J. (1978). *Secondary Metabolism,* Oxford University Press, New York.

Mothes, K., Schütte, H. R., and Luckner, M. (eds) (1985). *Biochemistry of Alkaloids,* VEB Deutscher Verlag der Wissenschaften, Berlin.

Nes, W. D., Fuller, G. and Tsai, L. S. (eds) (1984). *Isoprenoids in Plants. Biochemistry and Function,* M. Deunes Press, New York.

Porter, J. W., and Spurgeon, S. L. (eds) (1981). *Biosynthesis of Isoprenoid Compounds,* Vol. 1, J. Wiley, New York.

Porter, J. W., and Spurgeon, S. L. (eds) (1983). *Biosynthesis of Isoprenoid Compounds,* Vol. 2, J. Wiley, New York.

Preiss, J. (ed.) (1980). *Carbohydrates: Structure and Function,* Vol. 3. *The Biochemistry of Plants* (P. K. Stumpf, and E. E. Conn, eds), Academic Press. New York.

Robinson, T. (1983). *The Organic Constituents of Higher Plants,* 5th edn, Cordus Press, North Amherst.

Schütte, M. R. (1982). *Biosynthese Niedermolekularer Naturstoffe,* Fischer Verlag, Jena.

Thomson, R. H. (ed.) (1985). *The Chemistry of Natural Products,* G. Dummett, Glasgow.

Van Sumere, C. F., and Lea, P. J. (eds) (1985). *The Biochemistry of Plant Phenolics,* Clarendon Press, Oxford.

Weiss, V. and Edwards, J. M. (1980). *The Biosynthesis of Aromatic Compounds,* Wiley-Interscience Publ., New York.

Vickery, M. L. and Vickery, B. (1981). *Secondary Plant Metabolism,* Macmillan Press, New York.

Taxonomy of Medicinal Plants

Prior the establishment of the taxonomic system of Linnaeus, at the time of purely practical taxonomic systems, the medicinal plants formed one of the most important groups of the plant kingdom. A number of medicinal plants were also indispensable as food, while others were cultivated as ornamentals, and still others were gathered from the wild-growing populations.

Owing to the versatility of medicinal plants, they can be found in several classes of an econobotanical system (Terpo, 1981).

Class I. Food plants, e.g. spice plants.
Class II. Technical-industrial plants, e.g. timber industry plants, pharmaceutical industry plants.
Class V. Green environment and ornamental plants, e.g. *Tilia* species.
Class VI. Plants with detrimental effects, e.g. toxic species.

Phylogenetic systems classify plants according to their relationships, or, more accurately, according to their heredity. The basis for the phylogenetic systems applied today is the sexual system of Linnaeus, which, although an artificial system, contained a number of natural features. By taking all the morphological properties into consideration in taxonomy, the so-called natural (morphological) systems were established. Later on, the development of theories of Darwin facilitated the first development of phylogenetic systems, which are based on the principles of inheritance. Their effectiveness was highly increased by taking the results of sciences like geology, genetics, physiology, mathematics, and chemistry into consideration.

The introduction of chemistry to taxonomic methods enabled the recognition of new medicinal plants on one hand, and proved the unworthiness of several plants used in folk medicine on the other hand.

In a large part of Europe the medicinal plants can still be classified into two groups, namely the group of *collected* (wild-growing) and the group of *cultivated* plants. The latter group includes a number of *true* (ancient) cultivated plants (e.g. poppy) as well as the ever-growing number of species which have been taken into cultivation (naturalized or domesticated) recently (e.g. camomile, *Artemisia absinthum*).

The taxonomic treatment of medicinal plants is the same as that for other plants. A basic taxonomic category is the *species* (species = sp.). Related species form the *genus,* and the closely related genera are grouped into a *family*. The families which are close to each other are grouped into an *order,* and the related orders form a *class*. It is also a custom to group the classes into *divisions*.

In order to distinguish between the dissimilar populations of *wild-growing* species generally the intraspecific categories, *subspecies* (subsp.), *variety* (var.) and *form* (f.) are used.

Species (either natural or cultivated) in cultivation are divided into well-distinguishable infraspecific *varieties,* the classification of which is made on the basis of an econobotanical system. The cultivar is distinguished according to the features valuable to humankind.

For distinguishing and describing cultivars developed for medicinal plant cultivation, phytochemical methods are used more and more widely. The new cultivars, of course, may be described by any method.

If the description is made properly and the name corresponds to the International Rules of Botanical Nomenclature, then the new cultivar can be regarded as accepted.

For the taxonomic classification of natural taxa the decisive distinguishing factor is the relationship, while that of the cultivated plants is the *usefulness,* the *economic value*. Additionally, we cannot disregard the description of cytological, morphological and anatomical features either; sometimes even the ecological properties should also be described. For example, the species of camomile are distinguished by the amount and composition of active substances, as well as by the height of the plant, etc.

The used names of medicinal plants, according to the *binominal nomenclature*, are the same as those used for other cultivated or wild-growing plants. From January 1st, 1959 on, the species names may only be invented names. In ternary combinations the abbreviation 'cv.' is put just before the name of the variety, or the variety name is put between simple quotation marks ('—').

Example:

Papaver	*somniferum*	L.	cv. Kompolti or 'Kompolti'
genus	species	describer (author)	variety

The names of the drugs are also controlled by rules (Geneva Nomenclature). The Latin name of the medicinal plant (binominal or only the species name) in the singular genitive case is followed by the authorized name of the processed plant part in the nominative case. Examples: Vincae minoris herba, Daturae innoxiae folium, Daturae innoxiae semen, Lavandulae flos, Coriandri fructus, Belladonnae radix. In a number of cases, expecially for extracts (involving essential oils as well), the Latin name of the extracted part of the plant in the nominative case is followed by the name of the plant giving the extract in the genitive case. Example: Aetheroleum Menthae piperitae, Oleum Ricini, Amylum Ricini, Pix Juniperi. In European as well as Hungarian Pharmacopoeias the binominal name system is adopted.

Taxonomic review of medicinal plants

Medicinal plants can be found in most of the following categories of the plant kingdom:

A. SUBKINGDOM THALLOPHYTA
The Bacteria and Related Groups
The Algae
The Fungi
The Lichens

B. SUBKINGDOM EMBRYOPHYTA (The Vascular Plants)
The Bryophyta (Mosses and Liverworts)
The Pteridophyta (Ferns and Fern Allies)
The Spermatophytes (The Seed Plants): Gymnosperms and Angiosperms

The medicinal plants to be introduced here belong to group 'B' with one exception:

Division: Eumycota—The True Fungi
Subdivision: Ascomycotina
Class: Pyrenomycetes, Order: Clavicipitales
Clavicipitaceae—Ergot family

The main features of the family are the ascus, in which the initially one-celled, then divided string-like ascospores are located, and the lysergic acid derivatives. The most important wild-growing and cultivated species is *Claviceps purpurea.* Other *Claviceps* species are known as well. Area: Europe, Asia, North Africa.

Division: Gymnospermatophyta—The Gymnosperms
Class: Coniferopsida, Order: Pinales
Cupressaceae—*Cypress* family

These are densely branching evergreen, bracted, sometimes similar to pine-needle, opposite or three-whorled leaved trees or shrubs. The male cones are solitary, 6–9 mm long, with about 6 whorls of scales. The female cones are solitary, rather short at flowering time. The fruits are berry-like, globose, bluish-blake, prunose, containing up to 6 seeds.

Among the 60 or so species of *Juniperus* (Juniper), *J. communis* is a relatively common, wild-growing plant in Hungary. It is a circumpolar species, being found from the arctic sand plains to the alpine regions, in karstland brush forests, and in heath areas, and it is not sensitive to the quality of the soil. It is relatively rich in forms, both dioecious and diploid ($n = 11$). It is an important medicinal plant owing to its berry (Juniperi fructus), which gives the valuable essential oil (Aetheroleum juniperi). The species *J. sabina* is often cultivated; it is toxic (Sabinae herba, Oleum sabinae), and grows wild from southern Europe to central Asia.

Division: Angiospermatophyta—The Angiosperms
Class I: Dicotyledonopsida

Subclass 1: Magnoliidae
Order: Ranunculales
Ranunculaceae—Crowfoot family
(except Helleboraceae)

In several species of the family strong active ingredients can be found, as for example adonitoxin glycoside in *Adonis vernalis,* and alkaloid in *Ranunculus arvensis.* These plants are herbaceous; the leaves are alternate and parted, they have actinomorphic flowers, are apocarpous, and the fruit consists of numerous shortly backed achenes. About 2000 species are known in the family (*Ranunculus,* 250; *Adonis,* 20).

Order: Papaverales (Rhoeadales)
Papaveraceae—Poppy family
(Except Fumariaceae)

These are herbaceous plants (Th or H) sometimes shrub-like in size, and having alternate leaves. The flowers are hermaphrodite, and insect-pollinated; characteristic are the two sepals, which fall off even during flowering; inside the four-parted (2 + 2) corolla a number of stamens develop. The fruit is formed by lots of carpels, and is generally a capsule with porous dehiscence. In terms of phytochemistry, they can be mainly placed in alkaloid-containing families. The benzylisoquinolin-type alkaloids accumulate in divided laticiferous vessels; the latex is white *(Papaver)* or yellow *(Chelodonium)*. Almost all the genera have special alkaloids or alkaloid groups. According to the involved alkaloid (e.g. berberine) they are closer to or further from, the older families *(Lauraceae, Annonaceae, Berberidaceae, Ranunculaceae, Helleboraceae)*. Chromosome number: $n = 5$–11, 16, 19. A few natural hybrids are also known (e.g. *Papaver hungaricum*).

The species of the family grow mainly in the northern temperate region (about 200 species of the 26 genera); they are valuable ornamental, food and medicinal plants; several species are widely distributed weeds. The 50 species of *Papaver* are mainly native to Eurasia; several species are native to North America. Owing to its oil-containing seed, *P. somniferum* has been cultivated from the Neolithic ages. Chromosome number: $2n = 22$, 44. It is regarded as a native of Asia Minor, but its most close relative, *P. setigerum,* grows in the western Mediterranean region.

The Eurasian *Chelodonium majus* has a chromosome number of $2n = 10$, 12, 16; its orange-yellow milky juice is toxic, and contains a number of alkaloids (chalerythrin, sanguinarin, chelidonin, etc.). A recently cultivated medicinal plant is the Mediterranean *Glaucium flavum,* which also has yellow milky juice. In the related Fumiraceae family there are species which contain alkaloids *(Corydalis)*.

Subclass 2: Rosidae
Order: Rosales
Rosaceae—Rose family

This family includes several medicinal plants of minor importance. Most of the species are valuable ornamental and food plants. They grow mainly in the temperate region; their species can be found in almost all of the ecological life-form groups. The flowers are five-parted; the number of stamens is ten (or more). Being apocarpous, the developed fruit would be follicle or drupe theoretically, but a number of pseudo-fruits have developed which are generally characteristic of the family; Spiraeoideae (follicle), Pomoideae (pome), Rosoideae (nucularose-hips, etc.), Prunoideae (drupe).

Almost all of the species contain tannins; for this reason the family is regarded as a typical *tannin family*. Important drug plants are *Potentilla erecta* (Tormentillae rhizoma), *Rubus fruticosus* (Rubi fructicosi folia), and *Agrimonia eupatoria* (Agrimoniae herba). The seeds contain protein and fatty acid *(Amygdalus)*. The cyanogen compounds can also be found frequently *(Amygdalus),* except in the Rosoideae subfamily. Sorbite, mucilage and organic acid can also be found. The accumulation of ascorbic acid (vitamin C) is significant in the fruits of *Rosa canina*. Chromosome number: $n = 7, 8, 9, 17$.

Order: Fabales
Fabaceae—Broad bean or Pulse family

This family is exceptionally rich in species, and has a very significant role in the economy, but includes only a few medicinal plants of minor usage (12 000 species). The primary characteristic feature of the family is the typical papilionaceous flower structure, and the pod fruit, which develops generally from one carpel and opens into two parts. They mainly grow on firm ground and are distributed all over the world.

Among the materials which are important in terms of therapeutics, the alkaloids are frequent (e.g. Sophoreae, Genisteae), as are the saponins (*Glycyrrhiza glabra*—Liquorice). The species *Galega officinalis* (goats rue) contains the galegine glycoside. Liquorice ($2n = 16$) is a cultivated plant, and can be found in western, southern and central Asia (up to Iran); it is perennial (H), and its drug is Liquiritiae radix.

Order: Myrtales
Onagraceae (Oenotheraceae)—Evening Primrose family

These plants are mainly perennial, herbaceous, sometimes lignified stemmed, and grow in temperate and subtropical regions, mostly in America. The flowers are bisexual and four-parted. They are significant economically as ornamental, weed and medicinal plants. The red-spotted *Oenothera erythrosepala* belongs to the latter category, and has probably been established in Europe from American stock *(O. lamarckiana)*.

Order: Thymeleales
Elaeagnaceae—Oleaster family

These are woody, frequently thorny plants of temperate and warm regions, having alternate (Hippophae) or opposite leaves, which are silvery or covered with lepidote hairs. The characteristic metabolic products are polyphenols and alkaloids, as well as potassium oxalate. The orange – red fruits of sea-buckthorn *(Hippophae rhamnoides)* contain lots of ascorbic acid; this is already being cultivated in several countries. The flowers are unisexual. It grows wildly on dunes of sea shores and on sandy soils in Europe.

Order: Rutales
Rutaceae—Rue family

These are mainly woody plants having oil, balm or resin passages. The leaves vary in terms of shape and standing; they are frequently provided with transparent spots and with lysogene and scisolysogene passages similar to those of bark. The flowers are

4–5-parted and bisexual or unisexual, and the fruit is a capsule or berry. Important fruit-bearing and medicinal and ornamental plants belong here; these are very important in the economy of the world, and grow in the warmer (from the tropics to the Mediterranean) regions.

Several species contain alkaloids and coumarins (furano- and pyranocoumarin). *Ruta gravolens* originating from the eastern Mediterranean region is cultivated as a medicinal plant in Hungary.

Order: Rutales
Anacardiaceae—Cashew family

These are tropical and mainly Mediterranean woody plants; the leaves are generally alternate, and estipulate. The flowers are small and unisexual or bisexual. These are important ornamental and medicinal plants. The frequent occurrence of polyphenols and tannins is characteristic of the family. A rich tannin source is the leaf (Cotini folium) of the smoke-tree *(Cotinus coggygria)*. This species lives wild in south-east Europe and in the region up to and beyond the Caucasus.

Order: Rhamnales
Rhamnaceae—Buckthorn family

These are woody plants, with small flowers and simple leaves and stone-fruits. They grow mainly in warm regions. Their active substances are the anthroglycosides. The widely growing species in Hungary are the alder buckthorn *(Frangula alnus)* and the common buckthorn *(Rhamnus catharticus)*. The former accumulates its active substance in shell bark and the latter in the stone-fruit. The rhamnose sugar was also discovered here. Excellent drugs are prepared from the North American *Rhamnus purshiana* (Rhamni cortex) and the south-eastern European *Rh. fallax*.

Order: Araliales
Umbelliferae (Apiaceae)—Parsley family

This is a well distinguishable, so-called natural family. The species living in Hungary are herbaceous, and most of them have compound umbel inflorescences. The leaves are mostly deeply parted with sheath-like bases. The flowers are five-parted, from the two carpels the developed fruit is a special twin achene. Generally the whole plants contain essential oils in scysogen passages. They also contain coumarin materials (e.g. *Angelica*) (some of them are photosensitive), and, additionally, organic acids (e.g. malic acid, valeric acid), resins and alkaloids (hemlock—*Conium maculatum*). They are widespread mainly in temperate regions. Most of them grow on firm ground. Several species are giant. Some of them are well adapted to saline soils and to an aqueous lifestyle. They are significant food, ornamental, medicinal and weed plants. The species *Coriandrum sativum* (coriander) is diploid, $2n=22$, and originates from the eastern Mediterranean region. *Carum carvi* (caraway) is diploid, $2n=20$, 22 and a native of Eurasia. *Pimpinella anisum* (anise) is diploid, $2n=18$, 20 and originates from the eastern Mediterranean region; it is cultivated. *Foeniculum vulgare* (fennel) is diploid, $2n=22$, and a native of the Mediterranean region. *Anethum graveolens* (dill) is diploid, $2n=22$, and is a native of southern Asia; it is cultivated. *Angelica archangelica* (angelica) is diploid, $2n=22$, and a native of Eurasia; it is cultivated, and it has escaped cultivation to grow wild. *Levisticum officinale* (lovage) is diploid, $2n=22$, and originates from the eastern Mediterranean region; it is cultivated as a medicinal and spice plant. Cumin *(Cuminum cyminum)* is a spice plant which is also a native of the Mediterranean region. In the related *Araliaceae* family, species with similar active substances can be found.

Order: Dipsacales
Caprifoliaceae—Honeysuckle family

Most of these are woody plants growing in temperate regions. The leaves are opposite, the flowers are gamopetalous, and the fruit is a false berry. They are popular ornamental plants, and several of them are fruit-bearing and medicinal plants. The European elder *(Sambucus nigra)* is a shrub which is independent of associations, is generally distributed in Europe, and likes nitrogen. Its dried inflorescence contains the drug (Sambuci flos), and it contains cyanogen glycoside (sambunigrin).

Order: Dipsacales
Valerianaceae—Valerian family

These are mostly herbaceous, internoded plants with estipulate, opposite leaves. The flowers are small, grouped into dense paniculated inflorescences and may be unisexual monoecious or dioecious *(Valeriana dioica)*. The fruit is acorn-like (achene). The fragrant essential oil accumulates in the hypodermal cells of the roots (rhizome) of the plants. Among the nine genera and 300 species of the family, several *Valeriana* species are medicinal plants: *V. sambucifolia* (polyploid, $2n=56$), *V. officinalis* (diploid, $2n=14$) and *V. collina* (diploid, $2n=28$).

Order: Gentianales
Apocynaceae—Dogbane family

These are mainly the woody plants of tropical forests (liana), they are sparsely herbaceous, and the leaves are opposite, whorled or alternate, simple, entire-edged, and estipulate. The flowers are gamopetalous; the petals are fringed in the buds; the carpels may be free or gamopetalous; and the fruit is a follicle, capsule or berry. The species contains various alkaloids and glycosides. At least 300 genera and 1300 species live in tropical zones, while only seven species grow wild in extratropical regions. Two important medicinal species are exploited in Hungary; i.e. the wild-growing common periwinkle *(Vinca minor)* and the tropical red periwinkle *(Catharanthus roseus)*. The common periwinkle is diploid ($2n=46$), it is a native of central Europe in the Mediterranean region. It is and old ornamental plant, and a number of garden forms are known. The red periwinkle (Madagascar periwinkle) is also diploid ($2n=14$); it is cosmopolitan in the tropics. *Amsonia tabernaemontana* is a native of North America, is short-petioled, herbaceous, and has blue flowers.

Order: Gentianales
Gentianaceae—Gentian family

These are also herbaceous or shrub-like plants, with opposite leaves and bicollateral transporting bundles. They are generally distributed in every climatic region. For the preparation of the drug the aerial part of erythraea (*Centaurium minus),* Centaurii herba, and *C. vulgare* (uliginosum) and *C. pulchellum* can also be applied. They contain bitter substances (e.g. erythrocentaurin), similarly to several *Gentiana* species. *Centaurium minus* is a polytypic, polyploid ($2n=42$) species, avoids lime, and is often found at the borders of forests and on meadows and grazing fields.

Order: Lamiales
Labiate (Lamiaceae)—Mint family

These plants are very variable in habitat and life form (Th, TH, H, CH, N, MM); they are often characteristic subshrubs. The stem is generally four-edged, the leaves are opposite, the inflorescences are clustered in the axils of leaves, and the flowers are zygomorphic. The fruit falls into four partial fruits. They contain an essential oil which has a pleasant, often strong, odour. The essential oil accumulates in glandular hairs or in glandular bracts. The species of several families (e.g. Ajugeae, Stachyeae) do not contain or contain very little essential oil. The appearance of bitter substances (e.g. *Marrubium vulgare*), polyphenols and tannins is not scarce in the family. The alkaloids are infrequent; leonurin can be found in *Leonurus cardiaca*. The 160 genera and 3000 species can be found almost anywhere in the world, especially in regions with Mediterranean features. A number of very peculair and honey-producing plants belong to this family. The more important medicinal plants are as follows: lavender *(Lavandula angustofolia)* ($2n=36$, 54); the Mediterranean common balm *(Melissa officinalis)* is diploid ($2n=32$), cultivated, and can be found wild as well; the sage *(Salvia officinalis)* is a native of the eastern Mediterranean region ($2n=14$, 16); the clary sage *(Salvia sclarea)* is a native of the Mediterranean region ($2n=22$); the summer savory *(Satureja hortensis)* originates from the eastern Mediterranean region ($2n=46$, 48); these species are cultivated.

Hyssop *(Hyssopus officinalis)* is a cultivated plant which has many forms ($2n=12$), and is native to the western Mediterranean region. Marjoram *(Majoranna hortensis)* is cultivated in Hungary ($2n=24$), and originates from the Mediterranean region. The common thyme *(Thymus vulgaris)* is of Mediterranean origin ($2n=30$), and mother-of-thyme *(Thymus serpyllum)* grows wild, and is relatively common. The hybrid peppermint *(Mentha piperita)* does not grow wild ($2n=48$). The sweet basil *(Ocimum basilicum)* is of Indian origin ($2n=48$) and *Leonurus lanatus* is a medicinal plant native to Siberia and central Asia.

Order: Solanales
Solanaceae—Nightshade family

This is basically a herbaceous plant family including several subshrubs *(Solanum dulçamara)*, shrubs or small trees, the main area of which is Central and South America. The leaves are alternate, and the flowers are five-parted, the calyx characteristically persisting. The fruit is a capsule or a berry. The transporting bundles are bicollateral. The family is rich in alkaloids. The nitrogen-containing active substance of *Capsicum* species is capsaicin. The 75 genera and 2000 species can be found mainly in the tropics. As medicinal plants the genera *Hyoscyamus, Atropa, Datura* and *Solanum* are significant in Hungary; a rare mountainous species is *Scopolia carniolica*. Henbane *(Hyoscyamus niger)* is a native of Eurasia (Mediterranean) and requires lots of nitrogen; its usage is declining (diploid, $2n=34$).

Belladonna *(Atropa belladonna)* grows mainly on calcareous soils which are rich in nitrogen, and can be found frequently in clearing of beech and hornbeam forests (H; polyplolid, $2n=50$, 72). The Australian, originally shrubby perennial nightshade (*Solanum laciniatum*, Th; $2n=48$, 92) and the tropical American thorn-apple (*Datura innoxia*, Th; $2n=24$) are cultivated medicinal plants.

Order: Scrophulariales
Scrophulariaceae—Figwort family

These are herbaceous plants, subshrubs, shrubs or trees with alternate, opposite or whorled leaves. The flowers of the ancient types are actinomorphous or slightly zygomorphous. The fruit is a capsule containing lots of seeds. The glycoside content is high, and some species contain saponins as well. The 3000 or so species can be found in every zone of the world. Among the wild-growing genera in Europe, the gathered medicinal plants are certain *Verbascum* (mullein) species; mainly mullein *(V. phlomoides)*, which is independent of associations, likes nitrogen and is a native of south-eastern Europe (TH; polyploid, $2n=32$), and the species *V, densiflorum* and *V. thapsus* (common mullein). Two species of the genus *Digitalis* (foxglove) are cultivated: the purple foxglove *(D. purpurea)*, which is a native of the Atlantic side of the Mediterranean region (TH-H; $2n=112$) and the Grecian foxglove *(D. lanata)*, which originated from Balkanian-Pannonian, territories likes calcareous soil (TH; polyploid, $2n=56$), and can be found on dry, sunny hillsides.

Order: Plantaginales
Plantaginaceae—Plantain family

These are generally distributed annual, herbaceous, perennial, or sometimes shrub-like plants with conspicuous parallel veins on the leaves. The flowers are small, forming heads or spike inflorescences, anemophilus and the fruit is a capsule.

These are medicinal, weed and wild-growing plants. The dried leaves of *Plantago lanceolata* and *P. major* are the drug, which involves iridoid compounds (aucubin). Other species are also gathered, e.g. *P. altissima* and *P. media*. The epidermis cells of the seeds of *P. psyllium* contain mucilaginous substances.

Subclass 3: Malvidae Dilleniidae

Order: Malvales
Tiliaceae—Linden or Basswood family

Generally these are woody plants with leaves standing in two alternate rows. The actinomorphous flowers form a cyme inflorescence. In the case of the genus *Tilia* (linden), the peduncle has grown onto the bract. The stamens are free or bundled; from the pistil open or closed fruits (capsule, berry or stone-fruit) develop. The mucilaginous substance is accumulated in mucilage-holding cells. The flower or the linden contains mucilaginous substances. A small amount of essential oil, flavon glycosides and very little saponin. The 35 genera and 350 species can be found mainly in tropical areas. In Europe several species of *Tilia* can be found either wild or planted, such as the large-leaved linden *(Tilia platyphyllos)*, the small-leaved linden *(T. cordata)* and the silver linden *(T. tomentosa=argentea)*. All the tree species and their intraspecific taxa give valuable drugs.

Order: Malvales
Malvaceae—Mallow family

The plants of this family include herbaceous shrubs and trees. The leaves are alternate, and palmately lobed. The flowers have five petals, and the filaments of the stamens are joined to form a tube (Columniferae). The fruit is a multi-seed capsule, and consists of numerous carpels each containing seed. Some species contain mucilage substances, similarly to the linden family. For this purpose the root of marsh mallow *(Althaea officinalis)* is mainly used; this grows wild mainly in humid areas, is independent of associates, and is polyploid ($2n=42$). Other medicinal plants containing mucilage are the Mediterranean hollyhock *(A. rosea)*, which is a cosmopolitan plant, *Malva sylvestris* (polyploid, $2n=42$), and the Algerian subspecies of the latter, *M. sylvestris* subsp. *mauritiana*. 'Rama' *(Hibiscus sabdariffa)* is a tropical medicinal plant.

Order: Cucurbitales
Cucurbitaceae—Gourd family

These are mainly creeping, wild-growing plants of the tropics, generally annual. The laying stems cling with tendrils, and the nodes easily get rooted. The flowers are unisexual, and mostly monoecious. The leaves on the stem are united. The ovary consists of three locules, inferior, and the placentation is parietal. The seeds contain no endosperm, their shape is determined by the characteristic seminal leaves, and they contain lots of oil. The fruits are often bitter, which is caused by cucurbitacins and by the tetracyclic triterpene saponins. Several genera also contain alkaloids (lufanin, bryonicin).

These are ornamental, food (vegetable, fruit, oil) and odour plants. One of the cultivars, the so-called oil pumpkin is a cultivar of field pumpkin (*Cucurbita pepo* convar. *pepo* provar. *styriaca*).

Order: Capparales
Cruciferae (Brassicaceae)—Crucifers or Mustard family

These plants are widely distributed, mostly herbaceous, (Th, TH, H) and have a four-stamened flower structure and four sepals, which are very characteristic of the family; it is a so-called 'natural' family. The leaves are generally alternate; in the vegetative parts the two branched stellate hairs are also frequent. The fruit is a pod or silicle. The myrosin cells are characteristic, and contain myrosinase enzyme. Many basic food, dodder, ornamental, medicinal, weed and honey-producing plants belong to the family (3000 species). The seed of the black mustard *(Brassica nigra)* ($2n=16$) is a source of pharmaceutical raw material (sinigrine). By processing the seeds of the white mustard *(Sinapis alba)* table mustard is produced. The latter plant is a native of southern Eurasia, and is cultivated (Th; diploid. $2n=24$).

Lepidium sativum is cultivated for obtaining spice; the tips of shoots are consumed in a raw state.

Order: Primulales
Primulaceae—Primrose family

Comprising this family are herbaceous (Th-H) and sometimes shrub-like plants, which grow mostly in the northern temperate regions. The main characteristics are: the leaves are generally entire, the five-parted perianth is united, and the seeds are born on the central axis in the ovary. Usual active substances are the triterpene saponins. Saponin drug is prepared from the roots (rhizomes) of cowslip *(Primula veris)* and of oxslip *(P. elatior)*. Toxic saponin is involved in the wild-growing *Cyclamen purpurascens* (cyclamen). A number of species accumulate scented material in their roots.

Order: Asterales
Asteraceae (Compositae)—Composite family

Generally these are herbaceous (Th, TH, H), rarely lignified stemmy plants, the leaves of which are alternate or opposite. A characteristic part of the plants is the capitulum/pseudanthium inflorescence, which is surrounded by a group of bracts characteristic of the species, making up the involucre. The calyx is reduced, the corolla is united, five-parted, actinomorphous (Tubuliflorae) or zygomorphous (Liguliflorae), the filaments of the stamens are grown on the corolla, and the anthers are attached to each other, forming a tube. The pistil is formed by two carpels. The ovary is inferior, having one ovule. The fruit is an achene; the seeds contain no supplying tissues. The more important metabolic products can be applied in medical practice as active substances, such as oils, bitter substances, caoutchoucs, mucilages, waxes, glycosides, alkaloids, etc. Among the 20 000 species of the 900 genera, many live in Europe and are cultivated. Roman camomile *(Anthemis nobilis)* is of Mediterranean origin (H; $2n=18$); common yarrow *(Achillea millefolium)* is rich in forms, grows wild and is cosmopolitan (H; polyploid, $2n=54$); German camomile *(Matricaria chamomil-*

la=*M. recutita)* is a native of southern Eurasia, grows wild and is also cultivated in Hungary (Th-TH; $2n=18$); Dalmatian pyrethrum *(Chrysanthemum cinerariaefolium)* is a native of the Mediterranean region, and is cultivated in Hungary (H; $2n=18$, 36); tarragon *(Artemisia dracunculus)* is a Siberian spice and medicinal plant (H; $2n=18$); wormwood *(Artemisia absinthium)* is Mediterranean, contains bitter substances, and is often found in Hungary (Th-H; $2n=18$); pot marigold *(Calendula officinalis)* is an ancient Mediterranean, ornamental and medicinal plant (Th-TH; polyploid, $2n=28$, 32); blessed thistle *(Cnicus benedictus L.)* is Mediterranean, and is cultivated in Hungary (Th; $2n=22$); Variegated thistle *(Silybum marianum)* is a native of East-Mediterranean, in Hungary it is only in cultivation, it is herbaceous (Th-TH; $2n=34$).

Subclass 5: Eucommiidae = Hamamelidae
Order: Urticales
Cannabidaceae—Hemp family

These plants have bristly hairy shoot systems, and alternate or opposite stipulate leaves. They are dioecious, and wind-pollinated. The fruit is an acorn. Two genera are classified here: hop *(Humulus)* and hemp *(Cannabis)*. The European hop *(Humulus lupulus)* is human food, and a medicinal plant. The flowers are covered with bracteoles which accumulate resin and bitter substances (humulon, lupulon); the tips of the shoots are used for preparing vegetables. Hemp is a fibrous plant; its seed is pressed to obtain oil. Certain varieties are cultivated for the narcotic compounds; the resinous substance is hashish, and the active substance is tetrahydrocannabinol.

Class II: Monocotyledonopsida

Subclass 1: Alismatidae
Subclass 2: Liliidae
Order: Liliales
Liliaceae—Lily family

These are generally distributed perennial, herbaceous, rarely xerophyte tree-sized and liana plants. The characteristic parts serving vegetative propagation are, bulb, tuber, rhizome and viviparia. The flower is three-parted, the perianth generally is a coloured perigone, and the fruit is a capsule or berry. The most important active substances are the alkaloids (colchicine, veratrine), the glycosides and the saponins.

Many species of the family are medicinal plants, which are cultivated or gathered to various extents, as for example the species wild saffron *(Colchicum autumnale)*, hellebore *(Veratrum album)*, lily of the valley *(Convallaria majalis)* and onion *(Allium-sativum)*.

Subclass 3: Arecidae
Order: Arales
Araceae—Arum family

These are mainly terrestrial (tuberous, rhizomic, rosette, or giant, climbing lianas) or epiphytic plants of tropical regions (South America). The flowers are unisexual or bisexual, clustered in a dense spadix, which is surrounded by an insignificant or coloured spathe. The leaves are often provided with net-like vein systems. This family has lots of species; the 100 genera include about 1500 species. These are important ornamental, food and medicinal plants. An uliginal medicinal plant is the sweet-sedge *(Acorus calamus)*, which is a native of the East Indies and lives escaped in some places of Europe along the edges of water. Its rhizome contains oil.

The distribution of medicinal plants, gene centres

De Candolle was the first to process the information about the origin of domesticated plants. His main aim was to relate the studied plants to their ancestors. For this purpose he used two main principles: (1) the domesticated plant is cultivated at the location where its wild intraspecific taxa grow, and (2) the archaelogical, historical and language evidence for the relationship between the two types of intraspecific taxa. With this method he has processed 247 Old World and 45 New World cultivated plants. He called this method the 'phytological method'.

In the opinion of Darwin the number of ancient wild forms has significantly declined, and the relationship is impossible to reveal in a number of cases.

The study of the origin of cultivated plants was carried out by the expeditions and studies of Vavilov. The following aspects were taken into consideration for distinguishing the locations of origin: (1) the number of species and varieties; (2) the variations within the species (the concentration of varieties); (3) the features and level of morphological parameters; (4) the variation of areas of species; (5) the number of endemic species; and (6) the assumption that the genetically dominant features indicate the original centres, and the recessive forms indicate the borders of areas.

Primary and *secondary* locations of centres are distinguished by Vavilov. The secondary centres of intraspecific taxa are those areas to which the cultured plants have been taken from the original (primary) centre.

Although the wild intraspecific taxa of the plants did not originate in the secondary centres, they behaved in a very short time as in their original area (the establishment of heterogenous, resistant populations), e.g. the Chinese soy bean in the secondary centre in North America.

Vavilov named eight primary centres of establishment of cultured plant varieties, namely: (1) China; (2) India, (a) specific India, (b) India–Malay; (3) Central Asia; (4) Near East; (5) Mediterranean; (6) Abyssynian; (7) South Mexico and Central America; (8) South America, (a) Peru, Equador, Bolivia, (b) Brasil, Paraguay.

Two additional centres were suggested by the Hungarian RAPAICS; (9) Europe and (10) North America. Other classifications are also known (details can be found in the works of *Zeven-Zhukovsky*).

Medicinal plants are probably as old as human activity. The history of their origin and domestication has not been elucidated, as De Candolle and Vavilov investigated the centres of food and spice and the more important consumer goods (e.g. coffee and tobacco). Also, most of the medicinal plants were only gathered in Europe, so they have not reached the same level of domestication.

The areas of exploitation and distribution of medicinal plants depend on two factors: the first being the areas and locations of the establishment of plant families (or genera), and the second being the fact that the mentioned areas are identical to the gene centres, where the ancient culture centres have appeared, like China, India, the near-East and Central America.

If the distribution of the largest units (families) is considered in relation to the plant kingdom, then it can be concluded that families of the most important species (like Apiaceae, Papaveraceae, Rosaceae, Ranunculaceae, Helleboraceae, Cupressaceae, Cruciferae and Valerianaceae) are mostly temperate zone families (holarctic); their species grow in the holarctic flora regions.

Other significant medicinal plant families are widespread in warm zones, in the tropics and in the Mediterranean region, e.g. Solanaceae, Rutaceae, Apocynaceae, Labiatae, Rhamnaceae, Fabaceae, Tiliaceae, Malvaceae.

The East Asian (Chinese) gene centre is rich in medicinal plant species (e.g. *Panax pseudoginseng*). *Anethum* and *Sinapis alba* originate from the southern part of the Central Asian centre.

Ocimum is a native of India. It is remarkable that most of the cultivated medicinal plants originated from the eastern part of the Mediterranean centre and from the Near East gene centre, e.g. *Coriander, Pimpinella, Foeniculum, Ruta, Salvia officinalis* and one of the ancient cultivated plants, *Calendula officinalis. Hyssopus, Thymus* and *Digitalis purpurea* have reached Hungary from the western part of the Mediterranean area.

Relatively few tropical plants are cultivated in Hungary as yet. The domestication of tropical plants is closely related to the type of life form and the possibility of changing the life form (e.g. from perennial to annual). *Catharanthus roseus* is also a cosmopolitan plant, while the Australian *Solanum laciniatum* is more or less subtropical. Among the plants of the European centre, medicinal plants have not been cultivated for a long time; e.g. camomile, periwinkle (*Vinca minor,* southern and central European). Grecian foxglove (*Digitalis lanata,* originated from Pannonian–Balkanian territories) and *Artemisia absinthium.*

The origin of gathered (wild-growing) medicinal plants can be investigated according to their natural areas. As the areas are generally constant (except the areas of several species, like *Hyoscyamus niger*) area groups, the so-called flora element groups can be distinguished according to the geographical location, where a number of species have the same area.

A significant number of medicinal plants cultivated in Europe belong to the 'flora element groups' of Europe, Eurasia, and circumpolar and Mediterranean regions.

Environmental Factors

In recent years more attention has been turned to *enlarging yields through biomass production.* Although a considerable amount of information has been published (Lange *et al.,* 1983) the principles and practices characterizing and describing primary plant production cannot be directly adapted to the production of special plant compounds. The 'appearance' of special products in the production system implies the presence of special circumstances beyond general production considerations (Harborne, 1977).

In agreement with the general research strategy of modern ecology (Larcher, 1975), the starting point for work on secondary plant products is the study of the overall production system itself. The regulatory techniques developed depend to a great extent on the ecosystem where the plant will be grown. For the production of secondary plant products we can distinguish: first, natural or quasi-natural ecosystems being partially under human influence; and second, agricultural ecosystems 'producing' secondary products. The main differences between the two types of system are well demonstrated by the modified scheme of Gareth (1979) (Fig. 15).

As well as the ecological aspects, however, the following factors should also be taken into consideration for the interpretation of ecological influences, owing to the 'secondary' characteristics of special substances.

(1) The generation of active substances (special metabolic products) is basically connected to the metabolism; in this way the influence of ecological factors occurs via the universal processes. For this reason the harmony between the influence and the ecological requirement of the species cannot be neglected. The same active substances may be generated in ubiquitous and stenotopic species having completely different requirements (steroids), as well as in hydrophyte and xerophyte species (linalool).

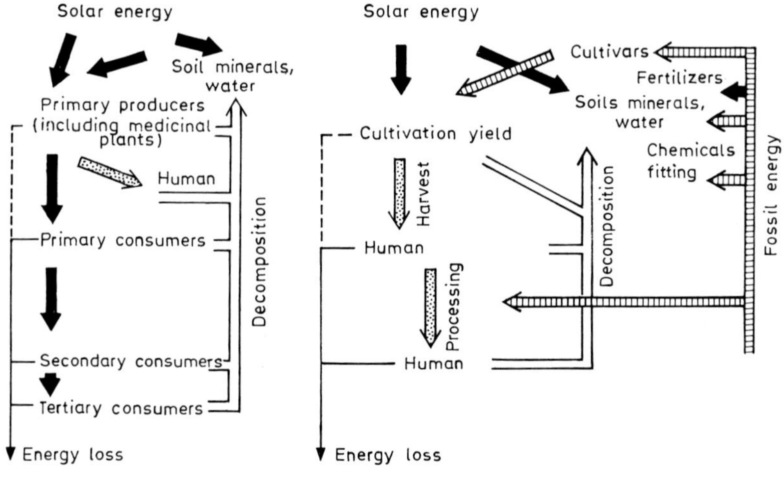

Fig. 15 Two main types of ecosystems producing secondary plant products characterized by the modified scheme of Gareth (1975). Under natural or quasi natural conditions (a) the harmonic utilization of solar energy is required. In agricultural systems (b) the optimal input of fossil energy has a great importance as well

(2) Chemotaxonomy present new requirements for the studies of ecological influences, especially for the recognition of chemical differentiation within the species (Tétényi, 1970). Even within the species, the differentiation of ecological requirements might be expected.
(3) The influence of ecological factors on the generation of active substances is complicated also by the fact that the amounts and ratios of active substances change during ontogeny (e.g. the amounts and ratios of supplementary alkaloids in the poppy). It is not the same if the tests are made before or after the differentiation time of active substance formation, which is characteristic of the intraspecific chemical taxon of the given species, i.e. different results can be expected in the initial period of ontogeny than in the phase when the plants contain the formed active substances.
(4) The influences can be evaluated only by the simultaneous application of general ecological and chemical methods. In this respect the detectability of special substances is of particular importance. Changes can be expected only within the limits determined by genetic conditions, and this often requires the measurement of very small differences in the amounts and quality of active substances.

According to the published results and the experiences of the authors, the external factors influence the medicinal plant from two different directions, i.e.

(1) As limiting factors, by their extent compared to the optimum;
(2) By their extreme values as stress effects.

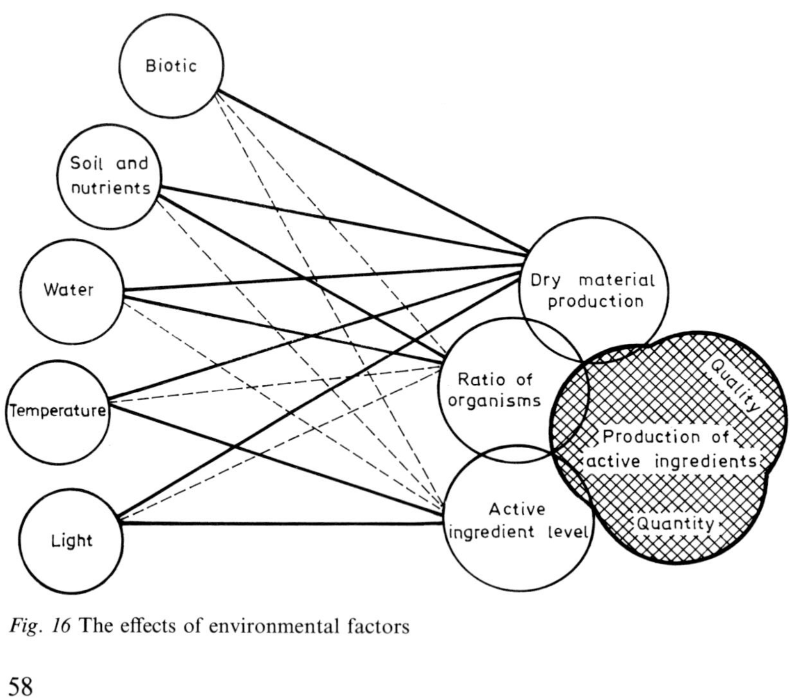

Fig. 16 The effects of environmental factors

The formation of special substances of medicinal plants and the production there of are modified by the extent of the factors relative to the optimum, which influence the growth and development, as shown by the scheme illustrated in Fig. 16. It can be seen that the external factor might succeed via the modification of dry substance production, as well as by the influence of the ratios of plant parts or by the level of active substance. Changes in dry substance production can be observed most easily. The increase in production is proportional to the increase in active substance yield. The difference between the 'medicinal' plants and other culture plants is less in this respect. As the formation of active substances is different in various parts of the plants, the changes in the ratios of plant parts, resulting from ecological influences, may also be significant modifying factors. With regard to the plant parts containing active substances, it is important how a certain factor influences these parts within the gross production. The level of active substance has been at the centre of attention since the start of medicinal plant ecological research. A very difficult problem is that the genetically determined limits and the extent of ecological variability are hard to differentiate. The problem is further complicated by the qualitative and quantitative differences depending on the growing stages.

The above changes are generated by the complex influences of environmental parameters among natural conditions, and in this way, for example, the increase of total yield is accompanied also by the modification of ratios between the plant parts. Similarly, the change in the level of active substance seldom occurs without the modification of total production and ratios of parts. The problem is still further complicated by the fact that it is not easy to determine the environmental factor which is responsible for the changes. In this field, however, significant progress has been achieved by the application of up-to-date methods, e.g. the phytotron.

The effects of the main environmental factors can be reviewed by the consideration of these results.

Light

In terms of ecological considerations, the quality, intensity and duration of light are important. As well as the general formative, photoperiodic and trophic effects, all of the three components influence the formation of special metabolic products.

Quality of light

After studying the effect of the quality of light it can be stated that unequivocal positive results have been achieved by applying short-wave ultraviolet light to the plants containing alkaloids (*Nicotiana tabacum, Atropa caucasica* and *Datura godronii*). In the latter species the higher alkaloid content is accompanied by increased dry material production. Studies made with the short-wave range of visible light (blue) confirm the above results. Kuznietsova and Hazanov (1973) have shown that short-wave blue light increases the alkaloid levels of *Stephania glabra, Rauwolfia canescens* and *Solanum laciniatum*. The same results were achieved in Hungary when studying

Solanum laciniatum and *S. dulcamara.* In this latter species the ratio of the main aglycone has also been modified, from soladulcidine in the direction of solasodine. As opposed to the above results, Pavlov and Ilieva (1972) emphasize the disadvantageous effects of short-wave light on the plants containing essential oils.

Intensity of light

According to the available literature data, the light intensity has a more pronounced effect on the ecotypes than the spectral composition of the light. In one study the light intensity was reduced to 1600 lux from 10 000 lux. In this case the amount of steroid aglycone has been reduced only in the heliophyte *S. laciniatum* (to a quarter), while the amount remained the same in the shade-enduring *S. dulcamara.*

Studies using the phytotron have demonstrated the role of light intensity in the formation of morphinic alkaloids (Fig. 17) (Bernáth and Tétényi, 1979, 1980, 1981).

By increasing light intensity—along the saturation curve—we can observe that the amounts are larger for both alkaloids. There is, however, a dissimilarity, which is the result of chemism, i.e. the morphine increases faster in the Hungarian cultivar 'Kompolti M', while the codeine increases faster in the English 'Reading' cultivar. This will give rise to a modification, which cannot be neglected in terms of chemotaxonomy (the cultivar 'Reading' is pronouncedly morphinic under a light intensity of 16 000 lux, while it assumes codeinic character if the light intensity is over this value).

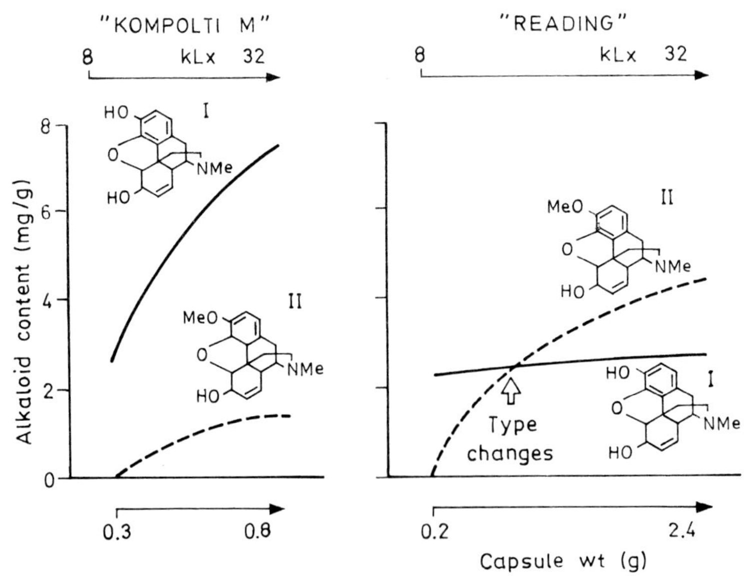

Fig. 17 Ecological and chemotaxonomical differences in morphine–codeine saturation curve of cultivars 'Kompolti M' and 'Reading' which are caused by simultaneous increases in capsule weight and light intensity. I, morphine; II, codeine

Table 3 Influence of light supply on the accumulation of essential oil in peppermint (Hornok, 1980)

Lighting	Essential oil		Content of essential oil (%)		
	Content (%)	Yield (kg/ha)	Menthol	Menthone	Menthone acetate
Full (natural) lighting	1.442	23.82	61.8	18.8	9.6
10 h natural lighting daily	0.647	8.46	49.2	24.2	20.2
Reduced (by 40%) light intensity	1.092	14.02	54.8	25.8	12.0
5% significant difference	0.267	5.12	3.1	1.7	2.5

Balbaa, Hilal and Haggag (1971) have also confirmed the connection between the increase of light intensity and the production of glycoside for *Digitalis lanata*. Camomile was studied in the phytotron by Saleh (1973), who has revealed that if the light intensity is decreased, the ratio of essential oil content and camasulene content decreases, together with the number and size of flowers. Similar results have been achieved by Hornok (1980) (Table 3) in the case of peppermint. There is, however, a number of dissimilar examples as well.

Duration of light (photoperiodic effect)

The naturalization of subtropical and tropical plants might be prevented by the short duration of light in temperate regions. The studies made by Cosson (1966) in the phytotron have also confirmed the direct influence of duration of light on special metabolic products.* Hyoscyamine, which is characteristic of the species, however, was formed only when the duration of light had been reduced to 9 hours.

Karnik (1972) has revealed the advantageous effect of long daylight treatment on the amount of sapogenins in *Dioscorea deltoidea*. The duration of light and the essential oil contents of various plants have been correlated by a number of investigations revealing positive effects, e.g. by Franz et al. (1975) in the case of camomile, Pavlov and Ilieva (1972) in the case of clary sage and Ruminska (1973) and Skrubis and Markakis (1976) in the case of thyme and common basil. This effect is also confirmed by the results obtained by the authors (Hornok, 1980) in the case of coriander and peppermint.

* When *Datura tatula* was under 16 hours of light in the flowering phase, scopolamine production was higher.

Temperature

The temperature is a basic influencing factor on the area and life form of plants, but it also has a substantial effect on the production of plants in given climatic conditions. In spite of this fact, exact data are scarce.

The connection between the heat requirement and the accumulation of active substances was investigated in the case of *Solanum laciniatum,* as well as in the case of *S. dulcamara*. Corresponding to results achieved in open-air fields, the content of aglycone in the phytotron shows a temperature-dependent and species-dependent relationship (Fig. 18). Saleh (1970) investigated the essential oil content of *Matricaria chamomilla*. The highest amount of essential oil was formed at a constant temperature of 25 °C, or when the night temperature was reduced to 15 °C. The ratio of camazulene was unquestionably the highest if the night temperature was 15 °C.

A number of publications deal with the investigations performed under natural conditions. For the interpretation of the results of these investigations one should consider the complexity and relativity of the investigations and the lack of knowledge of lower and upper limits. This is exemplified by the results relating to poppy. Brükin (1971) found a positive relationship between the temperature of the growing season and the formation of morphine, while Dános (1968) and Vágújfalvi (1968) present data questioning the validity of this relationship. According to Hotin (1968) the essential oil content of peppermint is produced proportionally to the daily mean temperature prevailing in the intense growing stage. In the opinion of Matusiewicz (1960) the essential oil content can be increased significantly by raising the mean temperature of the days prior to the harvest by 2–3 °C. Hornok (1980) has experimental results indicating that the essential oil content of dill increases proportionally to the temperature of the days prior to cutting (between 15–24 °C) up to 100 per cent. The essential oil content of coriander and anise in the stage of fruit development is less at a temperature over 19 °C than at one between 15 and 19 °C. It can thus be

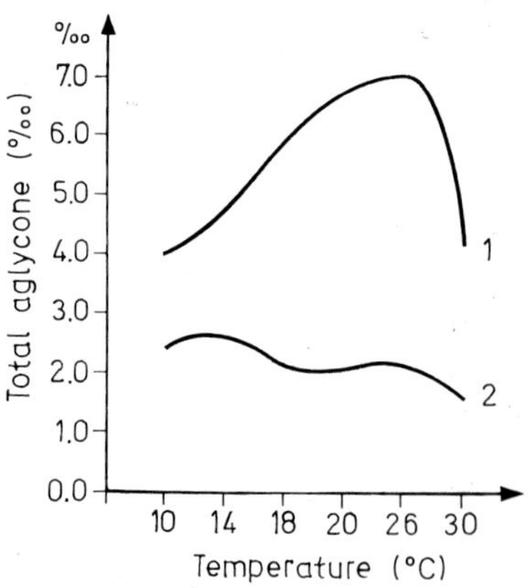

Fig. 18 Effect of air temperature on the solasodine content of leaves of species with different ecological requirements; 1, *Solanum laciniatum;* 2. *Solanum dulcamara* (Bernáth, 1976)

assumed that the influence of temperature is relative, the active substance content increasing or decreasing possibly according to the optimum temperature of the given species.

Water

A number of modes of adaptation to the water supply are known. The connection between the special metabolic products and the water supply depends strongly on the species. The effect of water supply on the essential oil content of the hydrophyte *Mentha* will be significantly different from that on the xerophyte *Lavandula* species.

Owing to the relatively low rainfall in the growing season and to the unreliable distribution of precipitation, supplementary watering is generally advantageous in Hungary, except perhaps for most of the xerophyte species. The watering mainly increases the production of dry substances, and it does not influence the amounts and qualities of special substances significantly, except in a few cases.

In the case of *Solanum laciniatum* and *S. dulcamara,* Földesi (1964), Kurszanov and Pikova (1971) and Máthé and Máthé (1973) have shown that optimal watering increases the yield of drug. The same conclusion was reached by Kerekes and Hornok (1972) for peppermint. In certain cases the essential oil content was increased by 0.3–0.5 per cent by an optimal water supply. According to Kerekes (1962), the changes in the drug yield are most significant in the case of camomile; however, they also presented data relating to the changes in amounts of essential oil and to the differences between essential oil compositions. Many problems are as yet unanswered. For example, Penka (1968) states that watering will increase the morphine content of poppy, while Dános (1968) thinks that the durable dry period is advantageous.

Soil

The soil has a complex influence on plants through its physical, chemical and biological properties. It is not easy to interpret the influences of separate factors. The soil types (even the base rocks) can cause macroclimatic or mesoclimatic differences, which presents a special problem when comparing the various plant locations. Most of the methodical investigations have dealt with the influence of nutrient supply, but these are not unequivocal either. The connection between the nutrient and the dry material can be regarded as being well understood. The results dealing with active substances are often controversial (slight decreasing and increasing influences were reported for the same species, and even for the same varieties). The reason for these controversial results is that the nutrient supply is a complex phenomenon; under open-air conditions the take-up and assimilation of nutrients is determined by the joint action of many physical, chemical and biological processes.

Vágújfalvi (1967) investigated the mineral nutrient supply of *Datura innoxia* and obtained a notable result; with the application of extremely different nutrient dosages, plant production and the change of alkaloid level occurred according to the optimum curve (Fig. 19). The optimum curves of phytomass and alkaloid level, however, do not coincide. Higher nitrogen dosages increased the mass, but the alkaloid content culminated at medium nitrogen levels. Nutrient supply slightly below or over the optimum level may cause an increase or a decrease. For example, the optimum for essential oil of *Valeriana officinalis* was reached with a medium amount of nitrogen (Fig. 20). The optimum for plant mass, however, is different for various parts of the plant (Bernáth, Földesi and Lassányi, 1973).

The research data of recent years indicate that the influence of nutrient supply on the accumulation of active substances is significant. As well the edaphic, ecological and agrotechnical factors, the extend and direction of the relationship between the nutrient supply and the production of active substances basically depends on the plant species, on the examined part of the plant, and, naturally, on the type of active substance. In certain cases the production of active substance is increased by raising the nitrogen level, but sometimes an increase in phosphorus or potassium level will give the same result. This is confirmed by the experiments performed on several plants containing essential oils (Fig. 21).

The results of investigations made in open-air fields are only valid for the test conditions and for the given soil type. The theoretical and experimental results will not relieve the cultivator from the task of adjusting their knowledge relating nutrition to the local conditions, according to which the nutrient level can be selected so as to produce the maximum amount of drug without a decrease in active substance content.

Among the chemical properties of the soil, the value of pH is the most significant. The adaptation ability of medicinal plants is very good for a wide range of pH values.

Hyoscyamus niger develops sufficiently and its active substance content does not change in the pH range of 6.4–7.9. The same is true for *Matricaria chamomilla* in the 7.3–8.1 pH range.

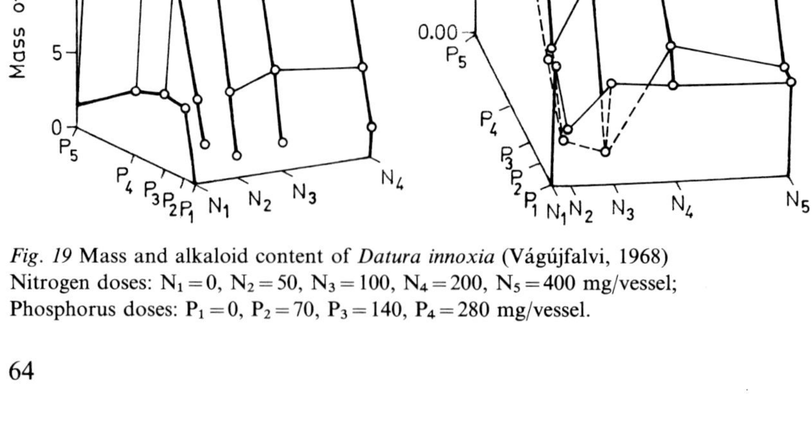

Fig. 19 Mass and alkaloid content of *Datura innoxia* (Vágújfalvi, 1968)
Nitrogen doses: $N_1=0$, $N_2=50$, $N_3=100$, $N_4=200$, $N_5=400$ mg/vessel;
Phosphorus doses: $P_1=0$, $P_2=70$, $P_3=140$, $P_4=280$ mg/vessel.

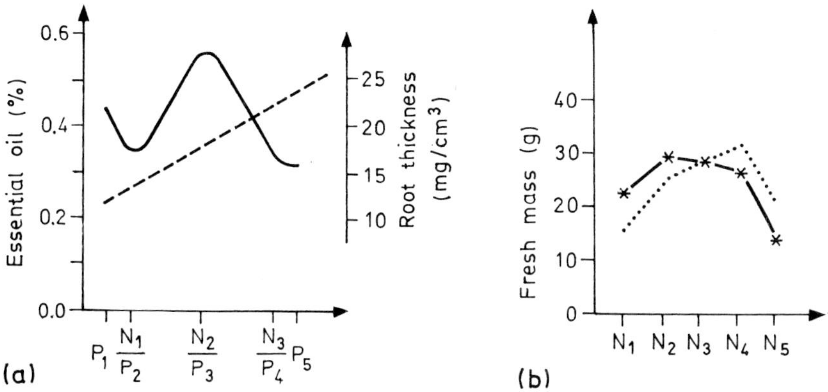

Fig. 20 Mass and essential oil content of *Valeriana officinalis* in culture vessel; (a) –, essential oil; ---, root thickness; (b) -, root; . . ., shoot (Bernáth, Földesi and Lassányi, 1973, 1975)
Nitrogen doses: $N_1=0$, $N_2=50$, $N_3=100$, $N_4=200$, $N_4=400$, $N_5=800$ mg/vessel;
Phosphorus doses: $P_1=5$, $P_2=25$, $P_3=100$, $P_4=200$, $P_5=400$ mg/vessel

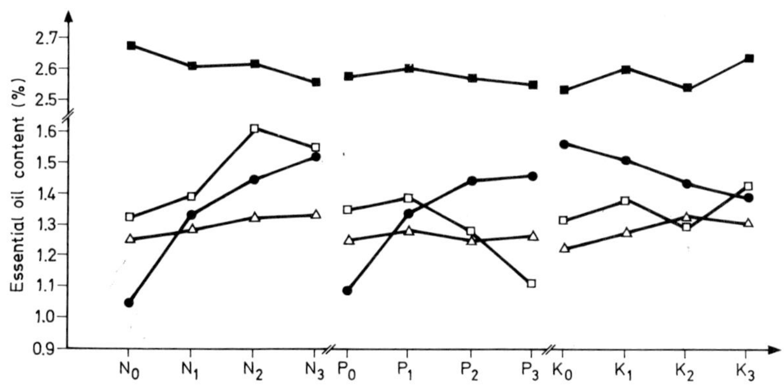

Fig. 21 Effect of nutrient supply on the essential oil content of several medicinal plants; ●—●, peppermint; □—□, dillgreen; △—△, coriander seed; ■—■, dill seed (Hornok, 1980)
Nitrogen doses: $N_0=0$, $N_1=150$, $N_2=250$, $N_3=350$ kg/ha nitrogen;
Phosphorus doses: $P_0=0$, $P_1=50$, $P_2=100$, $P_3=150$ kg/ha phosphorus;
Potassium doses: $K_0=0$, $K_1=40$, $K_2=80$, $K_3=120$ kg/ha potassium.

Growing location

Máthé and Máthé (1973) investigated the qualitative and quantitative aspects of the aglycone content of *Solanum dulcamara* as a function of geographical longitude (Fig. 22). Meyer (1973) studied *Digitalis lanata* divisions at four different altitudes (between 660 and 1840 m) and concluded that the amount of glycosides (mainly lanatozid A and B) decreases with the increase in altitude. The essential oil content of *Ocimum suave* was investigated by Rovesti (1975) in Ethiopia at altitudes between 900 and

1700 m, and significant differences in the compositions of essential oils were found. At lower locations the free or esterified linalool amount was more, while at higher locations the amount of eugenol was higher.

Other organic and non-organic factors

The plants are exposed to a number of influences externally—growth control materials, radioactive rain, industrial plant contamination—which might completely change plant associations. Most of the herbicides have physiological effects on the plants; the species differ only in their response reactions. Decreases in active substance occur only if the inhibition also decreases the production of dry material.

There is also an opposite example. Nastev and Jankulov (1975) observed an increase in scopolamine in the leaves of *Datura innoxia* under the action of Treflan, Dymid and Balan, although this was accompanied by a decrease in scopolamine in other parts of the plant.

Complex influences are also exerted on the plant by the surrounding organisms (other or same species, animals and humans). Cultivation can also be regarded as a competition taking place under controlled conditions. The competition between the same species is controlled by the allocated growing area, which is determined by considering the maximum active substance production, while the competition between different species is enhanced by the killing of weed plants. The allelopathy of special metabolic products is well known (alleopathic effect is the mutual effect of plants, although its role has not been investigated very much). Esculine, helleborine and allonine inhibit the growth of neighbouring plants through the soil, and colchicine might cause polyploidy even under cultivating conditions. Parasitism has a direct influence on the production of active substance. This relationship is a basic source of an important group of alkaloids (ergot alkaloids), although unfavourable effects are also known in the case of other organisms. According to Upadhyay (1976), cercospore infection decreases the essential oil content of common basil, and it also modifies the composition of the essential oil. Kerekes (1969) measured the solasodine content of leaves of *Solanum laciniatum* infected with virus, and found that the solasodine content was only half of the normal content.

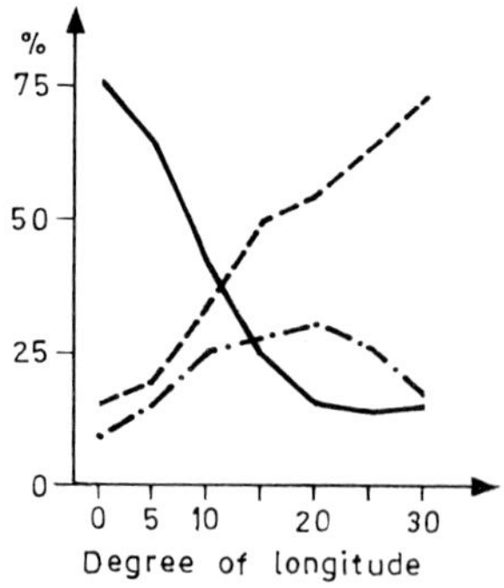

Fig. 22 Differentiation of chemism of *Solanum dulcamara* populations according to geographical locations (Máthé and Máthé 1973)

As well as the above influences, the extreme conditions of the environment, which cause stress in the plants, should also be taken into consideration in the case of medicinal plants.

These can appear either as biotic or physiochemical factors, but they influence the production of special substances via the modification of universal processes. In this respect, however, very few data are available as yet. In cultivation practice, well-known phenomena are the fast seed growing of *Papaver somniferum* in dry conditions, the shoot emergence of *Digitalis lanata,* frost affecting the generative processes of *Matricaria chamomilla,* etc. These phenomen are still to be investigated. The importance of the stress effects is emphasized by the investigations performed in connection with *Vinca minor* (Hubay, 1977). If plants are transplanted from forest to sunny areas (i.e. exposed to a stress effect), then the alkaloid content will increase by 50–200 per cent and this is accompanied by a continuous decrease in plant vigour. This measured alkaloid increase points to the possibility that the stress effects might be exploited beneficially to achieve higher production. This, of course, will not change the need to determine the role of the environment by taking the extreme and optimal values into consideration at the same time.

References

Balbaa, S. I., Hilal, S. H. and Haggag, M. Y. (1971). 'The effect of irrigation and nitrogeneous fertilizers on the growth and glycosidal content of *Digitalis lanata', Planta Medica,* **1,** 54.

Bernáth, J. (1976). Ecological factors influencing the development and active ingredients of *Solanum laciniatum* AIT. and *Solanum dulcamara.* Doctoral Thesis, Budapest, (in Hungarian).

Bernáth, J., Földesi, D., and Lassányi, Zs. (1973). 'Effects of nutrition supply and soil type on *Valeriana officinalis* L. ssp. collina (Wallr.). Changes in root organization, growth and drog yield of the plant' (in Hungarian), *Herba Hungarica,* **12** (1), 45.

Bernáth, J., and Tétényi, P. (1979). 'The effect of environmental factors on growth, development and alkaloid production of poppy (*Papaver somniferum* L.) I. Responses to daylength and light intensity', *Biochem. Physiol. Pflanz.,* **174,** 468.

Bernáth, J., and Tétényi, P. (1980). 'Alteration in compositional character of poppy chemotaxa affected by different light and temperature conditions', *Acta Horticulturae,* **96,** 91.

Bernáth, J., and Tétényi P. (1981). 'The effect of environmental factors on growth, development and alkaloid production of poppy (*Papaver somniferum* L.) II. Interaction of light and temperature', *Biochem. Physiol. Pflanz.,* **176,** 599.

Brükin, A. J. (1971). 'The morphine content of poppy-head as a function of harvesting time' (in Russian), *Bjull. Glavn. Bot. Sada Moszkva,* **78,** 27.

Cosson, L. (1966). 'Influence de eclairement sur la teneur en alcaloides des Daturas', *Herba Hungarica,* **5,** 157.

Dános, B. (1968). 'Studies on the yield of *Papaver somniferum* L. with special view on the production of alkaloid', (in Hungarian), PhD Dissertation, Budapest.

Földesi, D. (1964). 'Effect of fertilization and irrigation on shoot yield of *Solanum laciniatvm* AIT' (in Hungarian), *Herba Hungarica,* **3,** 2.

Franz, Ch., Fritz, D., and Schröder, F. J. (1975). 'Einfluss ökologischer Faktoren auf die Bildung des ätherischen Öls und der Flavone verschiedener Kamillenherkünfte. 2. Einfluss von Licht und Temperatur', *Planta Medica,* **27,** 46.

Gareth, I. (1979). *Vegetation Productivity,* Longman, London.

Harborne, I. B. (1977). *Introduction to Ecological Biochemistry,* Academic Press, London.

Hornok, L. (1980a). 'Effect of nutrition supply on yield of dill (*Anethum graveolens* L.) and the essential oil content', *Acta Horticulturae,* **96,** 337.

Hornok L. (1980b). 'Effect of environmental conditions on the production of essential oil plants' (in Hungarian), PhD Dissertation, Budapest.

Hotin, A. A. (1968). Effect of environmental factors on the accumulation of essential oils. In *Essential Oil Plants and their Processing* (Smolyanov, A. H. ed.) (in Russian), *Pishchevaya Promyshlenost,* Moscow. 35–43.

Hubay, R. (1977). 'Experiments on the cultivation of *Vinca minor* L'. (in Hungarian), *Herba Hungarica,* **2,** 101.

Karnick, C. R. (1972). 'The effect of photoperiod on steroidal sapogenins and other constituents in *Dioscorea deltoidea* Wa 11.,' *Annales of Botany* **36,** 305.

Kerekes, J. (1962). 'Effect of water on flower-yield and active substances of camomile (*Matricaria chamomilla* L.)' (in Hungarian), *Herba Hungarica,* **1** (1), 55.

Kerekes, J. (1969). *Medicinal Plant Producing* (in Hungarian), Mezőgazdasági Kiadó, Budapest.

Kerekes, J., Hornok, L. (1972). 'Data to irrigation and nutrition supply of *Mentha piperita* L.' (in Hungarian), *Herba Hungarica,* **11**(3), 39.

Kursanov, V. V., and Pikova, L. A. (1971). 'Productivity of nightshade under different conditions' (in Russian), *Sbornik Nauchnykh Rabot,* 20–40.

Kuznietsova, G. K., and Hazanov, U.S. (1973). 'Effect of light-spectrum on photosynthesis of some alkaloid-plants' (in Russian), *Fiziol. rast.* **3,** 554.

Lange, O. L., Nobel, P. S., Osmond, C. B., and Ziegler, H. (1983). *Physiological Plant Ecology* IV., Springer—Verlag, Berlin.

Larcher, W. (1975). *Physiological Plant Ecology,* Springer—Verlag, Berlin.

Máthé, I., and Máthé, I. jr. (1973a). 'Data to the European area of the chemical of *Solanum dulcamara* L.', *Acta Bot. Acad. Hung.* **1–4,** 441.

Máthé, I. jr. and Máthé, I. (1973b). 'Effect of some nutrition elements of soil on *Solanum dulcamara,* under glass-house conditions' (in Hungarian), *Herba Hungarica,* **12,** 29.

Matusiewicz, E. (1960). 'Effect of temperature on yield and leaf essential oil content of peppermint (*Mentha piperita* L.)' (in Polish), *Acta Polish Pharm.* **3,** 205.

Meyer, F. (1973). 'Der Einfluss des Gebirgsklimas auf die Cardenolidglykoside von *Digitalis lanata* EHRH.', University Doctoral Dissertation, Zürich.

Nastev, N., and Jankulov, I. (1975). Einfluss einiger Herbizide auf den Skopolamingehalt der einzelnen Pflanzenteile des Stechapfels (in Bulgarian), *Dokl. Sh. Akad. Bulg.* **3,** 11.

Pavlov, P., and Ilieva, S. (1972). 'Some biochemical changes in early developmental stages of *Salvia sclarea* effected by different illumination' (in Bulgarian), *Rasteniev Nauki,* **10,** 13–20.

Penka, M. (1968). 'Changes in water economy and morphine content in irrigated plants of *Papaver somniferum* L.' *Acta Universitat Agriculturae, Brno,* **16,** 579.

Rovesti, P. (1975). 'Ecological incidence in the composition of essential oils', (in Italian), *Rivista italiana delle essenze dei profumi e delle piante officinali. Aromi Saponi Cosmetica,* **6,** 321.

Ruminska, A. (1973). *Medicinal Plants,* (in Polish), Panst. Wydav. Nauk. Warsaw.

Saleh, M. (1970). 'The effect of air temperature and thermoperiod on the quantity and quality of *Matricaria chamomilla* L. oil', *Z. Landbauer—Hochenschule,* **15,** 17.

Saleh, M. (1973). 'Effect of light upon quality and quantity of *Matricaria chamomilla* L. oil', III. *Planta Medica,* **4,** 337.

Skrubis, B., and Markakis, P. (1976). 'Effect of photoperiodicity on growth and essential oil content of *Ocimum basilicum.*', *Economic Botany,* Bronx, **30,** 4. 389.

Terpó, A., and Bálint, K. (1981). Taxonomy of Horticultural Plants (in Hungarian), manuscript.

Tétényi, P. (1970). *Infraspecific Taxa of Medicinal Plants,* Akadémiai Kiadó, Budapest.

Upadhyay, D. N., Bordoloi, D. N., Bhapat, S. D. and Ganguly, D. (1976). 'Studies on blight disease of *Ocimum basilicum* L. caused by *Cercospora ocimicola* PETRAK et CIFFERI', *Herba Hungarica,* **15** (1), 81.

Vágujfalvi, D. (1967). 'The most important active substances in medicinal plants and their formation in the plants. Modifying effects of external factors and treatments' (in Hungarian), *Herba Hungarica,* 6 (3), 175.

Vágujfalvi, D. (1968). 'Studies on the physiology of plants containing alkaloids' (in Hungarian), PhD Dissertation, Budapest.

Plant Breeding and Variety Control of Medicinal Plants

The genetic improvement of medicinal plants includes two kinds of activity; the first one is dealing with the methodological side, and the second is how to ensure the necessary conditions in order to achieve any production. The methodological part means the maintenance of genetic diversity, breeding and biotechnological engineering. The second stage, by establishment of a complex system, might improve the supply of medicinal plants as well as raw materials originated and produced from them.

The aim of plant breeding is generally the improvement of production capacity, the large yield. Sometimes the intervention has the goal of improving the reliability of the yield, as for example, in the case of breeding for a more resistant or immune plant. A decrease in the yield is also sometimes allowed, if the yield can be harvested earlier, or if it is more valuable.

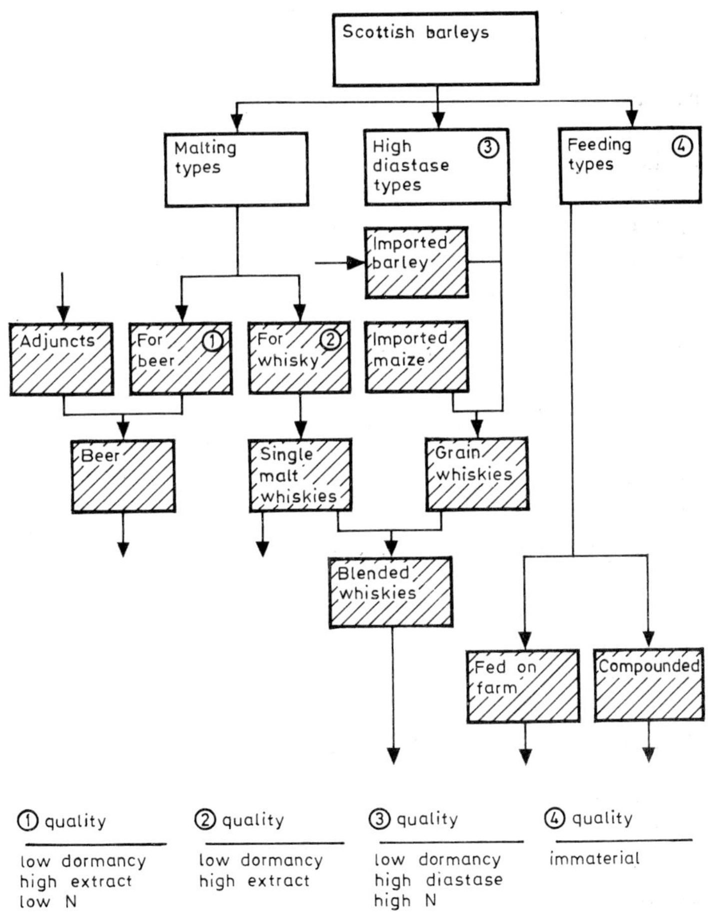

Fig. 23 Quality objectives in Scottish barley breeding (Simmonds, 1984)

Recent efforts in plant breeding do not simply aim at a greater yield, but the active principles of technically and biologically exploitable yield are also taken into consideration (Fig. 23). Thus the therapeutic value counts with medicinal plants, in addition to the amount of yield. This value, however, can be determined only by pharmacological methods, and chemical analysis is not used extensively. The recent development of phytochemical analysis allows chemical evaluation together with pharmacological measurements in place of these.

Biochemical differentiation, producing dissimilar taxa with respect to chemism, is the result of spontaneous processes in nature. The required target, i.e. the new medicinal plant cultivar, is achieved by the consideration and proper application of this process. The new plant may remain the same morphologically, but its chemical properties should be improved, and for this purpose breeders should monitor the inheritance and fixing of biochemical features.

The determination of the active principle content, however, is a very difficult task, because living plants and their metabolism should be considered and one should not make decisions on the basis of transient data. Therefore, differences resulting from the daily cycle of the active principles and from the changes during ontogeny should be excluded. It is a basic rule that the environmental effects should be compensated for and any misleading influences, which would lead to false conclusions, should be avoided.

In the evaluation of medicinal plant cultivars the primary parameter is the quality of the active principle, and in this respect the resistance, uniform ripening, thornlessness, freedom from seed spreading, and possibility of mechanized harvesting are secondary parameters. The same alkaloid or essential oil yield can be reached either by a large amount of plant material containing a small amount of active principle or by a smaller plant yield which contains a high percentage. The requirement is the achievement of the latter case, from the processing point of view, because the specific parameters of the extraction are better in the case of concentrated active substance content.

The total amount of active principle is frequently not so important as that of its constituents. The ergot strain containing ergocristine has less total alkaloids and auxiliary alkaloids than the ergot containing ergotamine. Despite this fact, the rye areas infected with the ergot strain containing ergocristine continuously increase in Hungary, because the alkaloids of this cultivar serve special pharmacological purposes. This is also the situation in the case of lavender: the French cultivars, having less yield and essential oil content, are cultivated instead of the high-yielding ones containing lots of essential oil, because their composition and ester content expressed in linalyl acetate better suit the requirements.

The active principle content, the main pharmaceutical constituent and the practically important chemical characters, should be considered most of the time when medicinal plants are bred. The chemical differences, however, can only be determined by rigorous serial analyses. We have high expectations from the improvement of the ETA and RIA methods, because they afford better possibilities for the determination in a greater number of individuals and for repeated controls of the effective substance either in *Digitalis* or in other plant species.

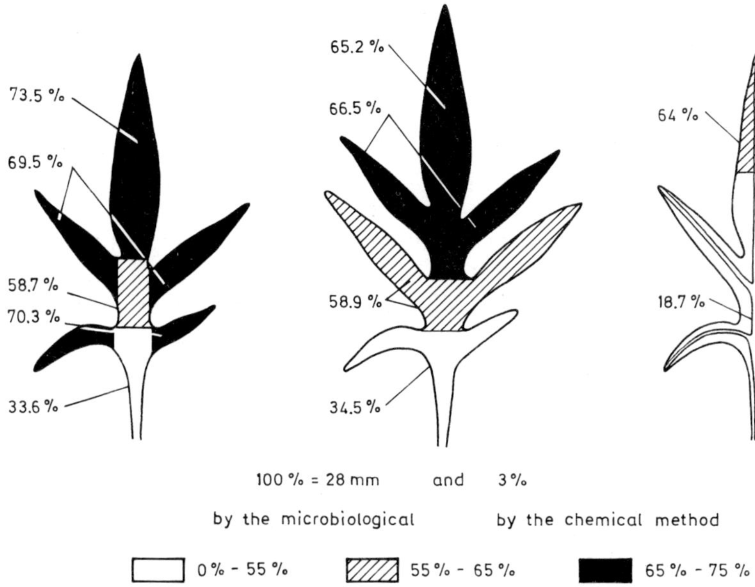

Fig. 24 Glycoalkaloid differences in the leaves of *Solanum laciniatum*

The efforts of the medicinal plant breeder are aimed at increasing the accumulation and yield of active principles by the special metabolic functions determining their content. The morphological differences (which might help in the breeding of other cultivated plants) can be insignificant or misleading in the case of medicinal plants. The breeders of poppy, for instance, wanted to establish an ovate capsule with a large number of valves, in order to establish a poppy cultivar of high morphine content, but the desired result was not achieved. The breeders of *Digitalis* assumed the correlation between the high glycoside content and the thick, wide leaves, but they did not obtain the desired results.

Sometimes it can happen that morphological features are determinant, especially in the case of direct consumption as spice. Thus, the large-fruited coriander is more popular than the high-content essential-oil-holding cultivars, which have small fruits. Among the large-fruited coriander cultivars, however, those with higher essential oil content are preferred. Sometimes morphological features are of significant help to the breeder; among essential oil plants, such as lavender, selection according to the number and size of follicles per leaf surface unit area have led to considerable success. However, if these morphological features are not correlated with biochemical characters, then they should be neglected; it happened that some lavender follicles were empty or only half full. The morphological characters of a new cultivar, having a more valuable active principle, can be ascertained in relation to the initial plant. Thus the breeder cannot be relieved of the obligation to describe these differences.

The variation of the active principle was initially studied quantitatively. In the rhizome of angelica, in the leaf of belladonna and in the root of rhubarb, differences have been observed. In the first plant the changing parameter was the level of essential oil content, while the alkaloid content was different in the second case, and the anthraquinone level in the rhubarb demonstrated a difference in quality also: in the meristem of the bud only rheumemodin is present, while the root also accumulates emodin, chrysophanol and aloeemodin. In the leaf of *Solanum laciniatum*, quantitative differences were also observed (Fig. 24); the tip and upper part of the leaf blade contained more alkaloid glycoside than the middle of the leaf blade, its underside and its veins. The variation of anthraquinone glycoside in the internode of the peduncle and leaves of rhubarb was called *morphogenetical variability* by Schratz; he considered it to be a general occurrence.

One of the difficulties in the breeding of medicinal plants arises when the active principles are not known exactly. Thus the aim of the breeding cannot be defined properly; as, for instance, in the case of camomile or catnip. Another situation may also occur where the content of the active principle and the effect itself are not proportional; the root of rhubarb contains anthraquinones, but their amount is not correlated with the effect.

Although the cultivation of medicinal plants is several thousand years old, certain species have not been bred at all. For this reason most of the species do not have any accepted cultivar; only populations exist. However, in the last quarter century, research has led to the establishment of almost fifty new varieties with higher contents of active constituents.

Despite the significant development of analytical methods, analyses of proper reliability were used only recently. They are rather expensive, and need special apparatus. Thus the breeding of medicinal and aromatic plants is very costly. To avoid this inconvenience, most breeding stations turned to the use of tissue cultures. The use of cell and tissue cultures in medicinal plant breeding and genetics is new.

Why use this method? The reply can be three-fold: to arrive at a higher degree of homogeneity in our plant material; conversely, to increase heterogeneity of heredity; and to induce mutagenesis.

To have more homogeneous plant material has long been practically exigent. Cell cultures provide us with the most potent alternative because they can produce plants of haploid state. There are several cases among medicinal plants where such types of result have been published, mostly in Solanaceae species. For a different example we mention the success with *Catharanthus roseus* (Apocynaceae) where the progeny contained more active material than their parents.

Having more genetically heterogeneous plant material may facilitate selection later on. We refer to only one example from medicinal plant breeding accomplished by this technique, producing protoplasts from the mesophyllum of leaves and of pollen sacs in *Atropa belladonna* as well as *Nicotiana tabacum*. These protoplasts were fused and afterwards generated plants. These varied in chromosomal number; there were haploids, triploids, and normal diploids, according to the authors. They also found higher variability in morphological and chemical characters, so their trial was judged not only to be successful in producing more heterogeneity, but also as being valuable in producing hybrids between two genera.

The third theoretical purpose of tissue cultures might be the inducement of mutagenesis. The most interesting results were published by Japanese researchers producing callus culture from germinating plant hypocotyls of *Papaver somniferum* (Papaveraceae). They could regenerate whole plants and were cultivating them in soil after treatment with a special combination of hormones. Plants flowered and produced fertile seeds. The most interesting physiological phenomena were detected in capsules containing no morphine alkaloid, in contrast to the mother plant, but instead of this the more valuable codeine.

Cell and tissue cultures can be used with high efficacy for practical breeding purposes, too. Embryo cultures, vegetative propagation, ensuring virus- and/or mycological infection-free conditions, as well as cryopreservation, are different fields where these new methods have been introduced.

We have many examples of multiplying positive factors of propagation through tissue culture. This technique is the most appropriate for cloning individuals just bred and for producing as homogeneous progeny as is practically necessary. We refer to medicinal plants through the case of *Urginea indica* (Liliaceae) where Indian experts produced, by tissue culture during 120 days from one bulb, 400 small bulbs, and 90 per cent of these produced normal plants when developed in soil.

In gene banks or for any other preservation purposes we can use the method of cryopreservation if we have any specially bred material or strains. Tissue cultures of *Digitalis lanata* (Scrophulariaceae) were tested after three years in liquid nitrogen, and it was found that whole plants could be regenerated and developed from them. Neither viability nor cardenolide production were affected, and plants had normal life cycles.

According to the literature, *Digitalis lanata* Ehrh. is a plant of great variability with respect to morphology, yield-increasing features and chemical composition. Breeding work with this plant involves many problems, the solution of which may be facilitated by means of vegetative propagation. At the Research Institute for Medicinal Plants in Budakalász (Hungary) we have been successful in propagating vegetatively more homogeneous seedlings grown from seeds of *D. lanata* 'Oxford' by tissue culture techniques.

The variability of medicinal plants is more extensive than that of other species. Populations found in natural growing locations are not at all balanced in terms of chemical characters and active substances (Fig. 25). At the first level of breeding the mixture of chosen infraspecific groups is already different from that of the spon-

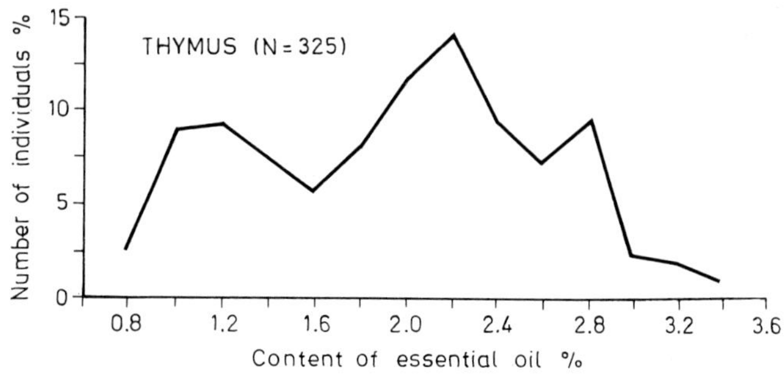

Fig. 25 Essential oil differences in a population of *Thymus vulgaris*

taneous populations of the species. In this way they are bred to some extent, although not purposefully. In spite of this the plants are extremely variable. The commercial propagating material taken from these types is selected randomly, and it is almost impossible to guarantee the inheritance of compositional properties. For example, peppermint retained for four years in the same plot contained various amounts of essential oil between 1.45 and 3.19 per cent depending on the weather.

The next level of breeding is represented by landraces. These are more or less also mixtures of infraspecific groups, but the regional features appear in chemical characters. In this case, the variability is somewhat restricted and the inheritance of chemical characters can be expected. A number of poppy cultivars of this level is grown in Hungary, including the 'Soproni' landrace, which is rich in codeine.

If the extent of variability is restricted by multiple self-pollinations, then an autogramous variety can be bred. The essential oil content curve of the French lavender cultivar bred in this way in Hungary has only one peak. The variation is minimal in the case of clones, as in English lavender. The essential oil content of clones of peppermint, for example, changed to a smaller extent (2.06–2.67 per cent) during the four years of study than that of the mixture of individuals belonging to the above-mentioned population. The same is true for *Thymus* clones (Fig. 26).

Breeding of medicinal plants should always be started from populations. According to population genetics the goal is not the panmictic propagation of every individual to as great an extent as possible. The extent of propagation should be controlled so that one of the genotypes is favoured, and the selection pressure is exerted in this way. This positive selection is successful mainly for the landraces. The selection of individuals is successful only if it is made according to chemical data. During the selection of an individual, care should be taken to prevent migration by the action of foreign pollen.

The new cultivar *Leucojum aestivum* 'Snejinka' as compared with the spontaneous population can ensure a two-fold amount of galantamine (0.4–0.6 per cent) and 13–21 per cent more drug production in Bulgaria. The plant breeders in India have reported

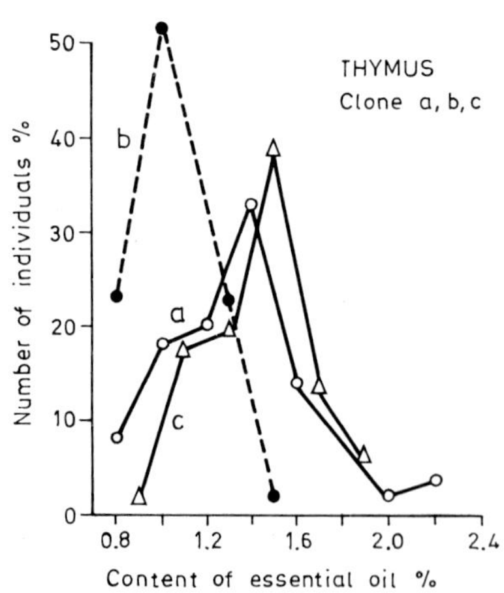

Fig. 26 Frequency curves of essential oil content in three clones of *Thymus vulgaris*

a new *Hyoscyamus muticus* cultivar with 4 per cent and an *Atropa* cultivar with 0.8 per cent alkaloid content.

The selection of individuals can be made more efficient by utilizing the drift. For this purpose the population is divided into parts, thus increasing the number of homozygotes, and the material will be more uniform. The breeding takes rather a long time as relatively small amounts of individuals can be analysed in each case. Generally a compromise is chosen, in which groups of 20 per cent are formed according to the active substance content, and the procedure can be hastened also by preselection.

In the case of medicinal plants, which have only recently been introduced into cultivation, it is generally regarded that the maintenance of a large mixture of individuals is necessary. The three levels of the selective, family-centred breeding system are based on the maintenance of genetic differences, while performing a restricted negative selection. The three levels ensure the continuity of the breeding process. The involvement of new material, on the other hand, retains the fitness of the plant. In this case the grouping with 10 per cent is the most appropriate for the selection of medicinal plants.

While breeding with the method of selection, it is very important to retain fitness, especially of medicinal plants. The uniformity can be improved by inbreeding, but the deterioration of cross-pollinated plants will increase, while vitality will decrease. Strict inbreeding is, for instance, harmful for *Digitalis;* the second-generation plants are more sensitive to diseases, and will not produce seed if inbreeding continues. The deterioration caused by inbreeding, as well as the heterosis effect of subsequent crossing on the total glycoside content, have been shown many times.

Breeding using the method of hybridization seems to be invariably a very long-term process. Indian researchers have produced hybrids of *Ocimum canum* x *O. basilicum* with a 47.7 per cent champhor content in their essential oil—thus differing from the linalool and methyl chavicol content, respectively, of the parents—but the synthetic allohexaploid is far from being stable.

Clone selection is regarded by many people as a procedure predetermining the uniformity of descendant plants. It was shown, however, that active substances of *Lavandula* and *Geranium* clones (the ratio of essential oil and linalyl acetate) are dissimilar for each individual plant. Clone selection is not an absolute method, as French researchers grafted cloned individuals on a *Citrus* clone base, but, in spite of this, the active principle content of the individuals was dissimilar. Neither is the body of the original mother plant homogeneous, and the organs are different in developmental stage or heredity. This is confirmed by chimeric and bud mutation phenomena, which are induced by environmental effects, mostly after cross-breeding, and appear as vegetative splitting, which is worth considering and registering during the breeding of medicinal plants.

The selection of clones has resulted in the greatest number of new cultivars: *Dioscorea* with 6 per cent and *Costus* with 3 per cent diosgenin content were selected in India. Bulgarian mint clones produce 110 kg/ha yield (and 0.6–0.8 per cent volatile oil content), with 46 per cent carvon from the cultivar 'Mehta' and with 90 per cent linalool from the cultivar 'Lina'. New *Nepeta transcaucasica* clones have been bred in

Bulgaria and new clones representing six chemotypes of *Thymus vulgaris* in France. Unfortunately, in the latter case the clones with thymol, linalool and geraniol content must be cultivated together, because customers demand the mixed compounds rather than the homogeneous ones.

Up to this time, more efforts have been made to loosen the heredity than to retain the vitality in the breeding procedures. A promising method is remote cross-breeding. In this process the different active principle contents of the parent plants might appear simultaneously in the descendants. It may also happen that the chemical characters of one parent suppress those of the other. A later generation of descendants can differ in the characteristics of active substances, as was observed in the case of the hybrids of poppy with *Papaver pseudo-orientale*. The advantage of remote cross-breeding may be heterosis (as in the case of the English-type lavenders), but sterility is a propagation disadvantage. This can be overcome, however, by vegetative propagation and polyploidization. A new method of producing polyploid varieties is valence cross-breeding: materials having different degrees of polyploidy are crossbred, which results in plants having advantageous active substances, e.g. the polyploid camomile has larger inflorescences and better characteristics of active substances than any of its parents. A German expert reported on trisomic mutants of *Datura stramonium* with an alkaloid content 155–227 per cent higher than that of the standard.

A popular method of plant improvement is radiation treatment. Its direct effect is accompanied by harmful biological consequences, but the larger number of variations facilitates breeding. The irradiation may also cause a heterosis effect. The treatment is successful only if the loosening of heredity is followed by continuous individual selection. In this way results can be achieved, and a new cultivar can be produced; e.g. the Hungarian cultivars of ergot, nightshade and poppy were bred. Mutants of *Capsicum annuum* in the third and fourth generation—following ethyl methanesulphonate (EMS) treatment—increased the alkaloid content by 20–60 per cent compared to the initial material.

To classify the bred plant material and to ensure its survey always requires meticulous work, with considerable trouble, but the results facilitate further progress for others. A Yugoslavian expert achieved, by several years of study, the chemotaxonomical system of the hop cultivars, based on the α-acids, on the proportion of α- and β-acids and on cohumulon content.

Although cultivation of medicinal plants has taken place for several thousand years, certain species have not been bred on a scientific basis at all. For this reason most of the species have no cultivars; only populations exist. The statement of Schratz from 1960 is still justified, i.e. that the breeding of medicinal plants is far behind the breeding of other plants. However, the breeding data in Table 4 showing results obtained during the last four years prove that development in this special field has accelerated: about 30 new cultivars were introduced into cultivation practice.

Hungarian breeders have had some remarkable results, too. Disregarding the previously arrived polyploid camomile, used also today there are recently registered, or patented, ones too, such as autumn-sowed poppy 'Cosmos', fruit-fixed caraway 'Maud', smoke-tree, mauritanic mallow and thornless safflower 'Pannonia'. We have to add that our 12-year-old mustard cultivar 'Yellow of Budakalász' was introduced

Table 4 Accepted new cultivars of medicinal and aromatic plants (Period 1981–1985: Hungarians not included)

Anethum graveolens 'Sari'	More aroma
Artemisia balchanorum 'Olimp,' 'Evrika'	70% citral or 65% linalool
Carthamus tinctorius 'Sidwill'	33% oil
Chrysanthemum cineriaefolium 'Hypy'	High pyrethrin and yield/ha
Coriandrum sativum 'Yantar'	High essential oil and yield
Cymbopogon jawarancusa—strains	Rich in piperitone, or in phellandrene or in elemol
Dioscorea floribunda 'Arka Upkar'	3.4% diosgenin
Foeniculum vulgare 'Feniks'	2.5% essential oil, 80% anethol
Glaucium flavum 'Glautsin Poli'	Higher glaucine yield
Hippophaë rhamnoides 'Zolotistaya'	More active principles
Humulus lupulus 'Nugget'	High α-acid
Hyoscyamus niger—strains	0.5% alkaloid
Lavandula angustifolia 'Irene Doyle'	Increased essential oil yield
Mentha arvensis—clone	0.6–0.8% essential oil and high yield
Mentha gentilis 'Hokuto'	Higher essential oil and yield
Mentha piperita '29'	3-4% essential oil, fewer constituents
Mentha spicata, from 'K 65'	52% menthone
Ocimum gratissimum 'Kelasuri'	40–70% eugenol
Origanum vulgare, chemotype '65/1'	High phenol
Papaver somniferum 'Sadovo 1235'	Higher morphine and yield
Pelargonium roseum 'Aist 4'	74% citronellol, high yield
Salvia sclarea 'Trakiika'	0.3% essential oil, 81% linalyl acetate
Sanguisorba officinalis, from 'II'	High polyphenol
Satureja hortensis 'Rzhevo Konare'	0.6% essential oil
Symphytum officinale, cytotypes	Higher phytosterol and triterpene
Valeriana officinalis 'Shipka'	0.5–1.2% essential oil, 1.3–3% valepotriate
Zingiber officinale 'Maram'	Higher oleoresin yield

in official testing experiments in both Germanies in 1987, and they obtained positive results.

We have to distinguish from these innovative researches the continous flow of raw material to be elaborated. For this very reason a real task exists: how to ensure germplasm of the quality and quantity needed for the large-scale culture of medicinal plants. This can only be achieved using a well-founded system. This system is the missing link between breeding and production, and therefore represents a *sine qua non* for the improvement of plant raw material.

With respect to the authorization of cultivars, we have to point out that many official steps are necessary in order to establish well-based evaluation methods. After 5–10 or more years of inside breeding work, the first official presentation is the registry sheet.

The next stage includes 3–5 years of regular comparative trials executed by the Hungarian Institute of Cultivar Testing. This governmental institute has different examining groups for the various plants, including the medicinal plants. It is well equipped and has a network of fifteen stations where trials are carried out; in the case of medicinal plants, six places are used.

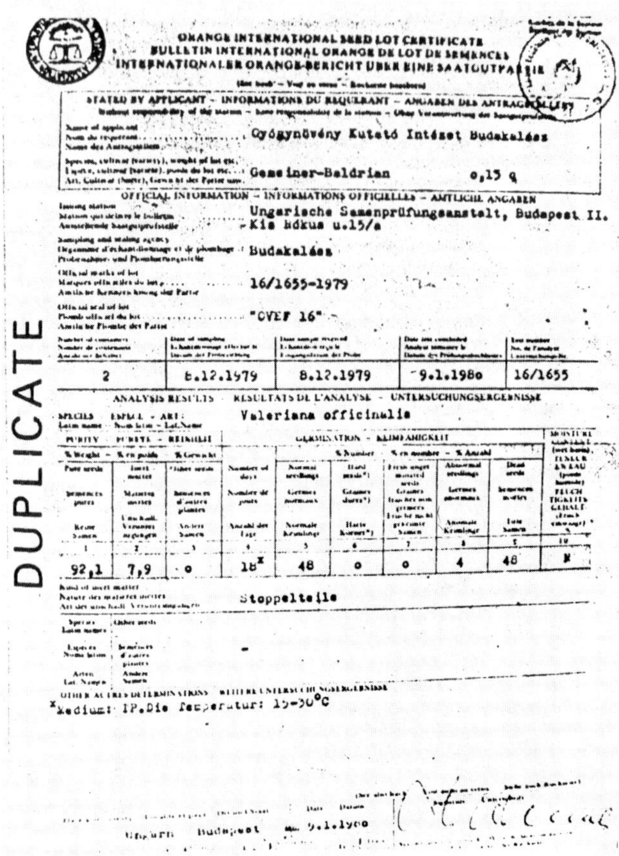

DUPLICATE

ORANGE INTERNATIONAL SEED LOT CERTIFICATE
BULLETIN INTERNATIONAL ORANGE DE LOT DE SEMENCES
INTERNATIONALER ORANGE-BERICHT ÜBER EINE SAATGUTPARTIE

STATED BY APPLICANT – INFORMATIONS DU REQUÉRANT – ANGABEN DES ANTRAGSTELLERS

Gyógynövény Kutató Intézet Budakalász

Gemeiner-Baldrian 0,15 q

OFFICIAL INFORMATION – INFORMATIONS OFFICIELLES – AMTLICHE ANGABEN

Ungarische Samenprüfungsanstalt, Budapest II. Kis Rókus u.15/a

Budakalász

16/1655-1979

"CVEF 16"

2	6.12.1979	8.12.1979	9.1.1980	16/1655

ANALYSIS RESULTS – RÉSULTATS DE L'ANALYSE – UNTERSUCHUNGSERGEBNISSE

Valeriana officinalis

1	2	3	4	5	6	7	8	9	10
92,1	7,9	0	16ˣ	48	0	0	4	48	

Stoppelteile

ˣMedium: TP. Die Temperatur: 15-30°C

Ungarn Budapest 9.1.1980

Fig. 27 Orange International Seed Lot Certificate

After completion of the comparative trials, the office evaluates all collected data and presents a report to the first level of qualification, namely the Professionals Committee. This includes experts on the field; in our case, medicinal plant cultivars are discussed by the Horticultural Committee. The report contains the reasons why the variety candidates have to be introduced, and the description and results of the trials. On the basis of a favourable view point of this Committee, the next level of judgement, the National Testing Council, authorizes the cultivar to be introduced into practice and orders its registration in the Hungarian Book of Cultivars.

The breeder's institute must regularly provide the users with the propagating material of the cultivar, and this can only be achieved by following regular preservative breeding practices.

The last step in the life of a cultivar is its withdrawal. The same procedure must be used—Committee and Council Meetings—as for the registration.

Returning to the second guarantee of the breeding system, it begins with the announcement of the seed.producer to the regional center of the controlling institute, at the start of vegetation. The announcing sheet contains data on sowing and seed origin, on one half-side, while the control half-side fixes—of course later—the result of official qualification and the produced seed material.

During the vegetation of seed-producing plants (beside the usual agronomical techniques) two reviews are produced by the experts of the controlling centre, each completed by registering all remarks on the health conditions of plants, protection against extraneous varieties, weeding, etc. These notes accompany the seed material until its commercialization.

However, the yield must be harvested earlier, and seeds elaborated (ridding them of powder, strange impurities, seeds of weeds and unviable own grains). Once the cleaned seeds are closed in containers or in sacks, the official laboratory controls for purity, germination power, health state, etc. begin. If everything is satisfactory, then the control institute gives out the certification of seed quality, but if something is wrong, they order the opening of the closed material and the renewal of elaboration.

The positive certificate makes possible the commercialization of seeds. For exporting the seeds there exists an international agreement on the uniformity of seed testing, and therefore a seed lot, as well as a seed sanitary certificate (Fig. 27), must escort the material. Hungary produces every year large quantities of seeds for use both at home and abroad, as more than 50 species including cultivars of medicinal and aromatic plants are cultivated here.

The guarantees provided by governmental rules and international agreements assure the quality of any cultivars and the testing of the seeds originating from them. The prescriptions are determined in standards, which pay attention to plant propagating material, to determination of active principles in the case of medicinal and aromatic plants, to sampling and to the rules of procedures.

Standards can be national ones, or international. The latter contain prescriptions for applications, apparatus operating conditions, etc., as well as an annex with typical chromatograms. Laboratories for quality control execute the determination of active principles according to these standards and to the Hungarian as well as international prescriptions of pharmacopoeias, which can also guarantee the requirements of health care, e.g. absence of pesticides or heavy metals.

The last part of the guarantee refers to the preservation of breeders' rights as well as those of their institute or country. The accepted form is the patent of a cultivar given out after a necessary examination by the office of inventions. The document delivered contains all characteristics of the cultivar and the points of claim for the patentee.

This system was elaborated slowly, but has been used in our country for a quarter of a century for improving the biological background of cultured plants and producing raw material of high quality for direct use or elaboration for medicinal and hygienic purposes.

Technical and Technological Conditions for Medicinal Plant Cultivation

Advanced medicinal plant cultivation and processing require many technical and technological conditions. Mechanization and the introduction of chemicals will be discussed in this chapter, as being basic technical conditions for efficient production. These two factors play the most important role in reducing labour requirements and in improving cost efficiency in medicinal plant cultivation as well as in other special branches of agriculture.

Possibilities and Conditions for Mechanization

The number of employees in agriculture has shown a declining trend all over the world. There has been a labour shortage, especially for operations which require hard physical work or which are detrimental to health. An efficient way of reducing manpower requirements is mechanization. Mechanization in medicinal plant cultivation has received more and more emphasis, as in other special branches of horticulture.

Apart from replacing manpower, mechanization also increases the economic and social efficiency of work. As a result, productivity, cost-effectiveness and profitability will increase, and working conditions will become more attractive. Results of progress in biology and chemistry cannot be utilized without advanced equipment. Mechanization has a crop-increasing and cost-decreasing effect as well. Using reliable equipment, the operations will take less time; therefore it will be possible to carry them out during optimal periods. In case of, for example, medicinal plants, it is especially important to do the harvesting quickly, because the value of the products depends on the quantity of the accumulated active substances.

If harvesting takes longer than the optimal time (which is very short for most of the medicinal plants) it will result in great losses of active ingredients, and consequently losses in value. The production of ergot can be mentioned as one of the best examples of the crop-increasing effect of mechanization. As a result of the application of advanced infecting and harvesting equipment, the volume of harvested crops has increased five-fold, and the costs of cleaning have decreased significantly. The level of mechanization in medicinal plant cultivation is relatively low compared to other branches of plant cultivation, due to the fact that this branch is new, covers relatively small area and has extended small-scale farming type cultivation. Besides, certain operations in medicinal herb cultivation—especially those in harvesting—are of a special character (e.g. the picking of inflorescence), and therefore they would require special, non-serial equipment. Designing and manufacturing any equipment is expensive, so it is cost-effective only if the equipment is highly utilized (i.e. larger area). Mechanical replacement of manpower without any increase in specific costs should be achieved.

In spite of these difficulties the level of mechanization has significantly increased during recent years. This development has been achieved partly by the increased use of advanced horticultural and field equipment, and the technical and technological progress in certain elements of cultivation (chemical weed control, variety, method of reproduction) has also contributed to it.

For large-scale medicinal plant cultivation, several operations have already been or will soon be mechanized. Many of the operations can be carried out by machines and implements belonging to the functional machinery system. Some other operations can also be mechanized if certain elements of the machinery system used in other branches are modified or converted or fitted with a suitable adapter in order to be applicable to medicinal plant cultivation.

The least mechanized operation is the harvesting of medicinal plants. Practically none of them can be mechanically harvested unless the normal machinery is reconstructed, and in most of the cases special means and special equipment are required (e.g. ergots, hop, camomile, foxglove).

Some of the implements that can now be utilized in medicinal plant cultivation have a power demand which can be met by applying power machines of 0.6–0.9 MPa force. In other branches of agriculture, due to the higher degree of concentration (i.e. larger field sizes), machinery of higher speed and larger working area is required. Some of these implements can be used in medicinal plant cultivation, too, and can be driven by power machines with 1.4–1.6 MPa force.

As far as the preparation of soil and seeding beds for medicinal plant cultivation are concerned, these are mainly identical with those in vegetable production. Therefore the elements and certain equipment belonging to the vegetable-producing machinery system can be applied without any modification for medicinal plant cultivation. High-power tractors provide better possibilities for subsoiling. Combined equipment completed with active tilling tools has a greater role in tilling and in the preparation of seeding beds. By using this equipment the number of operations can significantly be reduced. When using tilling machines of higher operating speed and of larger working area, the efficiency will increase and the operation time will drop significantly.

Machines for sowing and planting are identical to the units of the functional machinery system and they can be used for the sowing and transplanting of several medicinal plants without any modifications. In many cases it is necessary to use precision-type equipment and pip-sowing machines. Modified or special equipment is needed for planting peppermint, lavender and hop. The efficiency of sowing and transplanting machines can be improved by the introduction of adapters for spreading fertilizers, herbicides and soil antiseptics.

Plant protection, herbicide and nutrient spreading can also be carried out with the help of suitable units of the functional machinery system, without any modifications. It is advisable to do basic fertilization, involving large volumes, with trailers and automotive equipment, while top dressing and row fertilization, which involve small volumes, should be done with combined equipment.

In order to increase productivity and cultivation safety the irrigated areas of medicinal plant cultivation should be expanded. For the purpose of irrigation, which increases crop yield and decreases crop variations, it is advisable to apply machines which are easy to assemble and can be moved in an automatic or semi-automatic way, requiring less labour.

The harvesting equipment used in grain or hay production can be suitably applied, after some modifications, for harvesting several medicinal plants (mustard, coriander, caraway, fennel, peppermint, hyssop). For the harvesting of lavender, the hay-harvesting machines can be used after a substantial modification. For the harvesting of ergot, poppy, foxglove and camomile, Hungarian and foreign-made special-purpose machines are available. Efforts have been made in several places to design machines for the harvesting of other plant species (valerian, angelica, marjoram, etc.) but it is still a task for the future to construct highly efficient machines.

Within the entire machinery system for medicinal plant cultivation, an important role is played by driers. Nearly all of the medicinal plants should be dried immediately after harvesting. For this purpose the equipment used for drying grains or animal fodder is less suitable (in some cases cold-air driers, and in other cases hot-air driers, can be applied). For drying medicinal plants, plate-driers and belt- (conveyor)-type driers used in the food industry, and driers designed for drying rice, operating with either cold or hot air, can be efficiently applied; this latter is a grain-drying ventilation platform.

The cleaning of medicinal plant seeds can be done by sieves and separators belonging to the seed-cleaning machinery system, without modifications, in most cases. The cleaning of other plant parts (herb, folium) can be carried out by converted and special equipment.

Plants containing essential oils should be steam-distilled instead of being dried after harvesting. For this purpose, special mobile or immobile equipment is used. Its operation requires a lot of labour and its capacity is low. It lags behind the container-type equipment which has appeared lately.

Possibilities and Conditions for Introducing Chemicals

Chemicals are used in agriculture all over the world; this phenomenon is closely related to the development of economies. Its purpose is to increase the efficiency production and to decrease the need for labour.

In Hungary the application of fertilizers has developed rapidly during the last decade. The nutritional value introduced by fertilizers is two to three times higher than the volume of nitrogen, phosphorus and potassium introduced into the soil by stable manure. This has greatly contributed to the reduction of crop fluctuation and to the doubling of yields.

The application of plant-protecting chemicals has also developed significantly. Remarkable results have been achieved, especially in the introduction of herbicides. This has greatly reduced labour shortages in agriculture. Development has been most dynamic in the area of plant protection. The volume and assortment of plant-protecting chemicals applied have rapidly extended, and they are frequently changed. From day to day, new types of herbicide, defoliant and desiccant chemicals, soil disinfecting chemicals, insecticides and fungicides, and growth- and flowering-regulating chemicals are being tried and introduced.

Weed control. Through the introduction of herbicides, the conditions for the growth of cultivated plants will significantly improve, the harvest will be easier, the crop will be cleaner and of higher quality, and the yields will be higher. There are several medicinal plants (dill, coriander, anise, mustard) which do not require large cultivation areas and which are sown in wide row spaces only for the purpose of mechanical weed control. If herbicides are introduced, the row spaces will be narrower, the ratio of cultivation area to the number of plants will be optimal, and therefore the yield will be higher.

For several medicinal plants, sprouting is followed by a long period of initial growth, which creates favourable conditions for weed growth. Therefore the manual work demand for mechanical weed control is, for example, 600 work hours/ha for marjoram and 300 work hours/ha for peppermint.

If herbicides are introduced, the labour requirement will be lower. This is the reason for introducing chemical weed-controlling methods applying the latest herbicides in medicinal plant production. No special herbicides have been developed for medicinal plants, so herbicides used in other branches serve as a basis for research and practice. Their assortment is getting larger year by year.

A greater number of herbicides is required due to the introduction of chemical rotation and chemical combinations. Apart from the necessary crop rotation, this is an essential element of controlling resistant weeds.

The requirements for herbicides are strict in medicinal plant production, too. Besides efficient weed control and plant protection, the herbicides should not have an effect on either the formation or the accumulation of the active agents. Experience shows that herbicides causing no change in the external habitat of the medicinal plants will not hamper the accumulation of active agents either. In spite of this fact, the practical application of each new herbicide should be preceded by an analysis of the effects. Herbicides introduced with improper technology can do great harm. It is of primary importance to know the volume that can be introduced for a unit area, the spray dose and the degree of concentration to be applied. Chemicals should be introduced only during the optimal period (phase of growth, optimal soil conditions, degree of weed growth) and in an optimal way (an even volume of spraying or spreading on the soil or on the plant surfaces), or they will be inefficient. If introduced in other periods than specified by the technology, herbicides can cause significant damage. It should be emphasized that the efficiency of weed-controlling chemicals is considerably increased by proper soil preparation and order of crop rotation. Experience obtained during nearly twenty years has shown that in the case of some medicinal plants (e.g. marjoram, peppermint, lavender) the tilling row space used in the vegetation period cannot be replaced by introducing herbicides.

Herbicides are generally products with no indications of toxicity or they are nonviolent poisons. In spite of this, the safety regulations should be adhered to.

Pests and living organisms causing diseases do a lot of damage each year. The most effective weapon against them is chemical protection. Advanced plant protection, however, means not only the application of chemicals and the killing of pests. Chemical protection should be preceded by the creation of certain biological and agrotechnical conditions helping the growth of the cultivated plants and hampering the growth of pests. For the efficient application of plant-protecting chemicals, their properties, as well as the pests, biology and manner of living, should be known, because this is the only way of achieving effective protection.

Growth control by chemicals. Several biologically active compounds are used for the purpose of increasing the crop yields, and improving the quality and controlling the life processes of the plants. Natural plant growth hormones (auxins) and compounds with similar effects (regulators, growth-controlling agents, desiccants, defoliants) are widely used. Among these the synthetic auxins, gibberellins, phytoquinines and humic acids are worth mentioning, and among products, Ethrel, CCC and Wuxal should be noted. The defoliant Reglone is suitable for the defoliation of fennel, mustard and poppy (to facilitate mechanical harvesting), mainly for the purpose of sowing seed production. Wuxal can be efficiently applied for increasing the crop yield of woolly foxglove and other medicinal plants. Synthetic auxins can play an important role in the rooting of medicinal plant cuttings. In the near future these highly effective, biologically active materials might favourably influence the mechanisms of active agent formation and might increase the active agent content of the plant parts to be used.

Essential conditions for the effective application of chemical technology in producing medicinal plants are *health protection* and the production of drugs having *no harmful residue*. People come into regular contact with chemicals (fertilizers, plant-protecting chemicals, herbicides, regulators, etc.) during their application. Most of them are detrimental to health and many of them are poisonous. Therefore serious damage, even death, can be caused by not adhering to the precautions concerning their use. The fungicides and insecticides used for keeping off or killing pests have especially high poison contents. In order to prevent accidents, it is advisable to let only people with sufficient technical knowledge work with these highly effective chemicals, using the specified protective wear. The waiting periods should also be kept to, as otherwise serious trouble can be expected.

Further dangers lie in not applying the methods of chemical technology in conformity with the specifications. For example, the introduction of overdoses, not keeping to the hygienic waiting times, and polluting the air and soil, can easily lead to the accumulation of harmful residues. (The lead content of medicinal plants produced along highways, and the effect on soils infected with DDT–HCH-content plant-protecting chemicals are well-known facts.) Regarding all these, it has been a general requirement of increasing importance during recent years to produce and distribute drugs with no harmful residues. Such a residue (non-decomposing active agents and bioactive substances of decomposition) can be either superficial (cuticular) or sub-cuticular (penetrating into the plant tissues). Its volume is measured in mg/kg or ppm, and is related to the mass of the product, or drug. The volume of permitted (unharmful) residue has been controlled since 1968 in Hungary by decrees of the Ministry of Food and Agriculture and the Ministry of Health, and national standards have been in force. A drug containing any harmful residue (in any concentration) cannot be marketed either in Hungary or abroad.

There are no regulations especially concerning the residue content (harmless) of drugs. Limit values specified for food products seem to be adequate for medicinal plants as well. The danger of residues can be correctly judged only if the method of application of the drug is also considered. In the case of drugs used exclusively for pharmaceutical processing (e.g. ergot, nightshade), chemically clean compounds are produced (active agents), so an eventual residue in the plant would cause no damage (it is not consumed). Drugs in tea blends and spices, however, should contain no harmful residues, as they are directly consumed. Medicinal plants produced for processing essential oil should be regarded as having special aspects. It might be necessary to specify a new limit value which is lower than the previous one, because some chemicals (HCH, DDT) might get concentrated in the essential oil during the distillation in such a way that a dangerous volume of residue could get into the product.

In Hungary the introduction of chemicals is strictly controlled and the application of chemicals leaving harmful residues is forbidden. Regular analysis of residues is specified by strict regulations. The introduction of each new method of chemical technology is preceded by an analysis checking the residues. This means that by the skilful introduction of permitted chemicals no harmful residues or no harmful volume of residues are left in the drugs. The specified analysis of residues ensures that drugs and essential oils produced in an unskillful way and containing harmful residues are not sold and distributed.

Primary Processing of Medicinal Plants

The most valuable products can be prepared from the plant components which contain most of the active substances. One of the important conditions for improving the quality is gathering the plants when their active substances content is at a maximum. This time corresponds to the biological ripeness only for a few crops and seeds. For most of the medicinal plants the peak of accumulation of active ingredients is around the period of flowering, while it is around the end of the vegetation period for the parts of plants in the ground. In these cases the harvest, the time of collection is not the time of biological ripeness, but the period of technical or technological ripeness.

At this time the water content of the harvested plant parts (leaf, stalk with leaves, flower, root) is still high; consequently, these parts cannot be stored without deterioration even for a short period.

However, one of the common features of the medicinal plants is that they cannot be consumed in a fresh state. That is why the harvested plants should be prepared in such a way that they become suitable for storing, and should be preserved in order to prevent the occurrence of deteriorating biological processes.

Generally, the first preservation (processing) operations after harvesting are carried out by the producer himself. For this reason, these operations are called primary processing, so as to distinguish them from the subsequent industrial processing.

The most frequent (and at the same time, the simplest) form of primary processing is the drying and the extraction of essential oil.

Drying

After drying, the water content is reduced to a degree where storage is feasible. The water content of the plant parts is classified according to the mode of water binding, in terms of drying technique, as follows:

Chemically bound water has a high binding energy; its removal is possible only by destroying the material: consequently, it cannot be removed by drying.

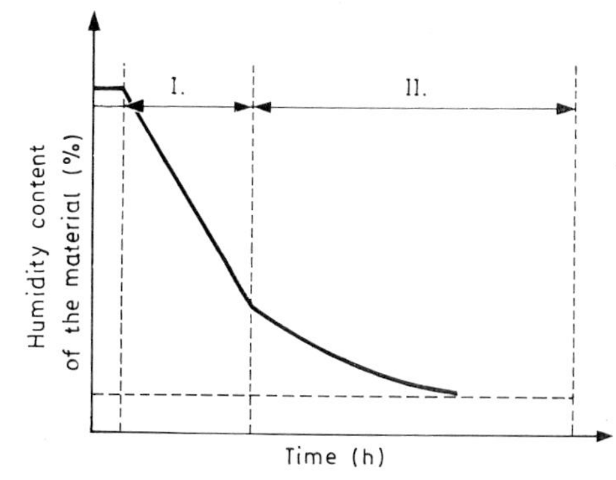

Fig. 28 Theoretical drying curve of medicinal plants; I, stage of constant rate; II, stage of decreasing rate

Physicochemically bound water is further subdivided into adsorbed and osmotically bound water, according to the strenght of binding. The binding strength of adsorbed water is high. This water is not removed by drying. The physicochemically bound water besides the adsorbed water is the osmotic water inside the cells.

Mechanically bound water is located in the macrocapillaries and on the surfaces. The rate and direction of water removal is determined by the concentration difference (concentration gradient) and by the temperature (temperature gradient). These two factors affect the drying jointly and simultaneously. Owing to the concentration difference, the water migrates to the location of lower water content from the area of higher water content.

The effect of temperature gradient is somewhat different, as the humidity migrates towards the cooler places from the warmer places owing to the thermal movement of water molecules.

The duration of drying is determined by the rate of water migration. The theoretical drying curve is shown in Fig. 28, as humidity content against time. For a short period of time, the humidity decreases proportionally to the time, i.e. linearly. The rate of humidity reduction slows after a certain elapsed time, gradually down to the limit value. The rate of drying thus consists of two main parts, the constant and the slowing drying stages. In the stage of constant drying rate, the whole amount of mechanically bound water and a part of osmotic water is evaporated. In the stage of slowing drying rate, the remaining water is evaporated at a significantly decreased rate, then the evaporation stops, depending on the conditions of the drying air, and the remaining water enters an equilibrium state. Removing more water than necessary is not justified, either from an economic or a quality point of view.

The rate of drying is dependent on the temperature and flow rate of the drying air. The higher the temperature and the higher the flow rate of the air, the quicker the drying. The temperature of the drying air is primarily limited by the heat sensibility of the active substances in the plants, and secondly by the rate of humidity migration. If the temperature and the flow rate of the drying air are too high, then the drying rate reaches a level where it is over the rate of water migrating outwards; consequently, the produced crust closes the capillaries, and this prevents the water from moving out of the inner portions. In this case the drying can only be continued by increasing the temperature, which may cause surface scorch and significant loss of active ingredients, i.e. it causes quality deterioration.

General conditions

The humidity content of harvested plant parts is generally high, around 60–80 per cent. The metabolic processes in the plants are not discontinued at the moment of cutting, but continue as 'starving' metabolic processes. This leads to deteriorating biological processes (such as bacterial activity, yeast and enzyme transformations, auto-oxidation, non-enzyme browning, moulding occurring within a certain period of time) unless the water content is significantly reduced.

The modified metabolic processes and the deteriorating biological processes both give rise to the transformation of the metabolic products. During this process, the active agents of the plant may be destroyed, and the exterior properties (colour, smell, taste) may become disadvantageous. It is a basic requirement to reduce the water content of the plants as early as possible in order to prevent the deteriorating processes.

A significant proportion of the drugs is directly consumed (tea blends, spices). That is why the exterior properties (colour, smell, taste) also need to be preserved, besides the active agents.

A major factor in the drying of medicinal plants is economics. The above requirements all imply the reduction of drying time. The simplest way of increasing the rate of drying would be to increase the temperature of the drying air. It is advisable to use the maximum temperature at which the active agent the exterior form of the plant (colour) do not change significantly.

Drying of medicinal plants should be so performed that the water content of the drugs at the end of the drying is 10–14 per cent. With this water content, most of the drugs can be stored for a long period without deterioration. Overdrying is not only harmful, but also uneconomical. The risk of harm increases with the water content. The ability of drugs to absorb humidity should also be kept in mind. This can be prevented by adequate storage methods.

Preparing operations

The harvested, collected, fresh plant parts should be prepared for drying. The aim of these operations—cleaning (by hand, by screening, by washing or by peeling), cutting up, stripping (folium)—is to increase the efficiency of the drying; unnecessary parts should not be dried, smaller portions may be dried with less energy, and the drying is faster.

Drying methods

In practice, the methods of drying are divided into two categories according to the heat source: natural and artificial. The artificial methods are further subdivided, according to the temperature of the air, into cold, warm and hot methods. All three methods are used for the drying of medicinal plants.

Natural drying

This is a method that has been used from ancient times. The simplest type is drying by sunshine at the growing area. It has the disadvantages of the risks of harm by extreme weather conditions. In order to exclude the influence of weather, airy barns and sheds have been used, and are also used today. This method is most frequently used for wild-growing plants. Another disadvantage of this method is the demand for large areas.

Medicinal plants are placed on drying stands, on replaceable drying frames, in order to better exploit area and speed up the drying, to reduce the demand for manual turning, and to prevent any harm resulting from the turning operation.

The plant components should be placed in layers of thickness corresponding to their size in order to achieve proper drying. On 1 m^2 area the amounts of fresh plants should be as follows: 0.5 kg of flowers, 1–2 kg of plants and leaves, 2–2.5 kg of roots. The duration of drying is mainly influenced by the weather. The air movement should be maintained carefully.

Artificial drying

Natural drying may take several weeks, depending on the weather conditions. The drying time may be reduced to hours by warm driers, and even to minutes by heat driers. For artificial drying the air is heated, its relative humidity content is reduced by intensive flow, and the demand for manual work is reduced.

Cold-air drying. In this method only the movement of the drying air is controlled. The air movement is maintained by machinery (fans). It is used for the drying of crops which are larger in volume than expected. The fan may only be operated during dry, warm weather, when the relative humidity content of the air is not above 50 per cent. The duration of the drying is 8–20 days, depending on the humidity content of the plant and the moisture content of the air.

A transitional method, between cold-air drying and warm-air drying, is drying with *forced ventilation*. In this method there is a possibility of warming the air as well as moving it. The requirement for manual work is also less. It is especially applicable to the drying of seeds and herbs without the risk of active substance loss.

In the equipment (Fig. 29) a high-capacity fan forces a large amount (20 000–30 00 m^3/h) of low-moisture air into the fresh herb bulk via the duct system with high pressure. The system may also be operated in the case of a high humidity content of the outside air, as mobile thermal blowers may be incorporated in the system. With

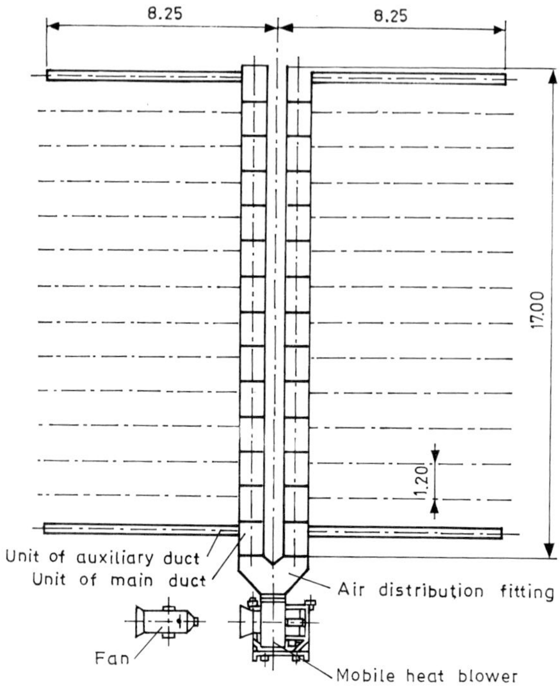

Fig. 29 Drier with forced ventilation

the thermal blower the temperature of the drying air can be increased, and the relative humidity content can be reduced to the level required for drying.

Warm-air drying. This method provides a much better product than the methods described above. For this reason, the main preservation process for medicinal plants is this method. The basic difference between this and the other methods is that instead of placing the plant outside the pipes, the plant is placed into the closed system, which also provides a means for the flow of warm air.

The most important drying parameters (air temperature, humidity content and flow rate) can be controlled with the warm-air drying equipment. The control range varies with the type of equipment. The duration of warm-air drying is a few hours as opposed to the several days necessary for the methods described above.

Both the plant and the air move in the presently applied warm-air drying method. Generally, the movements of the two materials are opposite to each other. This counterflow drying is effective and economical.

A variety of warm-air drier equipment has been developed during the past 10–15 years. The significance of tunnel driers is diminishing.

The *plate-chamber driers* have already been used for a long time for the drying of medicinal plants and vegetables. The most widely used types are the Dehidro and Shielde models (Fig. 30).

Their advantages are as follows: the surface area requirement is relatively small, they are effectively usable for drying flowers, leaves and other smaller plant parts, the plants are not broken, and they are economical owing to the recirculation of drying air. An amount of 3.0–5.0 t of raw material can be dried in 24 hours, depending on the conditions of the material to be dried. The disadvantage is that stalky plants cannot be dried, owing to the small drying space. The plates should be moved according to the process of drying in order to provide continuous drying.

The labour requirement is high for the placing of plants into the plates, and for removing the dried material after drying.

Conveyor driers are the most up-to-date equipment for the drying of medicinal plants (Fig. 31). The capacity of these is much higher than that of the two types above.

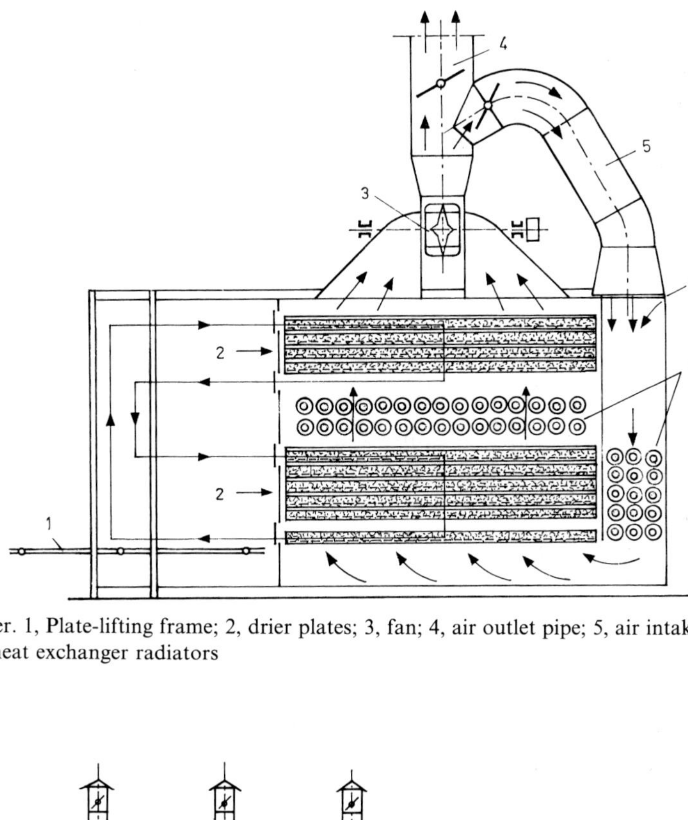

Fig. 30 Schielde–Favorit drier. 1, Plate-lifting frame; 2, drier plates; 3, fan; 4, air outlet pipe; 5, air intake pipe; 6, fresh air intake; 7, heat exchanger radiators

Fig. 31 Conveyor drier

They operate continuously, and the parameters influencing the process of drying can easily be controlled, or even automated. Their labour requirement is low.

The drying time for medicinal plants is generally 2.5–6 hours, depending on the water content of the plant components to be dried. The speed of belts, and the temperature and flow rate of drying air, can be individually controlled at each stage. This provides an economical way of drying, which is adjusted to the specific features of the plants in question, and controlled according to the progress of the drying. The conveyor drier has the following advantages: the operation is economical; high capacity; small labour requirement; and continuous preparation. Almost everything can be dried and high-quality products are obtained. In 24 hours the dried amount of raw plants is 8.0–14.0 t. The disadvantages are: investment expenditure is high; the specific energy requirement is high, owing to the double heat exchange (8.37×10^6 J); and it can only be used for cut plants, as the longer plant parts may stick between the conveyor belts.

Hot-air drying. This is the most widely used drying method in agriculture. It is mainly used for preserving hays. It can only be used to a limited extent in the drying of medicinal plants (nightshade, periwinkle). While the temperature of the drying air is 40–80 °C for warm-air drying, and the duration of drying is several hours, the temperature of hot-air drying is 200–1000 °C, and the duration is only several (2–5) minutes.

The capacity of fast driers is 1.8–2.0 t of dried drug per hour. The disadvantage is that the mixture of fume and air directly touches the plants to be dried. For this reason it is usable only for drugs intended for industrial processing (owing to the harmful sulphur compounds).

In several fields of the food industry the most up-to-date methods and equipment are used (such as dielectric freezing fluidization and spray drying). These methods are either too expensive for the drying of medicinal plants or they are not applicable for the careful drying of medicinal plants.

Effects of drying

A number of changes occur in the plant parts during drying. Owing to the removal of the water, the *weight loss* is significant. The extent of this weight loss varies over a wide range, depending on the properties of the plant components.

In terms of planning and economics, it is advisable to know the expected *drying ratio*. This number is the mass of raw plants in kg required to produce 1 kg dry drug. According to the experience of many years, for producing one kg of drug, 5–8 kg of fresh flowers required, 5–6 kg of fresh leaves, 4–5 kg of fresh herbs, 3–4 kg of raw roots, or 1.2–1.5 kg of fruit.

Operations after drying

After drying, the drugs are generally prepared as products for the wholesale or retail trades. The operations serving this purpose are called the manipulating operations. The most frequent operations after drying are as follows: sorting and cleaning, screening, stalk removal, crumbling, cutting up and grinding.

Extraction Processes for Essential Oils

The term 'natural essential oil' means or covers all aromatic extracts from vegetable matter (flowers, leaves, branches, roots, gums, fruits). The components of essential oils giving the scents of the plants are mostly terpenes and derivatives of terpenes (alcohols, esters, aldehydes, ketones, etc.). Among the other compounds the most important are sulphur compounds (in the essential oils of mustard, garlic, horseradish, etc.), and several aromatic compounds (phenols, aromatic aldehydes, coumarins).

The most widely used procedure for extracting the essential oil from the plants is *steam distillation*. With this method the volatile compounds and compounds insoluble in water, i.e. the essential oils, can be extracted. Steam distillation produces the essential oil in the purest form and requires relatively simple equipment and moderate labour even on an industrial scale. Three types of distillation are distinguished according to the method of execution; i.e. water distillation (Fig. 32a), water and steam distillation (Fig. 32b), and steam distillation (Fig. 32c). All three processes are based on the same principle, the so-called steam distillation or hydrodistillation; the different names refer only to the technological dissimilarities.

Solvent extraction processes may be used for plants which contain small amounts of essential oil or which dissolve in water very easily (mainly the scent materials of some flowers), or for plants which are not volatile, but whose soluble ingredients are valuable.

The third procedure of extracting the scent materials is cold *pressing*. This is used for extracting the essential oils from the fruits of citrus plants.

Classic steam distillation

The boiling points of the components of various essential oils differ significantly. For example, the boiling point of terpene hydrocarbons at atmospheric pressure is 160–180 °C, while that of terpene alcohols is 200–300 °C, and that of sesquiterpenes and their derivatives is 260–290 °C. Most of these materials have pleasant odours, as they evaporate even below their boiling points. Simple distillation is basically a process whereby the liquid is transformed into steam at its boiling point, and then the steam is condensed back into liquid with the help of cooling.

Steam distillation (hydrodistillation) is a simultaneous distillation of two different fluids, generally water and a material insoluble in water (e.g. essential oils). The physical effects of steam distillation are not the same as those of simple distillation.

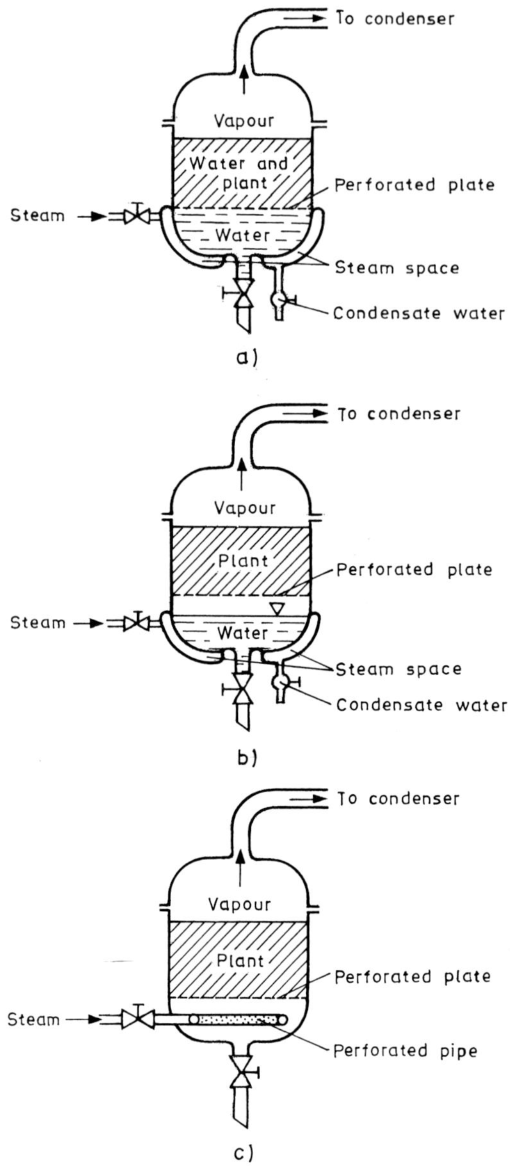

Fig. 32 Varieties of essential oil distillations. (a) Sketch of water distillation; (b) sketch of water and steam distillation; (c) sketch of steam distillation

Let us suppose that a vessel is being heated in which water and a liquid insoluble in water are being vigorously agitated. In this case the vapours of the two liquids generate a pressure in the air space which is related to the temperature. The mixture will start boiling at the temperature where the sum of water steam pressure and oil steam pressure reaches the external pressure.

$$P = p_v + p_o$$

(P is the pressure at which the distillation is taking place, p_v and p_o are the vapour pressures of water and oil at their common boiling point.)

This relationship is the characteristic feature of water steam distillation. The boiling point of such a mixture is below that of water or of oil. The steam pressures of the respective components are less than the external pressure; consequently they would not boil separately at the given pressure and temperature. This is why the essential oil components with boiling points as high as 300 °C can also be extracted from plants at ambient pressure and at a temperature below 100 °C. It follows from the relationship that the temperature will not rise during the distillation as long as both liquid phases are present. The composition of the steams, and that of the two phases of the distillate after condensation, will remain the same during distillation.

At the beginning of the distillation the cold plant contacts the hot steam, the temperature of the plants rises, and at the same time water accumulates on the surface of the plant. Because of the high temperature, the cell walls are ruptured and capillary gaps develop on the walls. The essential oil drops will flow through these capillary gaps. Owing to the capillary sizes, actual flow cannot develop; the movement of the essential oil is the result of diffusion instead. This diffusion is the most determining factor for the rate of distillation of a plant. Other influencing factors are the distance to the cell, the concentration of the essential oil component within the cell and in the water layer at the surface, and the extent of saturation of the steam space with essential oil.

The size of the cell capillaries depends on the preparation of the plant and the conditions of the distillation, as well as on the specific material properties. It is also influenced by the humidity content of the plant, the mode and extent of preparation (drying, cutting) and the pressure and temperature of the steam used for the distillation. By proper selection of these factors the rate of distillation can be increased and the essential oil extraction can be improved.

Chemical changes

At a distillation temperature close to 100 °C and in the presence of water and air a number of essential oil components might undergo chemical modification owing to hydrolysis, oxidation and polymerization. Some of these chemical reactions are advantageous, because they help to generate the essential oil (or its essential components). For example, the chamazulene content of camomile oil and yarrow oil forms from the prochamazulenes of the flowers at the distillation temperature. The distillation of these takes a considerable time, as the formation of chamazulene at 100 °C is rather slow.

Other chemical modifications caused by the high temperature are detrimental to the quality of essential oils, for the generated compounds are partly unstable, and are easily decomposed or polymerized.

The steam, and especially the superheated steam, might also cause the decomposition of other materials. If, for example, the hemicellulose content of the plant is disintegrated, then furfural is formed, which contaminates the essential oil. The chemical changes caused by the high temperature can be controlled, hastened or hindered according to the required goal, by adjusting the time of distillation, the steam speed and steam pressure, and the temperature.

The extent of ester hydrolysis can be decreased by shortening the time of distillation, i.e. by increasing the steam flow rate. The stabilities of various esters are dissimilar. Linalyl acetate, which is one of the main ingredients of lavender oil and clary sage oil, is easily hydrolysed, while other esters, e.g. menthyl acetate from peppermint oil, will not decrease during distillation.

Preparing operations

Essential-oil-holding cells and passages can be found on the surface of the plant and also in the tissues inside the plant. The aim of preparation prior to the distillation is to make these parts accessible to the steam, and to achieve certain conditions in which the diffusion of the essential oil can take place. Care should be taken to avoid restriction of the steam flow by the plant, or overpressure will occur in the distillation vessel.

In order to exploit the available distillation vessel space, the plants, which are distilled in a green or withered state, are cut up. The efficient extraction of essential oil from the roots and seeds can be ensured by cutting and crushing the roots and seeds. The extent of cutting depends on the raw material properties and the method of distillation.

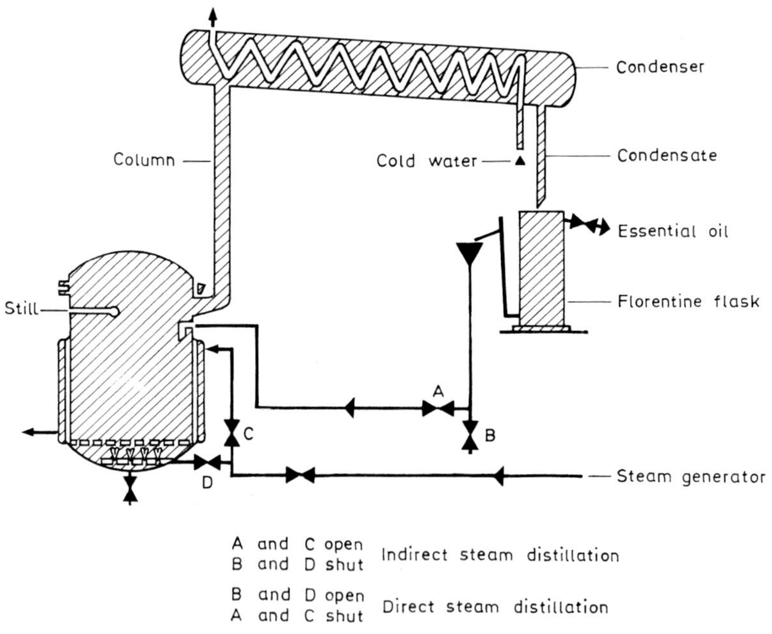

Fig. 33 Steam distillation diagram (Meyer-Warnod, 1984)

Fig. 34 Intermittently operated equipment of steam distillation

The procedure of steam distillation

The steam distillation of essential oil can be made with intermittently or continuously operated equipment. The operations are basically the same. The most widely used equipment is that for intermittent operation. The intermittent equipment (Fig. 33) consists of the following assemblies: *still* (evaporating vessel), *condenser* (cooler) and Florentine flask (essential oil separator).

The steam required for the distillation is produced in boilers, which are heated by either oil or coal, or both. The pressure of the steam in the boiler is 0.4–0.7 MPa and its temperature is 150–170 °C.

The still is basically a standing cylinder or a frustum cone widening upwards. Its material is metal (steel, copper), or sometimes wood. Its size varies according to its purpose, the upper limit being about 30 m^3. Owing to the attached condenser, the still constitutes an open system, i.e. the pressure is not over atmospheric in it.

The *intermittently operated vessel* (Fig. 34) is filled up manually by conveyor or by machine grab, or simultaneously with chaff cutting. It is very important to distribute the plants uniformly in the equipment; otherwise the steam would flow in the direction of less resistance, bypassing part of the plant material. Extensive compacting is also disadvantageous, because the resistance to the steam flow will increase, which will give rise to an increase in pressure, causing the plant material to move upwards, and in the meantime the steam will open passages in the plant material. In order to avoid this problem, when distilling roots, seeds or fruits, loose material (straw) or perforated metal plates are placed between the layers.

Table 5 The relationship between the steam speed and duration of distillation (intermittently operated distilling vessel of 4 m^3 capacity)

Name of plant	Steam speed (kg/h)	Duration (min)
Angelica root, raw	100–150	300–500
Angelica root, dried	100–150	300–700
Peppermint, raw	200–250	120–150
Peppermint, withered	200–250	90–120
Peppermint, dried	150–200	90–120
Common fennel, fruit	100–200	250–400
Hyssop, raw	250–300	60–90
German camomile flower, dried	100–150	300–600
Dill, fruit	100–200	250–350
Dill, plant, raw	150–250	120–180
Carrot seed	100–150	250–300
Lovage plant, raw	150–200	180–240
Lovage root, raw	100–150	300–500
Lovage root, dried	100–150	300–700
Lavender, raw	300–350	40–60
Clary sage, raw	350–400	40–60
Parsley leaves, raw	150–200	120–180
Parsley seed	100–200	250–400
Roman camomile, raw	300–350	60–120
Roman camomile, dried	300–350	60–90
Tarragon, raw	150–200	120–180

The specific volume of the plants (the amount which can be loaded into the vessel) depends on the humidity content, on the extent of cutting, and on the extent of compacting:

Green plant, without cutting	200–300 kg/m^3
Green plant, cut (8–10 cm)	300–500 kg/m^3
Withered plant, without cutting	150–200 kg/m^3
Withered plant, cut (8–10 cm)	200–350 kg/m^3
Root, without drying, cut (4–5 cm)	550–650 kg/m^3
Dried seeds, crushed in rolling mill	500–650 kg/m^3

In the closed vessel the essential oil is evaporated by the flowing steam.

The duration of the distillation is determined by the need to remove the essential oil as completely as possible while keeping the process costs within competitive limits. Determining factors are the properties of the plant, the type of distillation vessel, and the amount and rate of steam. The distillation may be stopped when the distillate flowing from the condenser no longer contains essential oil drops. In practice, the distillation is discontinued when a continuous oil layer is no longer formed on the surface of the distillate. This is because the essential oil adhered to the wall of the cooler is washed away very slowly, so the actual cessation of essential oil distillation will have happened 10–15 minutes earlier than the observable cessation. The data relating to the steam speeds and distillation times are compiled in Table 5 (for vessels of 4 m^3 volume).

In the *equipment operated continuously* the plant containing the essential oil is moved in the opposite direction to the steam flow. The speeds of the steam and the plant are controlled so as to achieve the time necessary for the distillation of essential oil while the plant passes through the equipment, and so that the necessary amount of steam will flow through the plant at the same time. The equipment is basically a standing cylinder, in which the plant moves downwards and the steam flows upwards. The continuously operated equipment has a better specific volume exploitation than the intermittently operated equipment, and additionally its steam requirement is lower (Fig. 35).

In *containerized equipment* the functions of harvesting and distillation are joined (Fig. 36). The right-prism shaped container is mounted on wheels, and has a volume of 25–35 m^3. It is towed by the harvesting machine, and the cut plant material is directly loaded into it. When the container is full, it is closed, the steam inlet and vapour pipe are attached to it, and the essential oil can be distilled just as in intermittently operated equipment. The distilled plant material is removed from the container in a separately allocated place, to where the container can be towed on its own wheels. The mass of plant material which can be placed into the container is 150–250 kg/m^3. The ratio of height to diameter of the container is less than that of a standing cylinder, so its specific steam requirement is higher. These disadvantages

Fig. 35 Distillation equipment of continuous operation

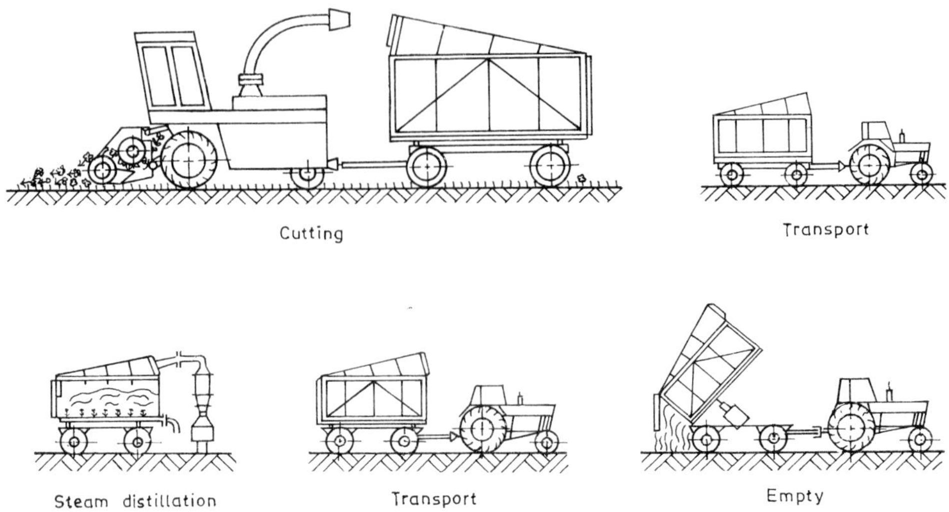

Fig. 36 Containerized distillation

are compensated for by the complete mechanization of the process and by the fact that the process needs very little labour.

The water steam and the vapours of the essential oil become liquids again when *cooled in the condenser*. The type of the condenser (cooler) is generally spiral-pipe or pipe-bunch (Fig. 37). Its material is metal: copper, aluminium or stainless steel. The cooling is effected by circulating water. The capacity of the condenser depends on the area of cooling surface and on the temperature and amount of cooling water. The amount of water is determined by the applied steam amount, and by the allowed temperature difference between inlet and outlet cooling water. If the inlet water temperature is 20 °C and the outlet water temperature is 50 °C, then about 20 litres.

From the oil–water mixture coming from the condenser the *separation* of essential oil is made in a so-called Florentine flask on the basis of differences in specific weights. These reservoirs are made of glass (the smaller ones), or of metal (copper, aluminium, stainless steel). The operational principle of the reservoirs is determined by whether the essential oil to be separated is heavier than the water or not. For the purpose of obtaining sufficient separation, the Florentine reservoirs are designed so as to provide straight water flow and to prevent turbulence which would hamper the sedimentation of oil. The size of the reservoir should be determined so that it would hold the amount of distillate generated in 40–50 minutes, i.e. the distillate and the water remain in the reservoir for this time. Sometimes a number of reservoirs are connected in series in such a way that the volumes of the reservoirs get larger step-by-step.

The separated essential oil should be *cleaned* to remove the floating impurities, and the water should be removed by filtration and clearing. In order to ease the operation it is advisable to perform a preliminary sedimentation, because the impurities are hard to filtrate and would block the pores of the filter very quickly. Simultaneously with the filtration, about 0.5 per cent of water-free sodium sulphate is mixed into the oil in order to remove the floating water drops and to reduce the amount of dissolved water. The filtration and clearing is carried out with a sedimenting centrifuge, with a Seitz filter or through filter paper placed in a funnel.

The water coming from the Florentine flask still contains some essential oil in a dissolved state or finely dispersed. The amount of this essential oil depends on the size of the flask, on the shape of the reservoir, on the rate of distillation, on the temperature of the distillate, and on the specific weight and chemical composition (solubility) of the essential oil. The composition of the essential oil of *cohobated water* differs from that of direct or primary essential oil.

A number of methods can be applied for extracting the essential oil content of cohobated water. In the case of water distillation and of water + steam distillation, the cohobated water is returned into the vessel, and in this way no loss occurs. In the case of steam distillation, the essential oil of the cohobated water is separated by another operation, redistillation (cohobation).

Other methods of steam distillation

Hydrodiffusion

In contrast to usual classical steam distillation, the steam enters from the top of the vessel. The resulting solution is condensed below the vessel, and oil plus water are separated in a Florentine flask (Fig. 38). The main advantages of this method are the energy saving resulting from reduced distillation time and therefore reduced steam

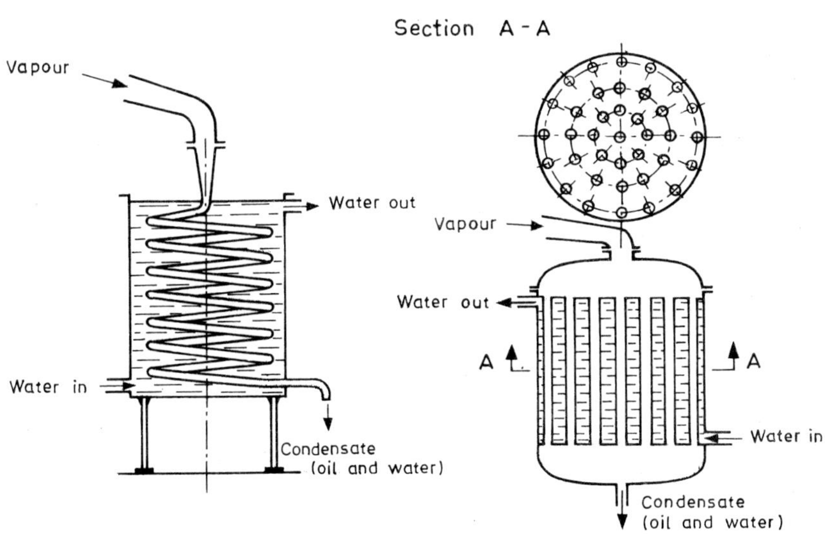

Fig. 37 Sketch of spiral and pipe bundle type cooler

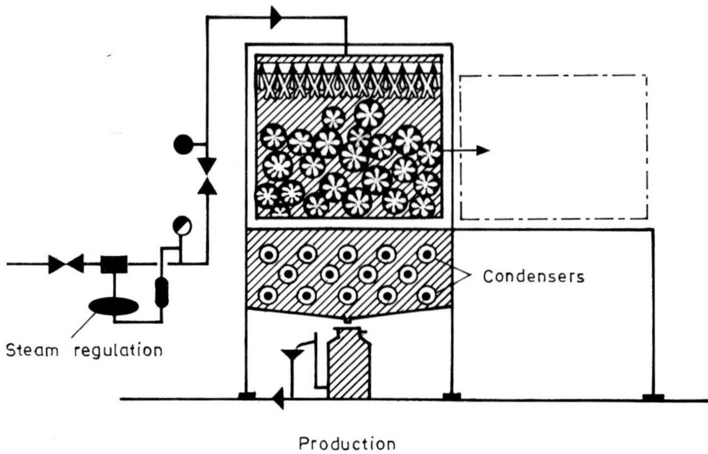

Fig. 38 Hydrodiffusion diagram (Meyer-Warnod, 1984)

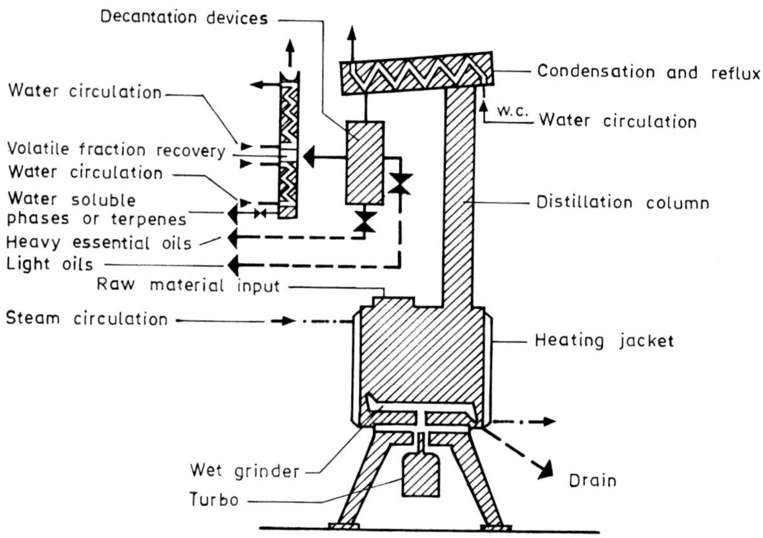

Fig. 39 Accelerated hydrodistillation diagram (Meyer-Warnod, 1984)

consumption, and the absence of hydrolysis because the raw material is never in contact with water (Meyer-Warnod, 1984). Hydrodiffusion is primarily used for seeds.

Accelerated distillation

This patented device (Aroma Process, 06 Mougins/France), shown in Fig. 39, accelerates distillation by rupturing the oil-containing plant cells with the grinder of a turbine. This process decreases distillation time and makes possible extraction of the most volatile components in the saturated gas at the top of the condenser, which is not possible with the usual steam distillation process. The yields may also increase (Meyer-Warnod, 1984).

Solvent extraction

The extraction vessel is filled with raw material and covered with an organic solvent, either hexane, benzene, toluene or other similar solvents (Fig. 40). The operation temperature varies with the solvent used and its boiling point (60–80 °C). After maceration, the saturated solvent-containing aromatic substances are pumped to a concentrator and heated at the solvent boiling point. The solvent is recycled and the volume of the solution is reduced to one-tenth. The product is then pumped into a vacuum finishing concentrator. The product obtained after final concentration is the *concrete,* which is usually waxy solids. *Absolutes* are highly concentrated perfumery materials obtained from concretes be repeated extraction with ethanol, followed by chilling of the extract (to precipitate waxes and other non-odorous matter), filtration or centrifugation of the remaining alcoholic solution, and finally removal of most of the alcohol by evaporation at moderate temperatures and under partial vacuum. Absolutes are usually liquids and are entirely soluble in alcohol.

Resinoids are obtained in a floating filter extractor by filling the vessel with ethanol, chlorine solvents or even benzene, strongly agitating and then filtering, and finally evaporating the solvent. When no solvent remains, the obtained product is the so-called resinoid, which is usually a viscous liquid or semi-solid.

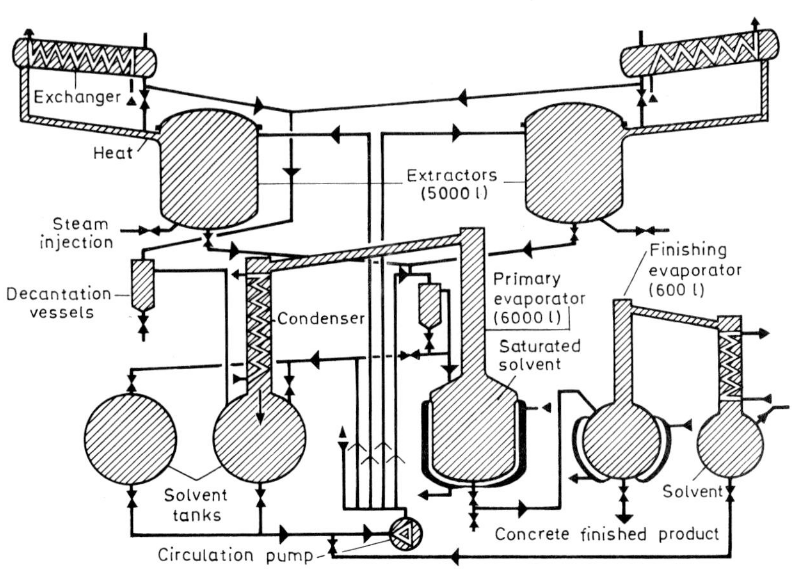

Fig. 40 Solvent extraction unit (Meyer-Warnod, 1984)

Vacuum distillation

In many cases the oil obtained either by solvent extraction or by steam distillation does not exactly fit the perfumer and/or the flavourist's needs (Meyer-Warnod, 1984.). It is sometimes necessary either to eliminate certain constituents, e.g. terpenes or heavy colourings, or to reinforce the percentage of major aromatic constituents such as irone, aldehydes, linalool or cedrol. Figure 41 shows a distillation column where, according to their boiling points, various components can be isolated. The temperature gradually increases during the process in order to separate the volatiles first, then the other components.

CO_2 extraction

According to Meyer-Warnod (1984) the advantages of this new method of extraction are as follows: no residual toxic solvents, low process temperature, important selectivity, non-flammability and energy saving. The method is based on the good solubility in CO_2 of most components of aromatic raw material and/or semi-finished products such as concrete or even essences.

The extractor (Fig. 42) is filled with raw material, and CO_2 is added. Pressure and temperature inside the vessel are regulated according to the manufacturing process.

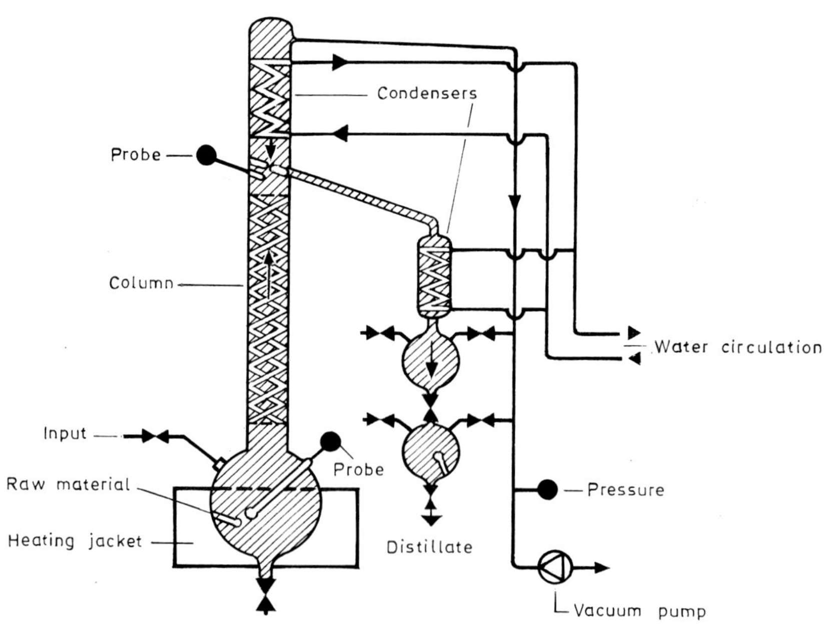

Fig. 41 Vacuum distillation diagram (Meyer-Warnod, 1984

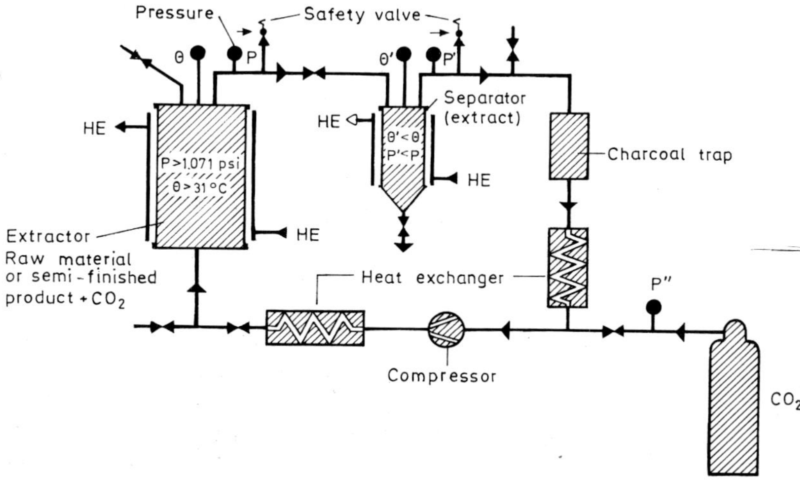

Fig. 42 CO_2 extraction diagram; HE = heat exchanger (Meyer-Warnod, 1984)

Saturated CO_2 is then pumped into the separator vessel, in which pressure and temperature are lower than in the extraction tank. The product is removed and the remaining CO_2 pumped into the charcoal trap tank to be purified and recycled.

References

Meyer-Warnod, B. (1984).'Extraction processes and application to some major oils', *Perfumer and Flavorist,* **9,** 93–104.

UNCTAD/GATT (1986). *Essential Oils and Oleoresins: A Study of Selected Producers and Major Markets,* International Trade Centre, UNCTAD/GATT, Geneva.

Storage, Packing and Certification

After drying and the extraction of essential oil, the drug should be packed and stored for various period of times in the production plant until delivery.

Storing drugs

Only well-ventilated dry rooms are suitable for storing drugs prepared by drying. The surrounding air is capable of taking up humidity (drying), and in certain circumstances it is also able to give up humidity. Most of the drugs are somewhat hygroscopic, and for this reason their humidity content might increase significantly in humid environments. Even if the storage is only temporary, in humid conditions the quality of the drug may decline to a great extent.

As the drugs are used in medicine directly or after processing, it is very important to keep the storage rooms clean, and to disinfect the rooms regularly. It is a very important rule that the poisonous drugs (Belladonnae folium, Daturae innoxiae herba) should be kept separately from other drugs, in a separate room. Pungent drugs (Valerianae radix) also require separate storage so that other drugs are not be contaminated by the unpleasant odour.

Storing essential oils

During the storage period essential oil may undergo chemical changes, leading to a deterioration in quality. The unsaturated compound of essential oils, especially the terpene hydrocarbons, are oxidized by the oxygen in the air, and the oxidation products become polymerized. These reactions are accelerated even in low temperatures by very low concentrations of heavy metal compounds, especially copper compounds. Chemical reactions causing the decomposition of certain compounds may take place during storage, and might also occur during distillation. The rate of these reactions is slower at low temperatures, but they could still be of significant extent during a prolonged storage time.

Dissolved water, even if its concentration is below 1 per cent, can cause the decomposition of ten times as much ester, thus increasing the acid content of the essential oil, which causes further decomposition.

These harmful processes must be prevented during storage. The rate of chemical reactions is reduced by decreasing the temperature; that is why storage should be at as low a temperature as possible. Oxidation may be prevented by keeping the air away from the product. For this reason the storage vessels should be filled up to their capacity and provided with airtight seals. In order to prevent oxidation, inhibitors may also be used. The effect of heavy metal ions can be compensated for by adding a very small amount of a complex-forming compound (citric acid), and by adding inhibitors at the same time. Essential oils, especially those stored in glassware, should be stored in a dark room, because polymerization processes may also take place under the influence of light.

A basic requirement is the accurate inventory of drugs and essential oils, even if the storage is only temporary. As well as recording the actual amount, the storage location should be mentioned. Every item of stored drugs and essential oils should be provided with an accurate inscription. The fire-prevention rules and the accident-prevention rules should be observed when storing drugs. The communication routes, which should be allocated according to the relevant rules, should not be blocked even for a short time.

Packing drugs

The mode of packing depends on the kind of drug, amount, mode of transport, distance to be transported and special requirements of the customer. The most widely used modes of packing are: bale, small or large suck, paper or plastic sack, wooden case, and cardboard box.

Generally, the drugs which can be compressed are packed into bales (herbs). The bales are made by high-capacity hydraulic baling machines. The mass of the bales is generally 60–100 kg. The bales are sewn into sackcloth (or plastics). Drugs which cannot be compressed are packed into large bale sacks (roots, barks). The bale sacks are closed by sewing after filling.

Drugs with high specific weights (fruits, seeds) are packed into small bags. The highly hygroscopic drugs (flowers, dog rose, juniper) should be packed mainly in paper or plastic (polyethylene) bags. Valuable drugs which are sensitive to handling are packed into wooden or cardboard boxes (Chamomillae vulgaris flos, Menthae piperitae folium).

Essential oils are mainly stored in metal vessels, canisters or tanks. The materials of the vessels are galvanized steel (or steel lined with lacquer) or aluminium. The best way of storing small amounts of oil is in glassware. Plastic vessels are not recommended for storage, because their material will chemically react with the compound of the essential oil.

Certifying drugs

Similarly to other products, drugs are certified prior to putting them into commercial circulation. The certification is made by considering the valuable properties of the drug (colour, odour, taste, cleanliness, content of active substance). In Hungary the certification (classification, permission to put into commercial circulation) is performed for every drug. This is made according to the latest edition of the Hungarian Pharmacopoeia, or to the national standard specifications, if the drug is not included in the Pharmacopoeia. The Pharmacopoeia specifies the strictest requirement for the drugs. National standards and standards of specific industries are obligatory for the drugs.

The certification of essential oils is made by trials using the senses, and physical or chemical methods, similarly to that of other drugs. The most important examination is made by the senses, with which the odour or the taste is qualified. Physical or chemical tests are not sufficient in themselves for the qualification of essential oil. Statistical treatment of checks with the senses will result in an objectively verifiable classification.

Charecteristic physical and chemical test results of essential oils are specified by the Pharmacopoeia and the relevant standards. The most frequent properties tested are: specific mass (density), refraction index, optical activity, and solubility in alcohol of 70, 80, 90 per cent concentration. The most important chemical properties are: acid content, ester content, alcohol content, aldehyde content and ketone content.

PART 2 THE CULTIVATION OF MEDICINAL PLANTS

Ergot (*Claviceps purpurea* (Fr.) Tul)

Ergot is mentioned in the works of Dioscorides and Galen. In obstetrics it was used by the ancient Chinese. In Europe, ergot became known only after the establishment of rye cultivation. It caused a serious and frequently fatal disease called ergotism, which was not initially recognized. As a medicinal plant it was first mentioned by Lonicer in 1582. It became an authorized medicinal plant in England in 1836, and since then it has become authorized all over the world.

Ergot is a very important plant material in modern medical practice and can be exploited in a very wide range of applications. The drug (Secale cornutum) is included in many pharmacopoeia of the world, but it is regarded today as a base material for medicines. Its active ingredients (alkaloids) affect the smooth muscles, the digestive and central nervous systems and the operation of the circulatory system. Its preparations are indispensable in gynaecology (at childbirth), and in treating high blood pressure, migraine and nervous system complaints. Many medicines are produced from ergot.

Ergot is cultivated in many parts of the world. High production is maintained by Switzerland, Czechoslovakia, Finland, Austria and Poland. In Hungary the first plan to cultivate ergot emerged in the second half of the last century. As early as the turn of the century, Práter *et al.* tried to produce ergot, with no success.

A method of growing ergot was first developed by a Hungarian scientist, Békésy. His method, which was patented in 1934, is being used all over the world, and Békésy was awarded the Kossuth price in 1952 for its development. The growing of ergot has been continuously developed since the end of World War II.

Characteristics

Ergot (rye-smut, spurred rye) is a parasitic ascomycete belonging to the family Clavicipitaceae, and having a number of developmental stages. According to present knowledge, the genus *Claviceps* includes 22 species and many varieties. Ergot itself was described as a number of separate fungus species (Secalis mater Thal., *Sphacelia segetum* Lev., *Cordyceps purpurea* Fr.) until 1853, when Thulasne showed that the forms regarded as various fungi are in fact the three developmental stages of the same fungus. The main host plant for the ergot is rye, so it can be found anywhere where rye is cultivated.

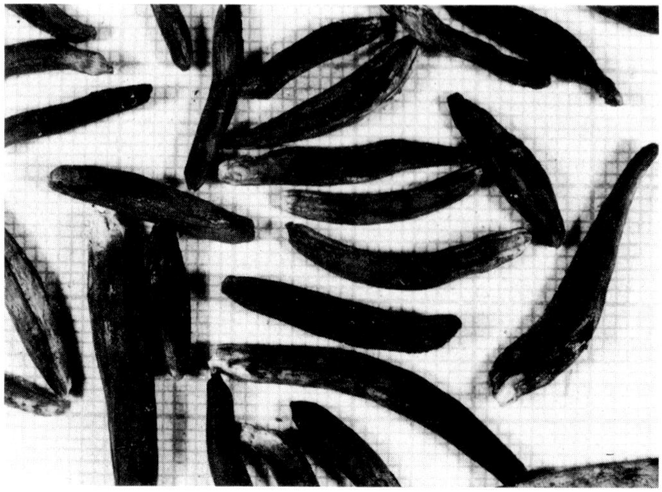

Fig. 43 Sclerotia of the ergot

The wintering form of the fungus is the sclerotium (Fig. 43), which is actually the drug. It has a spindle shape, straight or slightly arched. Its length is 10–50 mm and its thickness is 2–5 mm, depending on the host plant and on the taxon involved. Its surface is violet-brown or black, the broken surface is greyish-white, and the body is fresh and meaty, becoming horny after drying. The stroma develop on the sclerotium, and are purple-coloured and tiny. On their insides there are a number of peritheca, in which fibrous ascus spores develop. On the fruit of the rye long hyphal fibres grow from the ascus spores, forming a loose sphacelium. The vegetative conidia of the fungus are reeled around the end of fibres, and can be found in syrup (honey-dew).

The *active ingredients* of the ergot are alkaloids. No alkaloids can be found in ergot which is in the stage of sphacelia. The alkaloids are formed simultaneously with the development of the sclerotium, and the maximum content is achieved at the time of a completely developed sclerotium. The biosynthetic generation of alkaloids starts from tryptophan. The alkaloids are all true alkaloids, and belong to the indol-framed alkaloids. The alkaloids are classified into two groups according to their ability to dissolve in water, which plays a role in their physiological effects, and are also further divided into three groups according to their chemical composition and their pharmacological effects as follows:

(1) Water soluble:
Ergometrine group: ergometrine, ergobazine, ergine;
(2) Not soluble in water:
Ergotoxine group: ergokristine, ergokriptine, ergokornine;
(3) Ergotamine group: ergotamine, ergozine.

The total alkaloid content of the sclerotiums varies within a very wide range as a function of the host plant and of the given chemotaxon. Among the taxons growing spontaneously Békésy (1938) found ones without alkaloid content of the cultivated ergot is generally between 2.1 and 2.6 thousands expressed in ergotoxine.

Addition to the alkaloids, the ergot contains 18–40 per cent fixed oil (this is why the ruptured sclerotiums get rancid quickly), colour materials (antrakinone derivatives), amines, and other compounds containing nitrogen (hydrocarbons, protein).

The *process of development* of the ergot is shown in Fig. 44. The germination of the sclerotium (forming the productive body) starts at 9–10 °C after previous cold exposure according to Békésy–Garay (1960), and ascus spores ripen almost completely in three weeks in any case (actually, the optimum germinating temperature is 18–22 °C). In the opinion of Békésy (1960), the sclerotiums retain their germinating capacity for two years in aerob conditions, while they die in the first year in case of anaerob conditions. The ripe ascospores generated by sexual reproduction start to germinate at once as they get on the stigma of the rye or of other grasses, the hypha fibres penetrate to the bottom of the ovary and grow around it. The ovule is also weaved and spongelike sphacelia is formed. A large number of conidia reels down the hypha

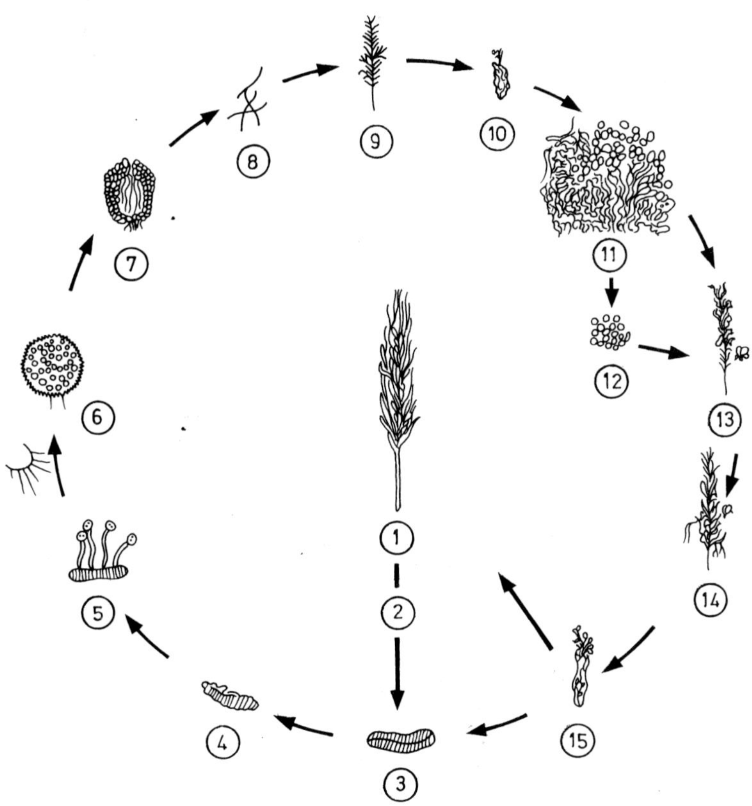

Fig. 44 Developing stages of the ergot. 1, ripe sclerotia; 2, drops to the ground; 3, wintering; 4, germination; 5, matured pistils; 6, pistil; 7, peritechium; 8, ascospores; 9, rye; 10, attacked ovulum; 11, sphacelia (mycelium); 12, conidia; 13, honey-dew on the rye; 14, secondary infection; 15, initial sclerotia

fibres, which leak out in the form of sticky drops from the among the glumella embedded in honey-dew. The formation of sphacelia (the appearance of honey-dew) takes generally 7–10 days.

The large number of conidia are transferred to other flowers in the vicinity by insects favouring the sweet honey-dew.

The stage of sphacelium (mycelium) and the generation of honey-dew lasts for 5–15 days, depending on the weather and on the developmental stage of the rye. After that the sclerotium starts to develop. The formation of the sclerotium begins at the bottom, where the dense hyphal web grows rapidly, forming sphacelium. The formation of alkaloids and of colouring materials is also started at this time. The duration of sclerotium development is 16–20 days. In this way, 4–6 weeks are necessary, on average, for the formation of ripe sclerotia from the time of infection.

Varieties

The alkaloids with physiological effects which accumulate in the ergot are not the same in terms of pharmaceutical exploitation and action mechanism. For this reason, the most appropriate biotypes (chemotaxons) have been practically sought from the beginning of ergot cultivation. Instead of the varieties containing high amounts of alkaloids, today the varieties containing single alkaloids are cultivated; these are more advantageous for the extraction of active ingredients and for medical purposes.

Environmental requirements

The environmental requirements consist of the ecological conditions of the fungus and its host plant, the rye. The optimal development of the parasitic fungus requires humid, rainy moderately warm weather, especially at the time of infection. Ergot can be successfully cultivated in areas where the annual amount of precipitation is over 500 mm. Especially advantageous for the development of the fungus is air having 65–90 per cent humidity during the period of infection, even when it is combined with drizzling rain and with moderately warm temperatures (16–19 °C). A vapour content near saturation gives rise to a prolongation of the flowering period of the rye, which is advantageous for the secondary infection. In dry, warm, or even windy weather, the conidia dry out. A high daily average temperature (over 20 °C) during the period of formation of the sclerotium is disadvantageous for not only the development of the fungus, but also for the synthesis of the alkaloids.

As a parasitic fungus, ergot is in close connection with the host plant. The most appropriate rye varieties are those which:

(1) Are at a uniform developmental stage at the time of formation of the spikes;
(2) Have their spikes almost at the same heights;
(3) Have their spike flowers in dense rows in the spikes, so that they will not lean away from the infecting needle prick;
(4) Have glumella which can be penetrated easily;
(5) Have stalk straw which is strong, short and resistant to the leaning effects.

For the production of ergot, a rye having uniform development and strong straw stems is necessary, which in turn requires a soil rich in nutrients and with good waterhandling capacity. It is not worth cultivating ergot on poor, sandy soil or on too firm soils liable to freezing. The plane or slightly sloping areas and hillsides are good for cultivation, where mechanized ergotizing and harvesting can be performed reliably. Ideal environmental conditions are provided by humid valleys and areas near water and woods protected from the wind.

Cultivation

Sowing the host plant

The working procedures for rye are mostly the same as those used when the goal of production is fodder. Rye is not especially sensitive to the kind of plant previously grown. However, as it is advisable to sow the rye early, the best previous plants are those which leave the field early, such as pea, mustard, coriander and green fodder mixtures.

The *supply of nutrients* for rye is similar to that for wheat. At the time of producing ergot, however, an increased nutrient supply is necessary, and the distribution of the nutrient is also somewhat different.

The total phosphorus and potassium requirement and one-third of the total nitrogen requirement (the actual value depends on the nutrient content of the soil) is introduced into the soil at the time of ploughing for sowing. The requirement is determined as the nutrient amounts required for the production of ergot + rye equivalent to 3.5–4 t/ha of rye. The balance of nitrogen should be spread in the spring in two portions.

For the purpose of producing ergot, the rye should be *sown* in the middle of September. The row spacing should be adjusted to the machines used in cultivation.

The 'Goldhamster' infecting and harvesting machines, however, require sowing in strips. Ten rows are involved in one strip, and the row distance is 120 mm. Unsown areas 700 mm wide are left among the strips; these are necessary to accommodate the width of the wheel of the bridged tractor. The optimum sowing depth is 40–60 mm. The sowing seed requirement is 100–110 kg/ha. The further nursing procedures of the rye (herbicide treatment, protection) are the same as those for fodder rye.

Infecting rye with conidia

The infection of the rye consists of the following operations:

(1) Producing the infecting material (conidia culture);
(2) Preparing the infecting material (spore suspension);
(3) Applying the infecting material.

The infecting material, the mycelium culture containing conidia, is produced from sclerotia which have been analysed for active ingredient content, in the laboratory under sterile conditions.

The mycelium cultures used to be produced in 'T' bottles (Colle vessel) for a fairly long time with saprofitone breeding (on agar–agar culture medium). The producer obtained the culture in 'T' bottles. The handling (transportation, storage) of 'T' bottles was troublesome, and the preparation of the surface culture for application was very costly. For this reason, only fermented and deep-frozen infecting material is used today; this has lower transport and storage expenses, and its preparation for application is very simple. The conidia cultures are produced in high-capacity fermenters on adequate culture media. The conidia concentrate is deep-frozen at −30 °C or −40 °C, then it is stored at −12 °C in serum bottles or in plastic bottles until is used. One disadvantage of the application of the fermented and deep-frozen infecting material is that a chain of cooling facilities (cooling chambers, cooling vehicles, cooling boxes) needs to be established between production and the place of usage. In spite of this fact, the application of the deep-frozen material is more economical than the use of surface cultures in 'T' bottles (one-third as expensive). In order to eliminate the need for cooling facilities, new methods have also been developed for the production of infecting materials (lyophilization, granulation).

The preparation of infecting material for application is made several hours prior to the infection. The method of preparing the spore suspension depends on the infecting

Fig. 45 Developing stages of the rye spike (Békésy, 1960), I, spike in sheath; II, half-emerged spike; III, completely emerged spike; IV, the stalk of the spike emerged from the leaf sheath; V, spike prior to flowering; VI, start of flowering; VII, in full flower

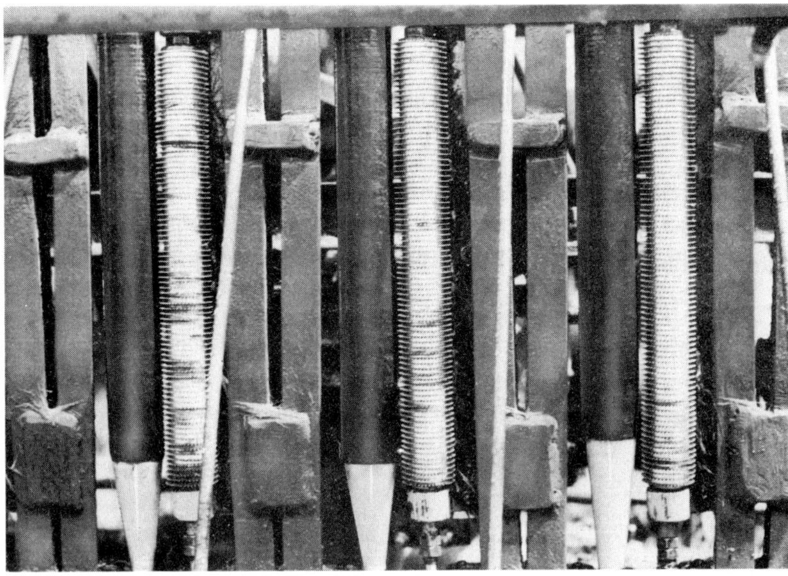

Fig. 46 The vertical rollers of the infecting machine (pinned and smooth roller pairs)

material. The fungus web is scratched from the surface of the culture medium in the 'T' bottle, and is then mixed with a little water.

After this process the material is filtered from the culture medium (perhaps using machines), and is then diluted with water obtained from a well or from the utility system. The degree of dilution is determined by the number of conidia capable of living. Thirty thousand conidia must be in 1 mm^3 of good suspension. The available infecting machines will spray 100–150 litres of suspensio on 1 ha of rye. The prepared suspension cannot be stored, because the conidia would lose their germinating capacity within a few hours in airless conditions.

The fermented deep-frozen infecting material is taken from the deep freezer 7–8 hours prior to application, melted at 1–5 °C, and then diluted with fresh water, preferably in the reservoir of the infecting machine. Generally, 1–2 litres of infecting material are necessary for producing the suspension containing 30 000 spores/mm^3 to be applied on 1 ha.

For the application of the spore suspension, the conidia are introduced into the ovule of the rye by needle pricking without damaging the spikes. Its success is influenced by the time and mode of application.

According to Békésy the infection should be started when the rye is still 'big bellied', as the spikes begin to emerge from the leaf sheath (Fig. 45, developing stages I and II) and may be made until the spikes emerge 10–15 cm over the leaf sheath (Fig. 45, developing stages III, IV and V). The infection should be discontinued 1–2 days prior to flowering, because otherwise fertilization would not take place.

The infection of rye today is done only with machines. The machines applied in Hungary are either self-propelled or can be mounted on basic machines. The operational principle of both types of machine is that the spikes are drawn between roller pairs consisting of one pinned roller and one rubber roller (Fig. 46), which perform the desired prickings, while the spore suspension is sprayed through spraying nozzles onto the pairs of rollers and just in front of the rollers. In this way the conidia are introduced into the ovulum with the help of pricking (Figs 47, 48).

The operation of the machines should be adjusted so that 70–80 per cent of the spikes would be pricked and the number of pricks in one spike would be 7–10. In the case of advantageous conditions, the success of the infection can be verified within 10–12 days on the basis of the amount of produced honey-dew.

Harvesting

The first ripe sclerotia appear 4–6 weeks after infection. The development and ripening of sclerotia are continuous, the actual process depending on the locations and origin (primary or secondary infection) of sclerotia. The ripe sclerotia are seated among the glumellae of the spike, and fall off easily.

The best method of harvesting ergot is the continuous gathering of sclerotia, which can be performed with the 'Goldhamster' machine. The machine lifts out the ripe sclerotia with the help of brush cylinder pairs in such a way that the immature sclerotia and the immature rye grains remain undamaged in the spikes. This careful method allows the harvesting of continuously ripening sclerotia in a number of stages. Loss can be minimized in this way (2–10 per cent), and the grown ergot can be gathered almost without loss. The first gathering of ripe sclerotia is made generally one month after infection. Depending on the weather and the rate of ripening, gathering is repeated for 6–8 days until the ripening spikes begin to break. Altogether four or five gatherings are made. The last gathering is made with a cereal combine harvester in one path. A large advantage of continuous gathering is that the ergot gathered on the first four occasions contains almost no rye grains (60–65 per cent of the total ergot yield), which means that the amount to be cleaned is less, and the cleaning costs are reduced. In Hungary the gathered amount of ergot is 220–300 kg/ha in the case of continuous gathering.

Fig. 47 Infecting machine 'PP-100-FA'

Fig. 48 'Goldhamster' infecting machine

Drying and separation

The ergot must be dried after harvesting. In order to ensure storage without damage, the water content of the sclerotia should be reduced to 9 per cent. The clean ergot may be dried using any methods, but the mixture should be dried in artificial driers, owing to its larger amount. The active ventilating equipment is very good for drying ergot mixtures. Do not use temperatures over 40 °C at the initial stage of the drying, when the humidity content of the ergot is still 30–40 per cent.

The separation of the ergot mixture starts with condensing. During this procedure the larger contaminating materials (spike and stem parts) are separated from the ergot with the help of ventilating screens and seed-cleaning machines, and most of the rye is separated. After this procedure the ergot content of the mixture is 60–65 per cent (the rye content is 25–30 per cent). Further separation is made with a fluid-bedded separating bench in two stages, then the final separation is made with a photoelectric separator on the basis of colour differences. At the end of the operation not more than 4 per cent rye is allowed in the ergot.

The alkaloids of the ergot are toxic, and for this reason health-protection and labour safety regulations should be complied with by anyone taking part in the working procedures.

Poppy (*Papaver somniferum* L.)

The poppy is one of the most ancient cultivated plants, having been known and cultivated for more than 4000 years. Its medical application was described by Hippocrates. The narcotic effect of this plant has long been recognized: it was first cultivated in Asia Minor, India and the neighbouring cultures. Arab merchants introduced the poppy to Europe and later to all parts of the world. Today it is cultivated all over the world.

The poppy serves several purposes. The immature poppy capsules and the opium obtained from these capsules have analgesic and narcotic effects. It is still cultivated in India and in China for the purpose of producing opium, which is a solidified latex obtained from the immature, slit poppy capsules. Opium is not only a pharmaceutical basic material but also a dangerous narcotic drug. This fact explains the importance of a process patented in 1931 by a Hungarian chemist, János Kabay, which opened the way for the industrial extraction of important alkaloids from ripe poppy capsules. This method has been widely used all over the world. Its importance lies in the fact that opium production, which is a dangerous and labour-intensive process, is eliminated, and therefore poppy can serve two purposes at the same time (capsules and seeds).

The active ingredients of the ripe poppy capsules (Papaveris caput maturis) form a multi-purpose basic material in advanced medicine for analgesic, antispasmodic and cough drugs. In Europe, however, poppy is produced for its seeds: poppy seeds are used as food and less frequently for oil production.

The poppy is cultivated in several countries of Europe, mainly in France, Bulgaria, Poland and Rumania. It is widely cultivated in Australia as well. The poppy has long been cultivated in Hungary. Its cultivation has been growing, partly because of the popularity of the poppy seeds, and partly because the production of poppy alkaloids and drugs has become a significant industrial operation in Hungary.

Characteristics

The cultivated poppy belonging to the family Papaveraceae and to the genus Papaver is an annual plant. This family includes more than 300 species, and the Papaver genus also includes nearly 100 species, and several forms and varieties. Some authors consider the poppy to be indigenous to Asia Minor, while according to others it originates from the Mediterranean countries.

Its primary root is stake-like 18–20 cm long, its diameter is 1–2 cm, and a slight lignification is characteristic. It develops relatively few lateral roots.

Its uptaking extremities can extend to a depth of 50–100 cm below the soil surface, depending on the water and nutrient content of the soil. Its stem is upright and can be 50–150 cm tall, depending on the variety and the environment. The upper part of the stem is cylindrical, smooth, silvery green and branched. Its rosette leaves are of long, elliptical shape, and they are sessile. Its stem leaves are scattered, and sessile; the lower ones are larger than the rosette leaves, and the upper ones become gradually smaller and of longish, elliptic shape.

As with the main axis, each of the branches terminates in a single flower. The petals of the cultivated poppy are white, and 6–12 cm long and the base parts of the petals are dark or light violet-shaded. Its fruit is a capsule containing a number of lacunar loculi; the shape, size and colour of the capsule is characteristic of the species. The

Fig. 49 Poppy in rosette-developing stage

sessile seeds on the septum fall to the bottom of the capsule on ripening; their colour and size is also characteristic of the species. Generally they are kidney-shaped, with a colour ranging from white to black. The seeds of the cultivated species are different shades of blue. The weight of 1000 seeds is 0.3–0.6 g.

The green capsule, as well as the ripe one, contains opium or phenanthrene structure alkaloids belonging to the phenylalanine group. The total alkaloid content of the ripe capsule is 0.4–1.5 per cent. More than 25 alkaloids can be isolated from the capsules. Its main alkaloids are morphine (9–10 per cent of total alkaloid content), narcotine (4–5 per cent) and papaverine (0.8–1 per cent). Its important secondary alkaloids are tebaine (0.4–0.5 per cent), codeine (0.3–0.4 per cent) narceine, cryptopine, pseudomorphine and laudanine. The poppy seeds contain practically no alkaloids. They have, however, a fatty oil content of approximately 40–50 per cent, which is suitable for human consumption and includes 20–25 per cent protein.

The vegetation period of poppy sown in the spring is 110–150 days. In its development, six stages can be distinguished, covering the period beginning with germination and lasting until the end of ripening.

These stages are as follows:

(1) *Embryo or dormant stage,* which can last several years, since the poppy seed preserves its germination ability for 4–5 years.

(2) *Germination period,* lasting from the rupture of the testa till the appearance of the first leaves. Under optimal conditions (+7 °C to +10 °C) this period lasts for 15–20 days.

(3) *Rosette stage,* from the appearance of the leaves till the emergence of shoots and the formation of reproductive carpels (see Fig. 49). This is the longest ontogenetic period of the poppy. Under average conditions it is 50–60 days in Hungary. It is shorter if the seeds are sown later.

(4) The period during which the *elongation* of the internodes and *branching* take

Fig. 50 Opium-ripe poppy

place lasts from the beginning of shooting until the blossoming of the flower on the main axis. This period lasts for 25–30 days.

(5) The period of *blossoming* and the *formation* of capsule and *seeds*. On average, this period lasts for 20–30 days, depending on the number of lateral branches, which is characteristic of the variety, and the number of flowers developing on these branches. Ten to fourteen days after the shedding of flowers on the main and lateral branches, the capsules and seeds assume their final shape and size and their growth is terminated. The poppy is said to be mature at this stage for opium harvesting, for this is when the capsules should be slit for collecting opium (Fig. 50).

(6) The period of *ripening* of the capsules and seeds lasts for 15–25 days, depending on the climatic conditions. In this period the capsules and seeds will assume their characteristic colour and the seeds will separate from the septum. Poppy seeds are ripe for harvesting when they give a rustling sound and when the foliage leaves have dried and the stems are brittle.

Cultivars

The main requirement for a variety is to have a capsule with high alkaloid content and to have a high yield of capsules and seeds. It is desirable for a variety to have seeds of vivid blue colour, to have a strong stem with a uniform height of capsules and a good resistance to disease.

At present, four state-approved poppy varieties (Figs 51, 52, 53, 54) are cultivated in Hungary (see Table 6). One of them is winter (autumnal) poppy ('Cosmos') and the others are spring poppy varieties.

Fig. 51 'Kompolti M' cultivar

Fig. 52 'BC-2' cultivar

Fig. 53 'Kék Duna' cultivar

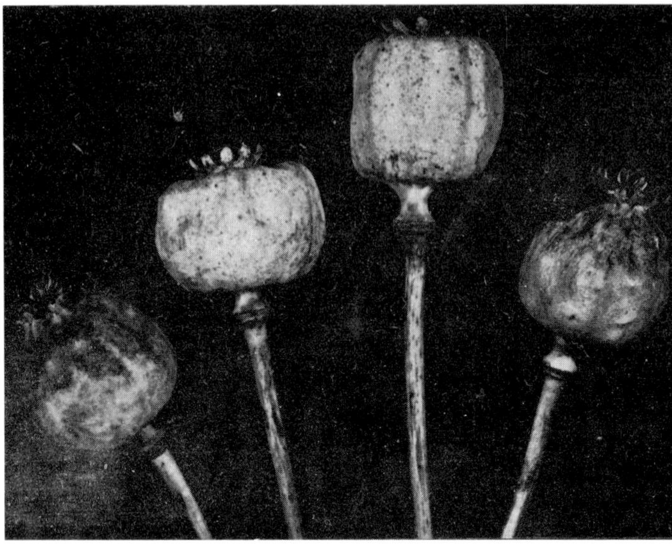

Fig. 54 'Cosmos' cultivar

Table 6 Features of the Hungarian poppy cultivars

Cultivars	Vegetation period in days	Seed yield (t/ha)	Capsule yield (t/ha)	Alkaloid content (in morphine) (‰)
'Kompolti M'	125–140	0.9–1.4	0.7–1.0	5–6
'BC-2'	110–120	0.7–1.1	0.6–0.8	4–5
'Kék Duna'	130–140	1.0–1.5	0.8–1.1	4–5
'Cosmos'	230–260	1.6–2.2	1.5–2.0	3–4.5

Environmental conditions

The poppy likes temperate climates, but its adaptation capability is high. It grows in Europe from the shores of the Mediterranean to Sweden. In Asia, cultivated poppies had been found by Pieper at an altitude of 2300–2400 m. According to observations made in Hungary, a heat amount of 2000–2200 °C in the vegetation period is necessary for the proper development of the poppy.*

The demand for heat is different in the various stages of development. The poppy begins germinating even at low temperatures (+2–3 °C), but 7–10 °C is necessary for sprouting. The high temperature is not advantageous for germination (20–25 °C) of spring varieties, but 18–20 °C is necessary for germination of winter poppy.

* The heat amount means the sum of the mean temperatures necessary from germination till seed ripening. (If the mean day temperature is high then the vegetation period is shorter.)

The optimum daily average temperature is 12–14 °C in the stage of rosette development, 16–18 °C from shoot emergence to the end of flowering, and 21–23 °C up to ripening. The sprouts can survive frost even with a temperature of -4.5 °C to -5 °C without damage. For normal development, a moderate temperature is advantageous; while a high temperature hastens development, the stems develop early and the yield is reduced. Dry, warm weather is most appropriate for flowering. In cool, windy weather, flowering takes place for a longer time and fertilization is not so effective. After the shedding of flowers the best weather is warm and slightly rainy weather. Dry, warm weather is required at the time of capsule ripening.

The poppy is a plant with a medium need for water. It responds actively to the amount of precipitation, and especially to the distribution of precipitation. The largest amount of water is required at the time of shoot emergence. Precipitation is not good at the time of flowering. The water requirement increases again during the time of capsule development, and if arid conditions occur during this period, then the seeds will be minute and the capsules will become forcedly matured. A large amount of rain is damaging in the period of capsule ripening, for the active ingredient content of the capsules is reduced.

The poppy prefers easily treatable soils with lots of extractable nutrients, and neutral (pH 7) or slightly alkaline reaction (pH 8). The acidic (pH 6) loose, sandy soils, and the firm, wet, clay soils, are not suitable for the cultivation of the poppy.

Cultivation

Sowing order

The poppy grows sufficiently on weed-free nutritious soils after almost any plant. The best previous plants to the poppy are those papilionaceous annual plants and the manured hoed plants which leave the field early and without weeds. Poppy should not be grown on the same field more than 3–5 years running.

Nutrient supply

The poppy requires lots of nutrients in its short vegetation period. Among the main nutrients, nitrogen and phosphorus increase the yield most efficiently. According to available data, the poppy extracts 102.5 kg of N, 192 kg of P and 11.4 kg of K from the soil to produce 3.5 t/ha of vegetative mass and 1.2 t/ha of fruit.

Preparing the soil

Owing to the small seeds, the poppy requires very careful soil preparation. Autumn deep ploughing is absolutely necessary, as the poppy is to be sown early in the spring. Unlike the usual procedure, the soil ploughed in the autumn should be smoothed in the autumn so that the seed-bed preparation to be made in the spring will not require much soil movement. The spring soil preparatory work should be performed as soon as the soil allows the movement of machines.

Sowing

The spring poppy can be sown early in the spring, and at the same time this is one of the conditions of successful poppy cultivation. In this way the sprouting is made more uniform and the probability of harming the poppy-head is smaller.

Sowing is carried out from February to the end of March, depending on the weather and soil conditions. The spring poppy is not sensitive to cold and moderate frost after sprouting. In general terms, the spring poppy should be sown when the frost in the soil has disappeared, the surface of the soil has dried, and the soil is not smeared after the furrows of the sowing machine.

The sowing of winter poppy should be done in the second half of September in Hungary, so that the poppy will develop into a stage having 4–6-leaved rosettes prior to the onset of winter. It is not advisable to sow the winter poppy earlier, because the more developed plant is more susceptible to the snow cover and cold.

According to examinations of cultivating areas and experiences, the optimal number of plants is 60–65 pieces/m^2, i.e. 600 000–650 000 pieces/ha. The poppy can be sown in different growing areas, while keeping the same number of plants. The row distance of 25–40 cm is advantageous according to Hungarian practice. In order to help the sowing and reliable sprouting, 0.6–1.0 kg of seed is sown on 1 ha. This

Fig. 55 Poppy field developed from radiation-treated sowing seeds and without thinning (Photograph by D. Földesi)

amount, however, gives rise to very dense sprouting, which makes thinning out necessary. For the purpose of reducing the labour requirement of thinning out, sowing with irradiated seeds is widely done in large-scale cultivation (Fig. 55) (Földesi, 1972). The sowing seed mixture consists of one-fifth of sound seeds and four-fifths of radiation-treated poppy seeds. The radiation-treated seeds will not lose their germination capacity, but the sprouts developed from them will die within a few days of sprouting. Only the plants germinated from the sound seeds will grow further, so thinning out is not necessary. In this case the radiation-treated seeds help the penetration of sound seedlings up to the surface.

Three kilograms should be sown on 1 ha from the radiation-treated seed mixture. This seed amount contains 800 000–900 000 sound embryos, which is sufficient for the development of the optimum 600 000–650 000 plants in 1 ha, taking the natural thinning out into consideration. The depth of sowing of the poppy is 10–15 mm, independent of the kind of soil. For the sowing of a relatively small amount of seeds, only the correctly adjusted small-seed sowing machines are adequate. The soil should be rolled immediately after sowing.

Care of plants

The protective works for the poppy should be started prior to the sowing. Against the Helminthosporia and the poppy peronospora, the sowing seed should be treated with TMTD or other treatment agents. The soil-disinfecting chemicals protecting against soil pests (grubs, wireworms, weevils) should be mixed into the soil during seed-bed preparation in the spring. For chemical weed control of the poppy, the Dicuran (chlortolurone) or Mezotox (nitrophene) herbicides can be applied (Földesi, 1975). These should be sprayed before sowing, Dicuran 1.5–1.7 kg/ha and Mezotox 8–10 litres/ha.

Herbicide treatment done before sowing will cease to have an effect within 4–6 weeks. The weeds among the poppies with three or four leaves may be destroyed by the application of 300 litres/ha of Reglone solution of 0.8 per cent concentration. Following this treatment, the poppy remains free of weed until it is at the stage of 8–10 leaves. At this time the plantation may be sprayed with 2 kg/ha of Dicuran or 2 + 8 kg/ha of Dicuran + Asulox (azulam), or 2 + 1.5 kg/ha of Dicuran + Dual (metolachlor) herbicide mixture, depending on the actual weed spectrum (Földesi, Bernáth and Pál, 1980). Herbicide treatment performed in this way ensures the weed-free state up to harvesting.

The poppy used to be attacked by a number of pests and parasites in the vegetation period. Protection should be provided generally against peronospore, weevils and sometimes against plant lice. Against peronospore *(Peronospora arborescens)* the protection may be provided partly by seed treatment and partly by spraying with fungicides (containing copper) of 0.4–0.5 per cent concentration, which should be applied when the symptoms appear. The most harm can be caused by the weevil *(Ceutorrhynchus macula-alba)*. The most important way of protecting against its damage is by early sowing. The protective operations should have already begun at the time of stem emergence. Efficient protection can be maintained up to the flowering stage against the parasites feeding on the stem part under the bud by the application of Wofatox powder (18–26 kg/ha) or with Wofatox (metileparation) spraying agent. At the time of flowering (when damage is most likely) only the bee-protecting Melipax (toxaphene) can be applied, or Thiodan (endosulphane) can be applied in the evening hours after the bees have left the field. To kill the plant lice *(Aphis sp.)*—which are also responsible for the spread of mildew—a solution of Pirimor (pyrimicarb) can be applied.

Harvesting

The poppy is harvested when the seeds are completely ripe, which is at the end of July or at the beginning of August. At this time the poppy capsules produce a rustling sound. Harvesting may be done in dry weather after the dew has dried. It is carried out by hand in small fields. The poppy heads are cut, together with 8–10-cm-long stems, and then the seeds are freed from the heads. In certain areas the poppy is pulled out and bound into sheaves, and then the seeds are removed from the heads with the help of threshing machines. For the purpose of harvesting in one path, a modified cereal combine harvester is used. The poppy heads with the 10–20-cm stems, are threshed by the cereal combine harvester together. In the case of harvesting by the one-path method, the seeds gathered in the holder of the combine harvester should be separated at once from the poppy capsule residues, so that the seeds do not become rancid because of the heating effect of the more humid capsules and green parts. After separation, both the seeds and the capsule residues should be dried. For drying, the seeds and the capsule residues are spread in a barn or in other suitable areas. Generally, no artificial driers are necessary. The seeds can be put into sacks with 9 per cent humidity, and the capsule residues may be put into sacks with 10–12 per cent humidity. The yield of the poppy is 0.9–2.2 t/ha of seed and 0.6–2.0 t/ha of dry capsule, depending on the weather and variety.

Rue (*Ruta graveolens* L.)

As a medicinal plant, rue (herb-of-grace) has been known in Europe for 1500 years. Its dried herb (Rutae herba) and leaves (Rutae folium) have been used in folk medicine to heal a number of diseases. Today it is a basic material in tea blends for healing gallbladder disease and high blood pressure, and it is a basic material for several medicines.

Rue is cultivated in several European countries, and the USA. It is also cultivated as an ornamental plant in a number of places. Rue has been cultivated in Hungary for a long time. It is also useful for exploiting poor-quality (sandy, stony) soils.

Characteristics

Rue is a perennial sub-shrub belonging to the family Rutaceae and originating from the eastern part of the Mediterranean (Fig. 56). It can be found on stony, lime soils with a warm local climate. The root is woody, branching and yellowish-brown, and

Fig. 56 Rue

penetrates deeply. The stem is erect, 50–100 cm high, and densely branching. The one-year-old shoots are greyish-green, glabrous and shiny, as are the leaves. The leaves are alternate, multiple, pinnate compound. The flower is dichogamous, and the petals are greenish-yellow. The fruit is a capsule covered with glands. The seeds are three-edged, reniform, dark brown or brownish-black. The mass of 1000 seeds is 2.05–2.19 g.

The rue belongs to that group of plants which are capable of accumulating a number of active ingredients in their bodies. The largest amount among these active ingredients is represented by the flavonoids, especially rutin, which has antispasmodic effects (1–2 per cent). The leaves contain 0.1–0.9 per cent coumarins (rutaringlykosid) and furanocoumarins (psoralene, bergaptene) as well. The unpleasant pungent odour of the plant comes from the essential oil accumulating in the skizolizigene passages to a level of 0.3–0.7 per cent. The components of this essential oil are methyl heptyl ketone (70 per cent) and methyl nonyl ketone (20 per cent). Even toxic alkaloids are formed in rue, involving acridone alkaloids, additionally to the 0.2 per cent quinoline and furano quinoline.

The seeds retain their germination ability for 2–3 years. The optimum temperature for germination is 12–15 °C. The duration of sprouting is 17–26 days. The initial development of the plants is very slow, and this is the reason why propagation is generally done with seedlings. Plants older than one year start to develop early in the spring (beginning in the middle of March). The first flowers appear in the middle of May, and full flowering generally lasts to the middle of June. The fruits ripen until the end of July or the beginning of August. The production capacity of rue starts to deteriorate from the sixth year of cultivation.

Environmental requirements

Rue requires plenty of warmth and sunshine. It develops insufficiently in cool, watery soils and in shaded places. The best locations for growing are moderately dry soils, which get warm easily and which slope in a southern direction. The most successful cultivation can be achieved on limey, nutritous soils, which have good water-exploitation capacity.

Cultivation

Nutrient supply

Rue requires plenty of nutrients. It develops 20–30 t of green mass in productive years. To develop this amount a soil rich in nutrients is necessary, and these nutrients should be of a type which can be easily taken up by the plants. Prior to the planting, 15–20 t/ha of manure 50–60 kg/ha of P_2O_5 and 40–50 kg/ha of K_2O active material in the form of fertilizers should be introduced into the loosely structured sandy soil. The regular replenishment of the nutrients in the soil should be ensured in the years following planting. Generally sufficient is a supply of 30–40 kg/ha of P_2O_5 and 20–25 kg/ha of K_2O in the autumn, and 40–50 kg/ha of N active material in the spring (prior to sprouting). Depending on the state of the plantation and the actual nutrient content of the soil, it may be advisable to spread nitrogen fertilizer after the first cutting.

Soil preparation

Soil preparation is done in the autumn or at the end of the summer, depending on the time of planting. The operations of soil preparation are the same as those used for other perennial plants.

Sowing and planting

The seeds are sown into open-air seed-beds to produce seedlings. Owing to the slow initial development of the plant, sowing in the final location is not widely done. Rue is cultivated mostly by seedling growing. The time of sowing depends on the time of planting. For the purpose of planting at the end of the summer or in the autumn, the seeds may be sown from the beginning of April to the end of May, and for the spring planting it is sufficient to sow the seeds in July or August. Sowing is made into well-prepared open-air seed-beds with a row spacing of 25–35 cm. The optimum depth of sowing is 1.5–2 cm. The field should be kept free from weeds after sprouting, and the correct amount of water for the proper development of the seedlings should be ensured. The seedlings necessary for planting 1 ha can be grown on an area of 200–250 m^2 with the use of 1–1.5 kg of sowing seed. The seedlings can be planted either in the spring or autumn. The well-developed seedlings of 20–25 cm height are suitable for planting. If the treatment of row spacings and the harvesting is to be done with machines, then the planting should be done with 60 × 20 cm or with 50 × 30 cm row and plant distances. The optimum number of plants is 70 000–90 000 on 1 ha. The seedlings can be planted with the machines which are generally used for planting vegetable plants. It is necessary to water the field after planting if the soil is dry.

Care of plants

The care of the rue plantantion consists of nutrient supply, regular weed killing, and soil-loosening operations between and in the rows. Weeds can be destroyed by mechanical methods (with cultivators) or by the application of herbicides. Hungazin DT containing Aktikon active material can be applied successfully from the second year, spread with a dosage of 3–5 kg/ha prior to the beginning of vegetation. In dry weather, watering might become necessary. In practice, protective measures have not been necessary until now, as the rue was not attacked by pests or parasites.

Harvesting

Rue is cultivated for the vegetative part which is above the ground when the plant is one year old. The active ingredient is most concentrated in plants beginning flowering. For this reason the optimum time for harvesting rue is this ontogenetic stage (end of May, beginning of June in Hungary). After harvesting, the rue starts to develop shoots again and flowers for the second time, provided that the nutrient supply is sufficient. In this case it can also be harvested in October before the first frosts. Only one harvest is done in the year of planting. The plants are cut just above the lignified part, which is generally 12–15 cm from the ground. The harvesting is done by hand (in small fields) or with windrowing machines. The out plant may be left in the windrows for drying, then windrow gathering machines pick up the plants. In the case of the second harvesting done in the autumn, the cut plants should immediately be transported to artificial driers.

The high-quality, greyish-green, strongly spice-odoured drug can only be prepared in artificial driers. The average yield of rue in productive years is 3.5–4.0 t/ha of dry herb. The ratio of green to dry material is 5 : 1.

Rue might cause slowly healing dermatitis if it touches the skin, especially on sunny days. This should be known by the people dealing with the harvesting, and protective gloves (and preferably clothes as well) and water for handwashing should be provided for them.

Coriander (*Coriandrum sativum* L.)

Coriander is one of the first-known multi-purpose plants. Its seeds were used for medical purposes and for food by the Egyptians as early as 1500 BC. The earliest known documents dealing with its cultivation date back to the sixteenth century. The seeds of the coriander *(Coriandri fructus)* are also regarded as a drug by the 7th edition of the Hungarian Pharmacopoeia. It is used for improving taste and for the general benefit of the digestive system. Coriander is a popular spice, and is mainly used for spicing pickles, drinks and meat products. Apart from medical practice and the food industry, relatively high amounts of the essential oil of the coriander (Aetheroleum coriandri) are used by the cosmetics and perfume industry. The main ingredient of its essential oil, linalool, is also widely used for the production of various scent and flavour materials.

Coriander is cultivated in a number of countries, the total yield being 200 000–220 000 t/year worldwide. More than 60 per cent of this is produced in the USSR. Significant amounts are also produced by Argentina, Yugoslavia, Romania, Poland, India and Marocco. Coriander is also cultivated for its green parts, which are consumed similarly to the green parts of parsley in most of the countries in Asia and Africa.

Characteristics

Coriander is an annual, herbaceous plant belonging to the umbellates (Apiaceae) family; its place of origin is the east coast area of the Mediterranean. Two species are currently known, i.e. *Coriandrum sativum* L. and *C. melphitense* T. et GUN. This latter is an endemic species, found in southern Italy; it has no commercial significance.

C. sativum is classified into two varieties according to the size of its seed. One of them has large seeds, *C. sativum* var. *vulgare* Alet; it has twin-achenial seeds of 3–6 mm diameter. The other, *C. sativum* var. *microcarpum* D.C., with small fruits, is

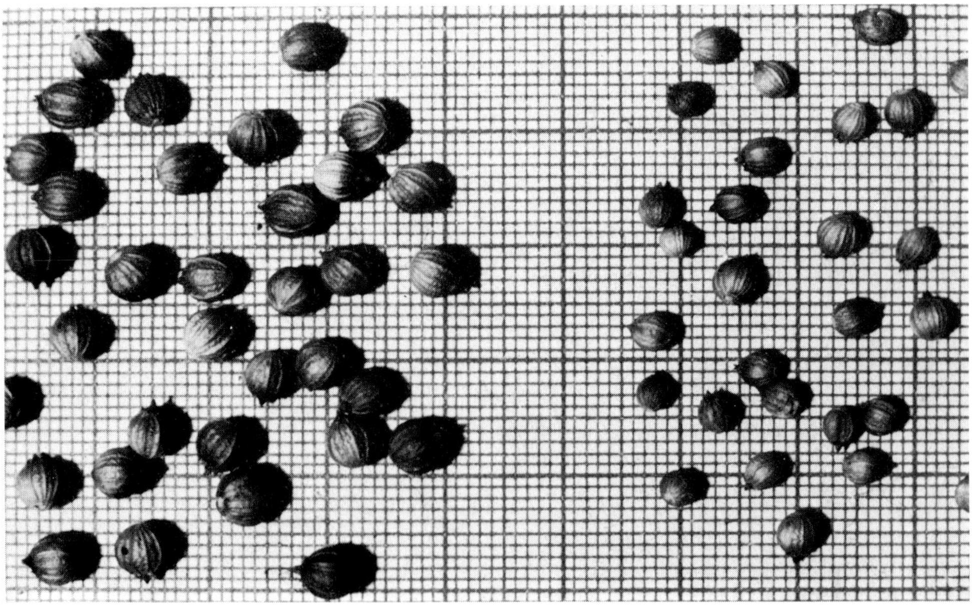

Fig. 57 Seeds of coriander varieties (left, var. *vulgare;* right, var. *microcarpum*)

Fig. 58 Coriander

cultivated in most of the European countries, including Hungary. The seeds of this variety have diameters in the range 1.5–3 mm (Fig. 57).

The root is slightly branched and post-like; the main part of the root develops in the 40-cm-thick soil layer. The stem is straight, erect, extremely branched, green and often reddish. Its height is 50–160 cm, depending on the variety and on the environmental conditions.

The main stem and branches terminate in flowers. The leaves are light green, and their shapes and sizes are variable. The bottom leaves have long stalks, their shape is round, they often form rosettes, and they become dry after the stem development has started (Fig. 58). The middle stem leaves are fringed and provided with stalks, and the upper leaves are fibrous and sessile. The flowers are tiny, and white or violet-pink.

The fruits are typically spherical, straw-yellow or light brown, and twin-achenial; the two seed parts are fused, and can be parted only with pressing; they are ribbed outside and have cavities inside. The mass of 1000 seeds is 5–7 g.

Every part of the plant contains essential oil. Owing to the accumulated essential oil in the stem and in the leaves, the whole plant smells like plant bug. The main ingredient (60–80 per cent) of the essential oil found in the vegetative parts is the unpleasantly smelly decylaldehyde. As opposed to the green parts, the ripe seeds of the plant have a pleasant smell and spicy taste. The essential oil content of the ripe seeds is 0.5–1.6 per cent or even over 2 per cent in certain varieties. The essential oil of the coriander contains more than 20 components, among which the most significant are α-linalool (60–80 per cent), geraniol (3–5 per cent), geranyl acetate (1–5 per cent) and l-borneole (1–4 per cent). The coriander seed contains 18–28 per cent fatty acid, 30 per cent protein and significant amounts of A and C Vitamins as well.

The vegetation period of coriander, calculated from sprouting up to the ripening of the seed, is 80–120 days. Its ontogenesis takes place in five phases: namely 15–20 days from sowing to sprouting, 30–40 days from sprouting to stem emergence, 15–20 days from stem emergence to flowering, 10–20 days from flowering to development of the seed, and 20–40 days from development of the seed to ripening. Coriander is generally in flower in the first part of June in Hungary. It is a plant of extraneous fertilization, mainly by insect pollination. It is also a good nectar-producing plant. Its seeds ripen unevenly late in July and early in August and shatter easily.

Cultivars

Cultivar 'Lucs' has a short growing season (88–90 days), a productivity of 1.1–1.4 t/ha, and an essential oil content of 1.3–1.6 per cent. Its seeds are medium-sized, the mass of 1000 seeds is 6.2–6.7 g. A Hungarian cultivar is 'Csillag', which has a growth season of 95–105 days, a productivity of 1.2–1.5 t/ha, and an essential oil content of 1.6–1.7 per cent. The linalool content of the essential oil is 80–82 per cent.

A recent cultivar in the USSR, 'Jantar', is especially useful for the production of essential oil. Its growing season is 110–120 days, it has a productivity of 1.3–1.6 t/ha, the essential oil content of its seeds is 2.4–2.6 per cent, and the linalool content of its essential oil is 72–74 per cent (Gluschenko, 1976).

Environmental requirements

Coriander requires a lot of sunshine, 900–940 hours of sunshine from sprouting to flowering, and 1400–1500 hours sunshine for the whole growing season. The germinating plants can survive at −8 to −10 °C, and the rosettes with 4–6 leaves can survive at −16 to −18 °C without any harmful effects. The heat requirement from sprouting to ripening is 18–20 °C. Growth, and as a consequence productivity, is affected detrimentally by a high mean temperature (18–20 °C) occurring in the period just after sprouting. The germination of the seed starts at 4–6 °C, the optimum temperature being 12–15 °C.

The coriander requires high humidity in the soil, especially in the period from germination to development of the rosette; then a high amount of precipitation is advantageous from stem emergence to flowering.

It can only be cultivated successfully on medium–firm, clay soils, having adequate structure, and provided with a proper water supply and nutrients.

Cultivation

Sowing sequence

Like other annual umbellates, coriander can be fitted into almost any sowing structure. Good previous plants for coriander are those which are harvested early, leave the field without weeds, and improve the physical properties of the soil. It is not possible to cultivate coriander after other plants belonging to the umbellate family, especially anise, owing to the common pests and parasites. It is not advisable to sow coriander for 4–5 years running.

Nutrition

Coriander requires a lot of nutrients. For producing 100 kg of seeds, it extracts 4.5 kg of N, 1.6 kg of P_2O_5 and 4 kg of K_2O from the soil. The nitrogen and phosphorus extraction is especially high in the stage from stem emergence to flowering, and more than 70 per cent of the potassium requirement is also taken up in the period of intense development. According to domestic and foreign data, it is sufficient to use, for 1 ha, a basic fertilizer consisting of 60–70 kg of N, 80–100 kg of P_2O_5 and 40–50 kg of K_2O, which should be worked into the soil to 25–30 cm depth.

As well as the above, it is advisable to introduce 10–15 kg/ha of N and 15–20 kg/ha of P into the soil during the spring soil preparation or at the time of sowing. The top dressing applied after sprouting has proved to be ineffective.

Soil preparation

Coriander can only be successfully cultivated in soil ploughed in the autumn. The humidity of the soils ploughed in the spring is generally not high enough for the germination of the seeds. Care should be taken to avoid the use of equipment causing drying when preparing the soil in the spring, and also operations requiring more runs should be avoided.

Sowing

The basic requirement for a good yield is to sow at the appropriate time. This is also verified by the experiments made in the Horticultural University (Fig. 59). The optimum sowing time in Hungary is the middle to the end of March, when the top 5–6 cm layer of the soil reaches a temperature of 7–8 °C. Coriander is also sown in the autumn (in September) in a few countries. It is advisable to sow it in rows spaced at 25–30 cm, according to the biological requirements. For sowing it is preferable to use a high-quality seed with a large mass (6–6.5 g per 1000 seed). In order to establish the optimum plant density (50–70 pieces/m), 16–20 kg/ha of sowing seed is necessary. The optimum depth of sowing is 4–5 cm.

Care of plants

The main work to be done is the elimination of weeds, as coriander has a long sprouting time, and the initial development is slow. For chemical weed control, Merkazin (prometrin) and Afalon (linuron) are sufficient. Both herbicides should be sprayed onto the ground before sowing in amounts of 4–5 kg/ha.

Plant-protection operations are made necessary by the so-called bacterial infection, which occurs in some years, especially when June is cool and rainy. Identified pests

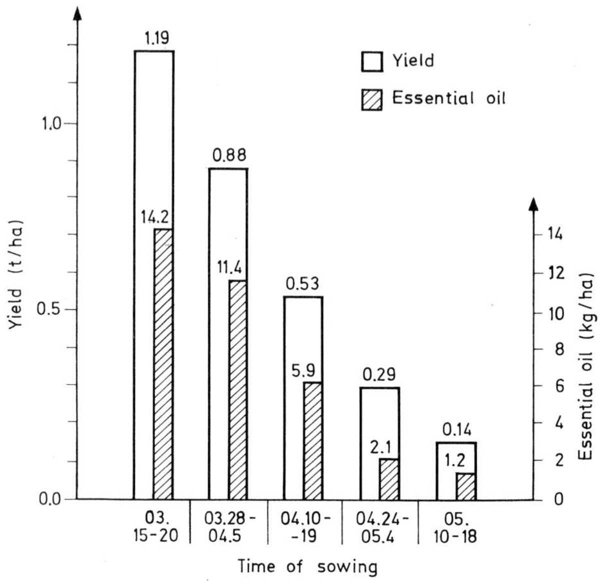

Fig. 59 Effect of time of sowing on the crop and essential oil yield of coriander (Hornok, 1980)

are *Ramularia coriandri, Erwinia carotovora* and *Xanthomonas heterocea* (Petrov, 1980). The diseases are spread by the seeds, and also by insects. Preventive protection is therefore very important. Most efficient is the TMDT treatment of the seed, but also successful is the combined application of fungicides and insecticides in the period from the appearance of the side branches to the beginning of flowering. With this treatment, rust *(Puccinia petroselini)* and the damage caused by the seed-consuming ichneumon *(Systole coriandri)* can also be prevented.

Harvesting

One run and two-run operations are both used for harvesting coriander. One-run harvesting with a combine harvester often leads to a seed loss of 20–25 per cent, according to Hungarian and other experiments. For this reason this one-run harvesting with a combine harvester is only suitable if most of the crops have already ripened (owing to the fast ripening or to other reasons), and the humidity content of the seeds has fallen below 30 per cent.

Two-run harvesting is more efficient. In the first run the coriander is cut in rows. The optimum time for this operation is when 30–40 per cent of the seeds are ripe, the stem of the coriander is still greenish, but most of the leaves have dried away. At this time the humidity content of the seeds is 40–50 per cent. Cutting can be done by any harvesting machine to a stubble height of 20–25 cm, if the speed of the pick-up reel is reduced.

The post-ripening takes 4–5 days in the avarage weather of early August. After this the coriander can be threshed directly from windrows with the help of a combine harvester, and this process causes practically no seed loss.

If the humidity content of the coriander is over 12 per cent, it must be dried after threshing, then cleaned by seed-cleaner machines up to 93 per cent of cleanliness. The average yield in Hungary is 0.8–1.0 t/ha of seed, or 10–15 kg/ha of essential oil; however, the 1.5–2.0 t/ha yield is not rare when good technology is applied.

Dill (*Anethum graveolens* L.)

The plant described as 'ammest' in the Ebers papyrus, which was used medicinally against headache and vein troubles, was dill, according to our present knowledge. It can be assumed that dill was cultivated in ancient Palestine, and it got into Europe during Roman times. Dill was a generally used and cultivated plant in the fourteenth century. As medicine, dill nowadays is used only in folk medicine for digestive troubles. Tea made with dill is used as a diuretic and antispasmodic. The essential oil of dill has antiseptic effects.

Dill is a well-known spice plant all over the world, owing to its pleasant taste, its appetizer effect and its wide usability. The food industry is the main user of dill. Its green parts are consumed in a fresh and a dried state *(Anethi herba)*, and its ripe seed *(Anethi fructus)* is consumed whole or in a ground state. The essential oil extracted from the dill *(Aetheroleum anethi)* is required also in large quantities. Dill oil consumption in the world is 60–70 t/year. The cultivated area is significant in India, England, the USA, Bulgaria, Spain and Italy. It has also been cultivated in Hungary for a long time, but its large-scale cultivation started in the last decade.

Characteristics

Dill belongs to the Anethum genus of the umbellates (Apiaceae) family. Three species are known in the *Anethum* genus (*A. graveolens* L., *A. chryseum* Boiss. et Heldr., *A. involucratum* Korov). In the literature, one of its kind is regarded as yet another species (*A. sowa* Roxb.), which is cultivated mostly in India, and has a dissimilar chemotaxon with respect to its essential oil. The cultivated dill (*Anethum graveolens* L.) is an annual plant with a characteristic smell and soft stem. Its ancient home was the eastern area of the Mediterranean, western Asia, Persia and eastern India; perhaps the regions of the Caucasus and Egypt can be regarded as the gene centrum of the species (Boros and Szujkó, 1970).

The root is spindlelike, thin, has few branches, and is 10–30 cm long. The stem is cylindrical, straight and erect, with a height between 50 and 180 cm, depending on the conditions of the cultivation area. The stem is provided with leaves along its whole length, and it has a flower axis end together with the side leaves. Its leaf is compound trifoliate, and its leaflets are thin and fibre-like. Its flower is a compound umbel, which is characteristic of the family, and has a diameter of 15–20 cm. The flowers are tiny, hermaphroditic, and their colour is bright yellow (Fig. 60). The fruit is twin achenial, and separates into two parts. The seed is dark brown; its shape is ovate. There are four essential oil passages on its back side, and two passages on the front side. The seed

Fig. 60 Dill

part has 5–6 cm length, 3–4 cm width and 0.5–1 mm thickness. The weight of 1000 seeds is 1–2 g.

All the parts of the dill plant contain essential oil. The accumulated essential oil in the various parts of the dill plant changes significantly both in quantity and composition during the vegetation period. At the time of the milky ripening the essential oil content of dill parts above the ground is 0.8–1.6 per cent depending on the weather conditions and the environment. The essential oil extracted from the dill herb cut at this time is generally called dill herb oil.

The essential oil content of the ripened twin-achenial fruits of dill is 2–5 per cent. The essential oil extracted from the fruits is called dill seed oil.

The dill herb oil contains 25–35 per cent *d*-carvone and a large amount of *d*-α-phellandrene (15–25 per cent), the seed oil contains 40–60 per cent *d*-carvone and 20–28 per cent *d*-limonene. As well as the above ingredients, both oils contain some dihydrocarvone, α-pinene, terpenene and other terpenoids.

Dill has a short vegetation period. Depending on the environmental conditions, 100–120 days are required from sprouting to ripening of seeds. The well-ripened and adequately stored seed retains its germination capacity for 2–3 years.

The optimum temperature for germination is 8–10 °C. The period of sprouting is 10–17 days.

The beginning of stem growth is expected within 35–40 days of sprouting. Depending on the weather conditions, the plant is in full flower 50–70 days after the appearance of the seed leaves. In cool, rainy weather the period of flowering is extended and the adherence of the seeds is very weak. On average, 40–45 days will elapse from the flowering of the main umbels to the ripening of the seeds in the main umbels. The crops ripen unevenly, and the ripe seeds drop away very easily.

Environmental requirements

Dill is a cosmopolitan plant; it can be successfully produced in almost any part of the temperate climate region. It requires warm, sunny weather at the time of seed development. Accordingly, growing it in the southern warm and sunny areas is more efficient if the purpose is the production of dill seed. It is not sensitive to cold; its optimum temperature of germination is low. Young plants bear the late spring frosts without any damage. It can be successfully grown on almost any soil, except for extreme types like sand, heavy clay and dry soil. Dill requires abundant water at the intensive development stage just prior to the development of umbels. For this reason dill can be most successfully cultivated on warm, medium–firm soils provided with a proper water supply.

Cultivation

Nutrition

It is demostrated by experimental (Table 7) and practical data that a crop yielding spices and herb oil is highly increased by improved nitrogen supply in the case of average soil characteristics. One-third of the nitrogen is introduced into the soil together with phosphorus and potassium as basic fertilizer, and the remaining two-thirds of the nitrogen is introduced into the soil with the soil preparation prior to sowing.

As opposed to the earlier assumptions, the essential oil content of dill does not decrease with improved nutrient supply, and in this way the essential oil production can also be significantly improved.

Table 7 Influence of nutrient supply on the yield of dill (Hornok, 1980)

Applied active material kg/ha			Mass of yield			
N	P	K	Spice (dry) (t/ha)	Herb oil (kg/ha)	Seed (t/ha)	Seed oil (kg/ha)
0	0	0	0.24	26.8	0.56	12.9
40	40	20	0.42	63.5	0.81	19.6
80	40	20	0.51	88.8	0.74	18.1
120	40	20	0.63	88.2	0.60	14,3
40	80	20	0.35	53.4	0.96	23.2
40	120	20	0.31	43.9	1.05	25.0
40	40	80	0.41	61.3	0.74	18.3
Least significant difference	5%		0.15	7.1	0.07	1.7

Table 8 Influence of sowing time on the yield of dill (Hornok, 1980)

Sowing time	Mass of yield		
	Herb oil (kg/ha)	Seed (t/ha)	Seed oil (kg/ha)
15–20 March	79.4	1.86	40.4
28 March—05 April	48.4	0.82	20.7
10–19 April	30.7	0.62	15.4
24 April—04 May	17.8	0.50	12.3
10–18 May	9.0	0.31	7.5
Least significant difference 5%	16.6	0.33	6.3

When the goal is the production of seeds, then the dill requires less nitrogen and increased phosphorus fertilizer. It is advisable to apply two-thirds of the phosphorus in the autumn, and one-third in the spring just prior to sowing. The increased nitrogen supply is disadvantageous if the dill is cultivated for seed production.

Soil preparation

There is no point in sowing dill in any soil except that which is ploughed in the autumn. It will sprout unevenly in soil ploughed in the spring. The seed-bed should be compacted by rollers prior to sowing. The rolling should be repeated after sowing.

Sowing

The time for sowing dill is determined by the purpose of cultivation. For the purposes of crop production and herb oil extraction, the dill should be cultivated as a main plant. With the average weather conditons in Hungary, the maximum yield is achieved if the dill is sown in March (Table 8). The dill is also sown in the period from early March to early April in order to lengthen the harvesting period, if the dill is cultivated for herb oil production. However, the differences in times of sowing are larger than those of harvesting, because the vegetation period of the dill is significantly reduced by sowing at a later date.

If dill is cultivated for spice (dry herb) production, then it can be sown in the spring, from middle March to middle May continuously, and then as a second plant from late in July to middle August. Sowing in June and early July seldom gives good results.

The sowing depth is 1.5–2. cm. The optimum row spacing is 12 cm in the case of spice dill, and 24 cm in the case of dill produced for herb oil or seeds. The required amount of seed is 18–20 or 12–14 kg/ha (100–200 seeds capable of germinating per metre).

Table 9 The essential oil content of dill during its vegetation period (Hornok, 1980)

Stage of vegetation period	Essential oil content (% of dry material)			
	Leaves	Stem	Umbels	Parts above ground
Emergence of stem	0.45	–	–	0.21
Umbel forming	0.51	0.09	–	0.46
Budding	1.12	0.16	1.71	0.75
Start of flowering	1.21	0.41	2.25	0.86
Complete flowering	1.28	0.70	3.12	1.27
Milky-wax ripening	1.05	0.58	2.81	1.84
Ripening	0.63	0.20	2.48	0.98
Full ripening	0.09	–	2.71	0.55

Care of plants

For weed control, Merkazin (prometrin) should be sprayed 1–2 weeks prior to sowing in amounts of 4.3–5.5 kg/ha. As well as Merkazin, Maloran (chlorbromuron) may also be used in amounts of 4–5 kg/ha. The advantage of Maloran is that it can be applied just before sowing or before emergence without the danger of harming the dill. As a post-emergence herbicide, Afalon may be used in amounts of 2 kg/ha.

The post-emergence treatment does no harm to the dill if the average height of the dill is 10–12 cm. (In the case of spice dill, the post-emergence herbicide must not be used.)

The most frequent and at the same time the most dangerous disease of dill is caused by *Megacladosporium depressum*. The dodin-type fungicides provide preventive protection against this.

Harvesting

The time of harvesting is determined by the goal of cultivation and by the mode of usage of the dill.

For producing spice the dill should be harvested prior to the emergence of the stem, or at the time of initiation of stems at the latest. In this stage the height of the plants is generally 25–35 cm. The plants still have abundant leaves, there are no thickened stems (which would decrease the quality of the herb), and the leaf/stem ratio is good. It is advisable to cut the plant with high-performance windrower or mower–loader machines. The cut part should immediately be transported to the drying facilities. No proper-quality drug can be prepared with natural drying. The cut plant should not be stored in bulk even for a short period, because it loses its original colour, and becomes unsuitable for use. The material dried at 40–50 °C is shelled, and then the stem parts are separated. The average yield is 400–600 kg/ha of cleaned shelled drug, which can be extracted from 4–6 t/ha of raw plant material.

If the purpose of dill cultivation is to produce *dill herb oil,* then it should be harvested when the essential oil content is at its highest. Experiments (Table 9) indicate that the essential oil content of the green plant gradually increases from the emergence of the stem, and the maximum content is reached at the end of flowering, when the green seeds have developed in the main umbels. The essential oil content of the plant continuously decreases after this time.

Dill cultivated for herb essential oil should be cut just above the bottom dry leaves, leaving stubble 20–25 cm in height. The height of the stubble is justified by the fact that the bottom part of the stem contains practically no essential oil, and consequently its transport and manipulation at the time of the extraction would only increase the costs. The harvesting may be done with the help of a mower–loader or, in the case of containerized extraction, by an ensilage harvester. The expected yield is 15–25 t/ha of green plant, or 40–80 kg/ha of dill herb oil.

When determining the time of the harvest, it should be taken into consideration that the seeds of dill do not ripen simultaneously, and that the ripe seed is highly susceptible to droping away. The most valuable seeds are those which are in the main umbels. The seeds of the second and third side umbellets are generally smaller, their essential oil content is less, and their biological seed value is also smaller. Consequently, for the determination of the harvesting time, the ripening of the seeds in the main umbels should be taken into consideration. One-path harvesting with the help of a combine harvester should be started when the seeds in the main umbels are already brown and do not smear. At this time the leaves are yellow and twisted. Harvesting in one path leads, in general, to a larger loss than harvesting in two paths. If the harvesting is done in two paths, then cutting into windrows shall be done when the plant is getting yellow, when the seeds in the main umbels are in the stage of wax ripening.

Fig. 61 The fruit of the star anise

Cutting to windrows with 30–40-cm high stubble can be done with any mowing machine. After several days of post-ripening and drying, the dill may be picked up and threshed by a combine harvester provided with a pick-up adapter. The expected yield is 0.7–1.8 t/ha of dill seed, or 20–40 kg/ha of seed oil.

Anise (*Pimpinella anisum* L.)

Anise was used in ancient times by the Egyptians. It is mentioned by Dioscorides as a seasoner and medicinal plant named anneson. Anise is still a popular spice today. Its seed (Anisi vulgaris fructus) is officially regarded as a drug by the 7th edition of the Hungarian Pharmacopoeia.

It has appetizer, digestive, and catarrheasing effects. As a spice it is mainly used by bakeries, confectionery and the liquor industry. The bacteriostatic effect of its essential oil (Aetheroleum anisi vulgaris) is widely known. Its pleasant odour and taste are also exploited by the medicine industry.

Owing to its wide field of applicability, anise is cultivated in many areas. The most important countries cultivating anise are Spain, Bulgaria, Romania, Italy, USSR, Turkey, India, Mexico and North Africa.

As a substitute for anise, star anise (*Illicium verum* Hook) is grown in Indo-China and in Japan, where it is native, and produces a crop known as Anisi stellati fructus (Fig. 61).

Characteristics

Anise is a species of the *Pimpinella* genus, which belongs to the umbellates (Apiaceae) family. There are about 150 species belonging to this genus in Eurasia and in Africa. The anise is native to the eastern coast area of the Mediterranean, to Asia Minor, and to Egypt. Spontaneously growing anise may only be found on a few isles of the Aegean Sea.

The cultivated common anise (Fig. 62) is an annual, herbaceous plant. Its height is 0.3–0.7 m, depending on the place of cultivation. Its root is spindle-like and thin, with sporadic branches, and it penetrates the soil to a depth of 0.2–0.3 m. Its stem is straight and erect, slightly hairy, circular, and grooved lengthwise, and has leaves along its whole length. The stem branches at its upper part. The stem and its branches terminate in a flowering main axis. The stem is weak, which is why the whole plant frequently leans on the ground. The rosette leaves have long stalks, are round, ovate or reniform, and form rosettes with variable dimensions. The middle stem leaves have short stalks, two or three folia, and are fibrously fringed.

The inflorescence is a compound umbel which is characteristic of the family, consisting of 10–15 umbellets. Its fruit is a twin achene with an egg-shaped or reversed pear form, the two remnants of the stigma. The crops tend to stick together. The peduncle frequently remains on the crop.

The twin achene has a greyish-green or brownish-grey colour. The length of the seed is 3–5 mm, and its width is 1.5–2.5 mm. The mass of 1000 seeds is 1–4 g, and it is similar to the poisonous hemlock seed (Fig. 63).

The characteristic odour and the sweet spicy taste of the seeds originate from the contained *essential oil*. The essential oil is accumulated in 30 small oil passages and 2–4 larger oil passages.

The ripened seed often contains 1.5–3.0 per cent essential oil (sometimes it can even reach 6 per cent). The main ingredient of the essential oil is anethole, the content of which may go up to 80–90 per cent. The essential oil also contains 5–10 per cent methilcavikol, anisaldehyde, aniscetone acid, and anise acid. The seeds contain 18–23 per cent fatty, 16–20 per cent protein and 3.5–5.5 per cent sugar after extracting the essential oil.

Fig. 62 Anise

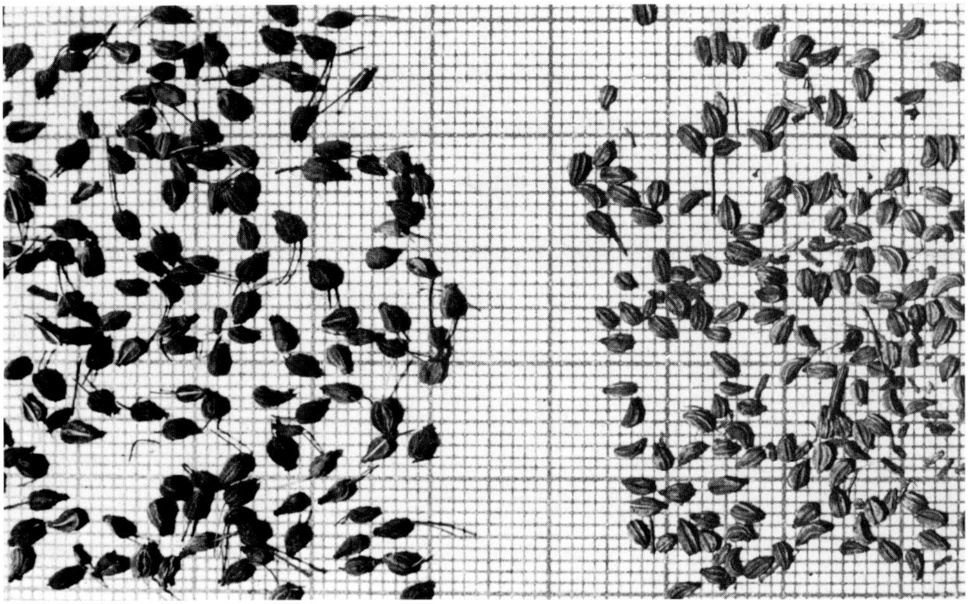

Fig. 63 Seed of anise (left) and of poison hemlock (right)

The total vegetation period of anise (from sowing to ripening of seeds) is 120–150 days. The germination (sprouting) time of the seeds is 17–25 days. The development of anise is slow during the initial period of vegetation; 35–40 days are required from sprouting to the emergence of the stem. Flowering starts at about 65–75 days after sprouting. The duration of flowering and seed development is 20–25 days. Anise is a cross-pollinating plant, pollinated by insects. Its tiny flowers produce abundant nectar. From the shedding of flowers of the main umbels to the ripening of seeds in the main umbels takes 25–30 days. The ripening of the seeds takes places unevenly, which is characteristic of the family.

Environmental requirements

The heat requirement of the plant is 2200–2400 °C according to Lukjanov (1976).*

Germinating plants are not sensitive to the cold; they are capable of enduring frosts of −7 °C without any damage. The seeds require abundant water for germination (130–140 per cent of the seed mass). The largest amount of water is taken up by the plant during the period from the emergence of stems to flowering. If the water supply is not sufficient in the soil at this period, then the anise yield might drop by as much as 150–200 kg/ha.

At the time of seed emergence and ripening, the plant needs warm, dry weather.

* The heat amount means the sum of mean temperatures necessary from germination to seed ripening.

Anise requires high-quality soil. It can only be cultivated on soils with proper structure, good water supply, abundant nutrients, and neutral chemical effects. Dry sandy soils and firm, clayey, cool soils cannot be used for cultivating anise.

Cultivation

Nutrient supply

The essential requirement for a suitable anise yield is an adequate nutrient supply. For the development of 100 kg seed of anise takes up 3.5 kg of N, 1.5 kg of P_2O_5 and 4 kg of K_2O. Most of this is taken up in the period from stem emergence to flowering. It is advantageous to distribute 20–25 per cent of the required nutrient at the time of the spring soil preparation. The top-dressing made during the vegetation period is rather ineffective, especially in dry weather.

Soil preparation

The soil preparation operations for the anise are identical to those described for dill.

Sowing

The determination of the time of sowing is of upmost importance. Although the germination of the anise seed begins at 4–5 °C, sowing too early may give rise to an extended germination period owing to the low soil temperature, and the sprouting will be uneven. On the other hand, if the sowing is too late, the soil humidity might be low, or the stem emergence might take place in the high-temperature period, and consequently the development of the umbels will not take place properly and the yield with be low. It is most advantageous to sow when the top 3–5 cm layer of the soil has an average temperature of 7–8 °C. In a number of countries the sowing of anise is done with calibrated seeds with a mass of 4–5 g for 1000 seeds. This results in a yield 20–25 per cent higher than the yield produced from the usual sowing seeds.

The optimum row spacing for anise is 25–30 cm. The seeds should be sown 2–3 cm depth. The optimum germinal number is 50–60 pieces/m, and the sowing seed requirement is 13–15 kg/ha.

Care of plants

The sprouting time for anise is rather long, and the initial development is also slow. For this reason, the biggest danger to anise comes from weeds. Efficient protection can be provided against weeds by spraying 5–5 kg/ha of Merkazin or 3–3.5 kg/ha of Maloran herbicides to the ground just prior to sowing. According to recent studies, the herbicide Afalon may be applied as a post-emergent herbicide in an amount of 2–2.5 kg/ha, without harming the anise.

Among the diseases attacking anise from time to time is bacteriosis (caused by the same pathogen as for the coriander), which might cause significant damage. It can be prevented mostly by agricultural methods (sowing sequence, sowing time).

Harvesting

As with the other umbellates, the ripening of the seeds of anise takes place over an extended period of time, i.e. the ripening is uneven. The most valuable drug in the main umbels can be extracted from the central twin-achenial seed. This is why the time of harvesting should be determined according to the ripening stage of this seed.

In accordance with the process of ripening, either the one-path or the two-path method of harvesting can be applied. If ripening is relatively rapid, then one-path harvesting is preferred. In this case harvesting should be done when the seeds in the main umbels are completely ripe. By this time most of the stem leaves have already dried. Harvesting may be done with a combine harvester generally used for cereals. It is advisable to raise the cutting equipment of the harvester slightly so that the twin-achenial seeds are not crushed. Two-path harvesting is applied to a lesser extent. In this case the anise is cut to windrows in the first path. Harvesting is started when most of the seeds in the main umbels are greenish-grey, and waxy ripe. Plants cut with a stubble height of 10–12 cm can be gathered without any loss and can be threshed after 4–6 days of post-ripening. The average yield is 0.8–1.5 t/ha of anise seed.

Angelica (*Angelica archangelica* L.)

This plant has long been believed to have healing effects, and its root drug is mentioned in a number of pharmacopoeias. Its root (Angelicae radix and rhizoma), and sometimes its leaves (Angelicae folium) and fruits (Angelicae fructus), are mainly used, although its leafy stalks (Angelicae herba) are also used from time to time. It is cultivated in a number of countries, mainly for the essential oil (Aetheroleum angelicae radix) extracted from its root.

It is an ingredient of a number of medicinal preparations, namely the spicy tea blend (Species aromaticae), fiorantibalsam (Spiritus balsamicus) and diuretic wine (Vinum diurecitum). It is used in both human and animal medicine, owing to its digestive, hunger-inducing and carminative effects. In folk medicine it is used against rheumatic and nerve diseases as a scrubbing agent, and it is used as an expectorant in the case of respiratory troubles. In tea blends it is used for blood cleaning, to induce sweat and as an antispasmodic, and for skin sweating baths in the case of muscle pains.

It is a most important flavouring agent for alcoholic liqueurs such as benedictine and chartreuse, and for gin and vermouth. Vegetables are also made from its young leaves and stems in some countries. Cakes are garnished with its crystallized thicker leaf stalks and stems, which are rich in liquids. Its essential oil is also used by the cosmetic industry. It is commercially cultivated in Belgium, Germany, France and several other countries.

Characteristics

Angelica archangelica L. (syn. *A. officinalis* Hoff.) is a biennial or short-lived, perennial plant belonging to the Apiaceae family. It is native to Greenland and Iceland, but it can also be found in Siberia and northern and central Europe. The southern limit of its range is in the Carpathians, where it grows in humid areas 700–1700 metres above sea level, along mountain rivers and alpine rivulets and in areas rich in clefts.

The root is like that of turnip with occassional branches in the first year. The thickened rhizome develops many auxiliary roots, which are 0.25–0.40 m long, covered with brownish cork, and ribbed lengthwise; their thickness is similar to that of a little finger. The plant develops only rosette leaves in the first year. The flower stems reaching heights over 2 m only develop in the second year (Fig. 64). Its bare cylindrical hollow stem has a thickness frequently over 50 mm. The underdeveloped plants (root thickness below 30 mm) will not develop flower stems in the second year.

The bifoliate leaves are large, dark green and light blue—green on the back side. The base of the leaf is an inflated sheath. Umbels are spherical, with diameters often above 0.2 m. Umbellets are collared, and have a spherical or half-spherical shape.

The hermaphrodite greenish-white petal flowers are in flower in June. Cross-pollination is by bees. The fruit is a detachable twin-achene, 6–7 mm in length, 4–5 mm in width, light yellowish-brown, and with winged margins. The mass of 1000 seeds is 2.0 g.

Roots, fruits and leaves contain essential oil. There is 0.5–1.0 per cent essential oil content in the root and rhizome, 0.6–1.5 per cent in the ripe fruits, and 0.2–0.3 per cent in the dried leaves. The main components are *d*-α-phellandrene, α-pinene, α-tujone and camphene. As well as the essential oil, the roots contain angelica acid, a number of coumarins and furocoumarins (osthole, osthonole, angelicin), and about 24 per cent sugar.

Fig. 64 Angelica

Angelica is one of the cultivated plants which sprouts earliest. Its leaves appear often as early as February. The vegetation period ends in the first year in the late autumn. It flowers durably, similarly to the other umbellates, generally from the beginning of June in the second year. The seeds ripen continously from early July and tend to drop away. The plant dies after seed ripening. The germination capacity of the seeds is best after ripening, and later on germination is effective only after a certain amount of exposure to cold.

Environmental requirements

Angelica likes cool climates. It is not sensitive to frost even in its vegetation period. Its optimal temperature is 5–19 °C. It develops sufficiently only in the lowlands, where the soil is humid, deep, and rich in nutrients and humus, with a pH of 4.5–7.3.

Soils originating from swamp are especially suitable for cultivation; here the gathering of roots is easy, and the yield is high compared to the yield in mineral soils.

The root system is in a relatively thin soil layer (0.4–0.5 m); water-absorbing roots do not penetrate the soil deeply. However, the large surface area of the foliage leads to the evaporation of significant amounts of water. The high water requirement of the plant does not cause too many problems in rainy climates (annual precipitation of 0.6–1.3 m) and in humid soils, but in dry areas the plant can only be cultivated with proper irrigation.

Cultivation

Nutrient supply

In the autumn, prior to planting or sowing, 60–70 kg/ha of N, 100–120 kg/ha of P_2O_5 and a higher amount of K_2O (150–180 kg/ha) active material should be introduced into the soil. Angelica is top-dressed with 70–80 kg/ha of N active material early in the spring.

Soil preparation

The field is prepared for autumn sowing with 30–35 cm deep ploughing in August, and then the usual treatments follow, consisting of cultivation, and elimination of furrows, lumps and weeds. For sowing it is necessary to prepare the soil so as get fine structure and a well-compacted seed-bed.

Sowing

Today, the cultivation of angelica by sowing in a permanent location is widely practised. In order to make it possible to gather the roots with machines, it is advisable to cultivate the angelica on ridges. In this case the seeds are sown in late

October or early November on the tops of ridges spaced 0.5–0.6 m apart. The sowing seed requirement is 18–20 kg/ha (80–100 seeds/m). In the case of autumn sowing, germination and sprouting start early in the spring, and the plants do not develop roots of a suitable size until the autumn.

Care of plants

Weed control is usually performed by cultivating between the rows with machines. Mechanical cultivation can be replaced by the application of herbicides. Chemical weed control in the sown angelica field can be accomplished sufficiently by the application of Merkazin (4–5 kg/ha) before sowing, or by post-emergent treatment with Afalon (3–4 kg/ha) after the plants have reached 0.15–0.2 m in height.

Fungus control treatments are generally started in May against peronospora, powdery mildew and septoria. Spraying is repeated every 10–12 days until the middle of September, if justified. According to experience, the best protection is provided with a Chinoin-Fundazol (benomil) solution of 0.1–0.2 per cent concentration. Against the plant louse, one of the following pesticides may be used: Pirimor (pirimicarb), Bi 58 EC (dimethoat), Wofatox (methyl parathion) and Phosdrin (mevinphos).

Harvesting

The roots are ploughed out in the autumn after cutting the foliage. In the case of ridged cultivation, machinery used for parsley roots can be applied. In the case of essential oil production, the removed roots are placed in a distillation vessel in layers separated by straw. Distillation is done with a slow steam speed; otherwise the steam would cut passages in the pulpy boiled material, and the extraction would be inefficient. The duration of the distillation is 8–10 hours.

For the preparation of the dry drug, the cleaned roots are cut up into 2–4 pieces, or may be sliced, and dried with an artificial drier. About 4 kg of fresh root gives 1 kg of dry drug. Sometimes leaf drugs are also produced. For this purpose the healthy leaves can be collected from May until the late autumn. Only the blade of the leaf is cut away. The leaves should be dried at a low temperature (30–40 °C) without delay, because the leaves easily get mouldy. Six kilograms of fresh leaves are necessary to produce 1 kg of dry drug.

The average fresh root yield is 7.0–10.0 t/ha. Considering the drying ratio, a yield of 1.7–2.5 t/ha of dry root drug is expected. The mentioned fresh root yields give 6–10 kg/ha of essential oil. The expected amount of leaf drug is 0.4–0.6 t/ha.

Sometimes the leaves or the flowering green plant with green seeds are also harvested, and essential oil is extracted from them; however, this essential oil is seldom required. In these cases the green yield is 5.0–20.0 t/ha and the essential oil yield is 2–6 kg/ha. Care should be taken during gathering and handling the leaves and roots, because the yellow liquid oozing from them can give rise to inflammation and blisters on the skin. Leather gloves should be used, and during work on the field, arm protection and skin guards should be used.

Lovage (*Levisticum officinale* Koch)

Lovage has been cultivated as a medicinal plant for a long time. It was called by Dioscorides 'libysticon' or 'lygisticon' (lybian). Columella also mentioned it. Its cultivation was encouraged by Charlemagne in his *Capitulare*. The plant is mentioned in the medical books of the Middle Ages and even the distillation of its essential oil is described.

All parts of the lovage have odours and flavours similar to those of celery. Its essential oil (Aetheroleum levistici radicis), root (Levistici radix), fruit (Levistici fructus), leaf (Levistici folium) and sometimes its herb (Levistici herba) are used.

The variations in aroma and fragrance among the different parts of the plant allow each portion of the plant to be utilized somewhat differently. The essential oil extracted from the roots is highly valued for use in perfumery, soaps and creams, and it has been used for flavouring tobacco products. The seeds and seed oil are used as flavouring agents in confectionery and liqueurs. The stems are used for candied products, and leaves are added to salads, soups and stews. Its root is authorized as a strong diuretic drug by a number of pharmacopoeias; many authors consider its name to be derived from the latin word 'levare' (lighten). The root drug is an important ingredient of diuretic tea blends (Species diureticae), but its extract and tincture are also used. It is used in folk medicine for digestive troubles.

Lovage is cultivated over large areas in central Europe and the eastern USA.

Characteristics

Lovage is a perennial herbaceous plant, belonging to the family Apiaceae (Fig. 65). It is native to the mountain regions of northern Europe and is naturalized in the eastern USA.

Fig. 65 Lovage

The rhizome is vertical, penetrates the soil to 0.4–0.5 m depth, and terminates in a taproot, which is ringed crosswise. The tubular stem, which branches and develops every year, is frequently over 2 m tall. The leaves are alternate, 0.5–0.6 m long, petiolate with stipules, shining, and bifoliate. The inflorescence is flat, compound umbel with 5–15 axes. The yellow flowers are hermaphrodite. The fruit is a flat, yellowish-brown winged twin achene. The mass of 1000 seeds is 3.7 g.

The roots, as well as the parts above ground contain essential oil. The essential oil content is 0.5–1.0 per cent is the rhizome and roots, 0.15–0.45 per cent in the leafy stem bearing green seeds, 0.08–0.24 per cent in leaves, and 0.8–1.5 per cent in ripe seeds. Among the above, the most valuable is the root oil. The essential oil consists mainly (70 per cent) of phthalides and terpenoids, including *n*-butylidene phthalide, *n*-butyl phthalide, sedanonic anhydride, *d*-α-terpineol, carvacrol and eugenol. The root also contains coumarins, sugar, apple-, angelica- and coffee acids.

Lovage lives for 6–8 years, but the plantation can only be maintained for 3–4 years, because later the stem and leaf development diminishes, and the roots become cavitated and rotten. The seeds retain their germination capacity for two years. The initial development of the germinated plant is slow; only rosettes are formed in the first year. In the second and later years, sprouting begins early in March. The emergence of the stem begins at the end of May, and flowering lasts from the end of June to the middle of July in central Europe. Characteristically for the umbellates, the fruits ripen unevenly from the end of July, and complete ripeness is expected in the early days of August. The seeds are susceptible to dropping away.

Environmental requirements

Lovage can be cultivated in any temperate climate. Its range is 6–18 °C with 0.5–1.5 m annual precipitation and a soil pH of 5.0–7.8.

Lovage likes medium–firm soils which are rich in nutrients and humus and thick-layered. It can also be successfully cultivated on swampy soils, where intensive development takes place and the roots are easy to remove. In these areas, however, the essential oil content is somewhat lower.

Cultivation

Nutrient supply

Organic fertilizers are preferable for previous plants. Prior to sowing, the following active materials should be introduced into the soil in one hectare: 100–120 kg of P_2O_5, 140–150 kg of K_2O and 60–70 kg of N. Plants containing a high mass of green parts extract a large amount of nutrients from the soil. For this reason, a sufficient supply of nutrients is also necessary during later years.

Sowing

The optimum sowing time for lovage is in the late autumn (November) or early spring (first ten days of March). The row distance is 0.5–0.7 m, and the depth of sowing is 10–20 mm. The sowing seed requirement is 10–12 kg/ha, and the number of seeds should be 70–80 per m.

Care of plants

In the case of autumnal sowing, Maloran (chlorbromuron) can be applied sufficiently with pre-emergent treatment (2.5–3.0 kg/ha) for chemical weed control. In the spring, Merkazin (prometrin) can be applied before sowing in amounts of 4–5 kg/ha. Maloran is applied in a dosage of 8–10 kg/ha on lovage plantations in their second or later years, in the early spring (pre-emergent).

Insect and fungus control should be achieved as described for angelica.

Harvesting

The roots are harvested generally in the third or fourth year of vegetation, in the autumn, depending on the development, after cutting the herb with silage machines. Accordingly, the roots are turned out from the soil with a plough from which the steering plates have been removed and which cuts to 35–60 mm depth. The soil is shaken off the roots and then prior to processing the roots are washed. The sprout residues are removed from the washed roots, and then the roots are split into 2–4 parts of 0.1–0.15 m length, or are sliced and dried in shaded open-air or in artificial driers. Four kilograms of fresh root give 1 kg of dry drug.

For essential oil production, the roots are washed and sliced, and placed into a distillation vessel in layers which are separated with straw. The distillation and extraction should be performed in a manner suitable for an essential oil having a specific gravity higher than that of water. A 3–4 year-old plantation yields 6–8 t/ha of fresh root, which gives 5–6 kg of essential oil.

For the production of vegetable oil, the plantation is maintained for 4–6 years. The base leaves of the first-year plant are harvested in the autumn just prior to the frosts, because the too early cutting would hamper the strengthening of roots.

From the second year on, the plants are harvested together with the base leaves at the time of main umbel waxy ripening with the help of machines. Following the main harvesting, a secondary crop consisting of base leaves develops until October.

The average leaf yield is 4–6 t/ha, from which 2–4 kg of vegetable oil can be produced. In one year 10.0–20.0 t/ha of seedy plant can be harvested, from which 8–20 kg/ha of vegetable oil can be extracted. The ripe seeds are harvested only if specifically requested by a customer to produce seed essential oil or seed drug. The expected 0.4–0.6 t/ha of seed gives 3–6 kg of seed oil.

Caraway (*Carum carvi* L.)

Caraway is one of the most popular and widely used spices in the world. Its excellent properties are even mentioned in the Bible. In the first century BC it was proposed as strengthener by Dioscorides. Caraway has been cultivated in Europe, from Sicily to northern Scandinavia, since the Middle Ages. It is mentioned in German medicinal books as early as the twelfth century as 'Cumich' and 'Kummich'.

The achenial fruit of the caraway (Carvi fructus) is regarded as an authorized drug by almost all pharmacopoeias, including the 7th edition of the Hungarian Pharmacopoeia. It is used in human medicine mainly as an antispasmodic, carminative and stomachic. The essential oil extracted from its fruit (Aetheroleum carvi) has antibacterial properties; it hinders the multiplication of *Escherichia coli, Streptococcus fecalis, Streptococcus aureus* and other infectious bacteria.

Caraway and caraway seeds, however, are still mainly used by the food industry and in the household, for spicing various baked products, meats and other foods, and salads and pickles and drinks, in almost all areas of the world. Its essential oil is a welcome natural base material, not only in pharmacology, but also in the cosmetics industry. It is cultivated on a large scale in many areas, owing to its popularity and its wide applicability, mainly in countries with temperate climates. The most important countries cultivating caraway are the Netherlands (main supplier of the world market), Poland, Spain, Egypt, Turkey, the USSR, the USA and Morocco.

Characteristics

The caraway species belong to the family Apiaceae. More than 20 species are known in the *Carum* genus, including forms developing both small and large fruits, and also biennial, annual and perennial forms. The species which has been cultivated for the longest time, and which is also the most widely spread in the world, is the biennial caraway (which has a large seed). The annual caraway has been cultivated in a few countries only (Poland, Egypt, Israel, the USA, Hungary) since not more than 10–15 years ago, over a relatively small area.

Eighty per cent of the cultivated area of caraway in Hungary is used for the biennial caraway. Its pot-like root penetrates deeply into the soil; the root thickness is about equal to that of a human finger; slightly fleshy, white inside and light reddish-brown on the outside.

The stem, which develops in the second year, is erect, 1–1.5 m in height, and generally branched at ground level. The rosette leaves develop in the first year and are petiolate, with lengths sometimes over 0.2 m (Fig. 66).

The stem leaves are sessile in cuticle coats. The leaves are trifoliate compound. The stem of the annual caraway is 0.5–0.7 m tall, and its leaves are yellowish-green, lighter than those of the biennial caraway. The inflorescence of both forms of caraway is a compound umbel with 5–10 rachises; the colour of the corollate leaf is white or pinky. The flowers are monoclinous. The fruit is a twin achene with five ribs. The seeds of both caraways are 3–7 mm long and 1–1.25 mm thick; the mass of 1000 seeds is 2–3.5 g.

Fig. 66 Rosette of the biennial caraway

The characteristic spicy taste and flavour of the caraway originate from the essential oil accumulated in the passages of the achenes. The essential oil content of the biennial caraway seed is 3–7 per cent, and that of the annual caraway seed is 2–3 per cent. As well as the essential oil, the fruits contain 20 per cent protein and 14–22 per cent fixed oil. The main constituent of the essential oil is carvone (50–70 per cent). Significant amounts of *d*-limonene (25–30 per cent), dihydrocarvone, carveol and dihydrocarveol are also included in the essential oil.

Other parts of the plant–roots, leaves, stem, flower and inflorescence–contain essential oil only in traces, and its composition differs from the composition of the seed oil.

According to Gluscsenko (1976), the accumulation of essential oil in the seed begins just after seed binding, and the peak amount is reached at the time of milky ripening. The essential oil accumulation decreases somewhat after that, and it stabilizes at the ripening of the fruit. The essential oil of milky ripe seed contains less carvone than that of ripened seeds.

The seed of the caraway retains its germination ability for 2–3 years. Germination lasts for 18–25 days in advantageous conditions. The initial development of the biennial caraway is slow; the first rosette leaves appear 11–14 days after the development of seed leaves. After this the rosette, characteristic of the biennial plants, develops continuously, and consists of 7–18 leaves when fully developed (in the autumn). The plants start developing rapidly rather early in the spring of the second year (first part of March), and the formation of the flowers starts within 30–37 days counted from sprouting (middle April) (Fig. 67). If the first-year development of the plant is inadequate (owing to shading or dryness) and the diameter of the root collar is less than 5 mm, then stem emergence does not occur and the plant develops fruit only in the third year. In average conditions the biennial caraway flowers in May. The average duration of flowering is 40–50 days. The fruits ripen within 40–50 days

Fig. 67 The flowering biennial caraway in the second year

counted from full flowering. Ripening is continuous, and the ripened seeds shatter easily.

The total vegetation period of the biennial caraway is 440–460 days from sowing to the ripening of fruits. Unlike that of the biennial caraway, the vegetation period of the annual caraway is only 140–160 days. It develops rapidly after sprouting; the stem emerges in the middle of June and it flowers from the first days of July for 20–25 days. Its fruit ripens from the end of August to the first part of September, and the seeds are less susceptible to shattering.

Environmental requirements

The biennial caraway is a native even of hillside pastures and humid hay fields. It prefers a rainy climate. It requires relatively low amounts of heat; its germination begins at soil temperatures of 7–9 °C. High temperatures are not even required in the consequent phases of development; a daily mean temperature of 16–20 °C is advantageous at the time of flowering and seed binding. It is sensitive to the availability of sunshine, especially in the first year. According to experiments, it develops no fruit even in the second year if the shading is with (intercropped) vegetables. The water requirement is high. The plant will not develop fruit if the water supply is not sufficient, even in the second year. The water intake is highest at the beginning of stem emergence and at the time of flowering. The precipitation requirement is 600–650 mm per year.

Caraway can be cultivated on almost any soil. The highest yield can be reached on deep, caulky, humid (although without ground water), medium-firm soils, which are rich in nutrients. Dry, sandy soils, areas where water accumulates and arid soils are not suitable for the cultivation of caraway.

The annual caraway has a somewhat higher heat requirement, and the water requirement is also higher owing to the fact that it ripens later.

Cultivation

Associated (intercropped) cultivation

The biennial caraway is frequently cultivated together with annual plants, in order to exploit the cultivation area more efficiently, as it does not develop fruit in the first year. The plants usable for association are those which have short vegetation periods and relatively small shade areas. Previous publications have suggested a number of plants for cultivation in association with caraway in small-scale production. The most suitable plants for associated cultivation are dill and the annual caraway. These have very similar requirements in terms of agrotechnique and biology; but sowing needs to be done in two paths, owing to the dissimilar row distances.

Dill associated with caraway should preferably be cultivated only for producing vegetable oil, because it may be harvested earlier (90–95 days) and thus not hamper the development of the caraway. The annual caraway causes shading to an even lesser extent than dill does, and the biennial caraway is able to strengthen satisfactorily under it.

Caraway may only be sown on the same field after four years.

Nutrient supply

Caraway requires high amounts of nutrients. It does not require direct fertilization with organic fertilizer, and it can also exploit the effects of fertilization after two or three years. A sufficient nutrient level in the soil can be realized by supplying fertilizers, the amount of which is should be determined taking the needs of the associate plant into consideration.

Inadequate nutrient supply would cause slow development, and consequently the caraway would not be able to develop stems which bear fruits even in the second year. As far the observations made by Schröder (1964), caraway extracts 72 kg of N, 32 kg of P_2O_5 and 80 kg of K_2O from the soil for the production of 1 t of seeds. Hungarian experiments indicate that it is justified to introduce 50–70 kg/ha of P_2O_5 and 50–80 kg/ha of K_2O active material into the soil at the time of deep ploughing in the autumn. Naturally, the actual amounts of these depend on the nutrient content of the soil at the time of ploughing. A nitrogen amount of 50–60 kg/ha (depending on the needs of the associate plant as well) should be introduced into the top layer of soil simultaneously with the soil preparation operations performed in the early spring. In the case of soils having poor nutrient contents, the yield can be increased if 30–40 kg/ha of P_2O_5 and 35–40 kg/ha of K_2O are spread on the ground in the autumn of the first year. This is especially required if cultivation is performed with associate plants. It is also advisable to perform a top-dressing with nitrogen (50–60 kg/ha) in the early spring of the second year.

Soil preparation

Caraway requires a wet seed-bed which has a fine structure.

In order to avoid desiccation of the soil, the early spring soil preparation should be made with equipment which does not move the soil excessively or penetrate deeply into it.

Sowing

The basic factor in cultivating the caraway successfully is sowing, which is performed at the appropriate time. The optimum sowing time for caraway in Europe is in March and early April. As a frequent result of late sowing, the rosettes would not develop until the winter, provided that sprouting was sufficient, and consequently the plants would not develop their generative organs even in the second year.

The method of sowing depends on the mode of cultivation. When the cultivation is performed without associate plants, then the row spacing should be 0.30–0.40 m. The number of germinating seeds should be 80–100 per m. In this case the requirement for sowing seed is 10–12 kg/ha. The annual caraway may be sown with a 0.2–0.25 m row distance owing to its lower space requirement; 12–15 kg/ha should be used in order to get the same number of germinating seeds in 1 m. The sowing depth should not be over 20 mm, because the seeds that are deeper than 20 mm sprout inefficiently and unevenly.

In the case of cultivation with associate plants, sowing is done in two paths. Initially the densely cultivated associate plant (dill) is sown, then the biennial caraway is sown. The dill is generally sown with a row spacing of 0.2 m and with half the amount of sowing seed, and then the caraway is sown with a row spacing of 46–50 cm by the application of 8–10 kg/ha of sowing seed. The seeds of the two plants may also be mixed and sown simultaneously if the masses of 1000 seeds and the sowing row spacings are similar, as in the case of biennial and annual caraway. The soil should be rolled after sowing to prevent the formation of cakes, especially if the weather is dry.

Care of plants

The care of the caraway consists of spreading herbicides in the first year, cultivating between the rows with the appropriate machinery if necessary, and spreading fertilizers in the autumn. In the second year, herbicide spraying, and sometimes cultivation between rows and plant-protection spraying against pests, should be done in the spring. Merkazin, 4–5 kg/ha, will not harm the caraway if it is sprayed before sowing.

It is a safe procedure to apply 3–4 kg/ha of Maloran before sowing if cultivation is done in association with dill. At a later date, Afalon may be sprayed onto the 0.10–0.20 m-high plant in amounts of 1–2 kg/ha. After harvesting the associate plant, it is advisable to loosen and aerate the soil by cultivating between the rows as necessary. If weed development is significant, then Afalon herbicide treatment may also be repeated in amounts of 1–2 kg/ha. In the spring of the second year (early in April), chemical weed control should again be performed with 1–2 kg/ha of Afalon. There is also a need for regular plant protection spraying in the second year. Among the fungus diseases, powdery mildew *(Erysiphe umbelliferarum)* and peronospora *(Plasmopora nivea)* most frequently cause damage. In the case of extended infection, proper protection can be realized by the application of fungicides.

Regular preventive protection has been required in recent years against *Eryophyes pencendani* f. *carvi,* which can cause deformation of the inflorescence and sterility. For this purpose it is advisable to combine the herbicide with acaricid chemicals, and in this way the first protection can be applied together with the spreading of herbicides in one path. The protection should be repeated at the time of bud development. If the symptoms appear again at the time of flowering in spite of the first protection, then a spray may be applied with acaricid pesticides, which do not harm the bees. The caraway moth is also a dangerous pest *(Depressaria nervosa)*. Its caterpillar damages the stems and leaves, and later the crop is annihilated by their feeding on flowers and seedy umbels. Pest control should be started early when larvae emerge from the eggs layed in the rosettes. An efficient protection may be realized by spraying or powdering with insecticides; however, this should be repeated after 8–10 days.

Harvesting

The fruits of the caraway should be very carefully harvested, owing to their uneven ripening and high susceptibility to dropping away. The manual processes which used to be applied (mowing, binding into sheaves, post-ripening in cones) may nowadays only be used in garden cultivation.

The biennial caraway is generally harvested in two paths. In the first path the caraway is cut by windrowing machines, and in the second path (after 5–6 days) it is threshed with cereal harvesting combines. Cutting should be started at the beginning of ripening (prior to the autumn harvesting of cereals), when the umbels are yellowish-brown and the seeds are just starting to harden (waxy ripe state). In order to reduce the drop-away loss, work should be performed early in the morning or late in the afternoon.

The threshed seeds should preferably be immediately cleaned and dried with artificial driers, on which the seeds are spread in thin layers. The seeds get mouldy rather easily, and for this reason careful handling is necessary. The annual caraway is less susceptible to losing its seeds, and therefore it can be harvested in one path with the help of cereal combine harvesters when the seeds are completely ripe.

The seed yield of the biennial caraway is 0.5–1.0 t/ha, and that of the annual caraway 1.0–1.6 t/ha.

Cumin (*Cuminum cyminum* L.)

Cumin was well-known to ancient civilizations, for it has been cultivated since the earliest times. It was included in the list of medicinal plants used in Egypt and recorded in the Ebers Papayrus. Cumin seed (Cumini fructus) has been utilized as an antispasmodic, carminative, sedative and stimulant. Its essential oil (Aetheroleum cumini) has been reported to have antibacterial activity. In indigenous medicine in India, cumin seeds have traditionally been employed as a stimulant and stomachic, useful in diarrhoea and colic. In Holland and Switzerland cumin is used to season some kinds of cheese, while in France and Germany it is used to flavour cakes and bread. Cumin seed, with its strongly aromatic, hot and bitter taste, is an essential ingredient of curry powder and chili powder. It is also used commercially in the preparation of meats, pickles, cheese, sausages and chutney. Cumin makes an excellent seasoning for soups and stews, for which it is widely utilized throughout Latin America.

Fig. 68 Cumin

Primary cultivation of cumin is in Europe, Asia, the Middle East and North Africa, with India and Iran being the largest cumin exporters.

Another aromatic herb, *black cumin, Bunium persicum* B. Fedtsch., has smaller and sweeter seeds than *Cuminum cyminum*. This plant, which grows wild in the Middle East, especially in Iran, is not commercially significant.

Black cumin, Nigella sativa L., which is sometimes called *black caraway,* is a herb native to the Mediterranean region and is not related to cumin.

Characteristics

Cumin, also known as *Cuminum odorum* Sahib., is a small, herbaceous plant species belonging to the Apiaceae family (Fig. 68). It is native to upper Egypt, Turkestan, and the Eastern Mediterranean region.

It grows to 0.2–0.4 m. Its root is spindle-like, thin, has sporadic branches and penetrates the soil to a depth of 0.1–0.2 m. Cumin produces a stem with many branches bearing long, finely divided, deep-green leaves and small flowers that are white or rose-coloured. The fruit consists of two united carpels, each carpel containing a single seed. The fruits are elongated-oval in shape and yellowish-brown in colour, frequently with a pedicel. The length of the seeds is 3–7 mm. The mass of 1000 seeds is 1.2–3 g.

The distinctive odour and flavour of cumin seeds are derived largely from the essential oil. Cumin seeds contain about 2.5–5.0 per cent essential oil, which contains cuminaldehyde as the main constituent. Other components of the oil are dihydrocuminaldehyde, *d,l*-pinene, *p*-cymene, β-pinene, dipentene and cuminyl alcohol. The fixed oil content of cumin is about 10 per cent.

The three major styles of cumin seed on the market, Iranian, Indian an Middle Eastern, differ in seed colour, quantity of essential oil, and flavour.

Environmental requirements

The total vegetation period of cumin is 100–110 days. The reported life zone of cumin is 9–26 °C with an annual precipitation of 0.8–2.7 m and a soil pH of 4.5–8.3. Cumin is a tropical plant and it can be cultivated in areas in which atmospheric humidity is low during the flowering and fruiting period. It thrives on rich, well-drained, sandy loam soil. The plant is intolerant of long periods of dry heat.

Cultivation

In the Middle East and in India, cumin is grown as a winter crop.

Well-rotted farmyard *manure,* 15–20 t/ha, may be applied uniformly in the field before ploughing. In addition to manure, 25–30 kg of urea may be applied with irrigation after 40–50 days of sowing to ensure proper growth of the crop (Pillai and Nambiar, 1982).

The land for sowing should be repeatedly ploughed to bring the soil to a fine tilth. If soil moisture is not adequate, a pre-sowing light irrigation should be given preparatory to tillage operations.

Sowing

In India and the Middle East, sowing is done between 15 November and late December. According to Pillai and Nambiar (1982), late sowing results in poor yield. Normally, the temperature before November is too high to permit survival of germinating seeds. Regularly, cumin is sown broadcast. The proper method would be to first divide the field into seed-beds and then broadcast the seed uniformly. Immediately after broadcasting, the seeds should be covered by the soil. Care should be taken that the seeds are not covered deeper than 15–20 mm. A seed rate of 12–15 kg/ha is sufficient to obtain an optimum plant stand.

Soaking of seeds for 24–36 hours in running water has given good results in improving germination percentage.

Care of plants

The first hoeing and weeding is done at the time when the plants are 50 mm tall, to facilitate proper growth of the crop. At the same time thinning is also done to adjust the plant spacing to 150 mm. After this one or two more hoeings may be done to get better crop growth.

Cumin is severely damaged by powdery mildew *(Erysiphe polygoni)*, blight *(Alternaria burnsii)* and wilt (*Fusarium oxysporum* f. *cumini*). Powdery mildew can be effectively controlled by spraying the plant, at the initial stage of infection, with wettable sulphur. The spray should be repeated after 10–15 days. Blight can be effectively controlled by spraying with a solution of copper fungicide. Dithane can be mixed with Karathane LC to control both the blight and the powdery mildew. For wilt control, the seeds should be treated with any mercurial fungicide (2 g/kg seeds).

Harvesting

Seeds are usually collected and threshed by hand, since the small, tender plants are difficult to harvest mechanically. The seeds are ready for harvesting when the plants begin to wither, the seeds become hard, and the fruit changes colour. The plants are uprooted by hand and small bundles of uprooted plants are stacked in the sun for drying. The seeds are separated by beating the plants with light sticks and cleaned by winnowing. The average yield is 0.8–1.0 t/ha of cumin seed.

Fennel (*Foeniculum vulgare* Mill).

Fennel is an ancient spice plant. Its agreeable flavour and useful medicinal effects were known to the ancient Greeks and Romans. Its seed has a pleasant, sweetish, slightly hot odour, which is similar to that of anise. It is authorized by most pharmacopoeias

Table 10 Main characteristics of fennel *(Foeniculum vulgare)* subspecies and varieties

Name	Subsp. and var.	Taste and odour of seed	Composition of seed essential oil (%)		
			Anethole	Fenchone	Estragole
Piper fennel	subsp. *piperitum*	Strong spicy	Trace	55	46
Bitter or wild fennel	subsp. *capillaceum* var. *vulgare*	Pleasant sweetish	60–70	12–22	2–5
Sweet or Roman fennel	subsp. *capillaceum* var. *dulce*	Pleasant sweetish	85–90	Trace	4–8
Indian fennel	subsp. *capillaceum* var. *panmorium*	Slightly strong taste	40–60	7–20	Trace
Florence or finocchio fennel	subsp. *capillaceum* var. *azoricum*	Unpleasant taste, brown Thickened base of leaves can be eaten as a vegetable	Trace	1–2	2–3

as a drug (Foeniculi fructus). In modern medical practice it is used as a cough medicine and antispasmodic, and is also an ingredient of hunger-inducing tea blends. It is the basic material of a number of Galen preparations. Chinese herbal medicine includes the use of fennel for gastroenteritis, hernia, indigestion, abdominal pain, and to resolve phlegm and stimulate milk production. Its essential oil (Aetheroleum foeniculi) is used in condiments and liqueurs, and it is the basic material for a number of cosmetic preparations. The seed is also a popular spice; it is used for flavouring vegetables, salads, fish, sauces, tea cakes and liqueurs. The word 'foeniculum' comes from the Latin word for fragrant hay, reflecting the plant's odour. It is cultivated throughout the world. Large areas are used for its cultivation in central and southern Europe, in Asian countries (India, China, Japan, Indonesia), in Egypt, and in Argentina. It has also been cultivated in Hungary for a long time.

Characteristics

Fennel is a perennial, herbaceous plant belonging to the Apiaceae family. It is native to the Mediterranean area and southern Europe. A number of its subspecies and varieties with dissimilar properties and active ingredients, are cultivated. The most widely known ones are summarized in Table 10. Generally the bitter fennel (var. *vulgare*) and sweet fennel (var. *dulce*) are cultivated (Fig. 69.)

The root penetrates into the soil relatively deeply; its shape is spindle-like, it is bulky and its colour is raw white. The stem is erect, 1.5–2.0 m high, abundantly branched, rounded, and ash-green. The leaves are palmate, shallow lobed, light greyish-bluish-green. The inflorescence is a doubly compound flat umbel. The flowers are minute and yellow. The fruits are 6–10 mm long, 2–3 mm wide, cylindrical, and twin achenial, they

Fig. 69 Fennel

tend to thin at the top, and are greenish or brownish-grey. The mass of 1000 seeds is 4–8 g.

All parts of the plant contain *essential oil,* and consequently the whole plant has a characteristic sweet odour. The essential oil accumulated in the parts above ground (leaves, stem), amounting to 1–1.5 per cent has a composition similar to that in the seeds.

The essential oil that can be extracted from the root (0.6–0.7 per cent) has a variable composition (90 per cent is dilapiole).

The ripe seeds contain 2–6 per cent essential oil, 18–20 per cent protein and 12–18 per cent fixed oil. The main component of the essential oil is anethole, and it also contains fenchon, methyl chavicol, camphene, camphor, limonen and several other compounds.

The seed of the fennel retains its germination ability for 2–3 years. The seed germinates within 14–20 days in advantageous conditions. The initial development is slow; the period from germination to the emergence of the stem is 2–2.5 months. Flowering takes place from early July to late August in the first year in temperate zones. The seeds ripen rather late in the year. Sprouting in the second and third years occurs rather early, in middle March, and consequent development is rapid. The plants reach a height of 1.50–1.80 m as early as the beginning of June. Flowering begins in the middle of June.

The life expectancy of fennel is 4–5 years, but significant amounts of seed are only produced in the first and second, or sometimes in the third, years. For this reason it is cultivated for two, or sometimes three years.

Environmental requirements

Fennel likes warm conditions, due to its Mediterranean origin. Although the germination of the seeds starts at 6–8 °C, a soil temperature of 15–16 °C is necessary for rapid sprouting. Fennel is not able to survive every winter in central Europe. The complete freezing of the soil frequently occurs, especially in loose soil (which cools down significantly to a great depth), and in winter when no snow layer protection is available. It requires a rather high temperature (20–22 °C) during its vegetation period, especially at the time of flowering and seed ripening. Its water requirement is unusually high at the time of germination and emergence of the stem. Fennel grows well in most mild climates, but thrives in sunny situations in limy, well-drained loam soil.

Cultivation

Nutrition

Fennel develops an enormous vegetative mass (40–60 t/h), for which relatively high amounts of nutrients are necessary. In spite of this, it is not advisable to cultivate in soils which have been recently provided with organic fertilizers, because the roots of the fennel become brown-spotted, and tend to rot. An abundant nitrogen supply is to be avoided, because it would result in a disproportionately high vegetative mass and poor seed development. However, soils with a poor nutrient content should be provided with fertilizers. Undernourishment would not only diminish the yield, but would also endanger the cultivation as the undernourished plants are less capable of surviving the winter. The capacity to survive frost is also decreased by nitrogen overdosing. The plant requires a uniform, harmonic nutrient supply; its phosphorus need is especially high.

According to our experiments and observations performed over a number of years, it is advisable to apply 80–100 kg/ha of P_2O_5 and 40–60 kg/ha of K_2O active substances to the soil at the time of the autumn soil preparation. In the case of soils with a poor nitrogen content, good results are obtained by introducing 40–50 kg/ha of N into the soil during the preparation of the seed-bed in the spring. If sprouting of the plants is slow in the spring of the second year, then it is wholly justified to perform a top-dressing with a 20–40 kg/ha N dosage.

Soil preparation

The operations of soil preparation for fennel are identical in every respect to those for caraway.

Sowing

Owing to its long vegetation period in the first year, fennel should be sown earlier than any other umbellates, generally at the beginning of March.

Fennel is generally sown today with 0.36–0.48 m row spacing. The optimum sowing depth for fennel is 20–30 mm, and the sowing seed requirement is 8–10 kg/ha (60–80 pieces of seeds capable of germinating/m).

Care of plants

The development of the common fennel is rapid, provided that the nutrient supply is adequate, and it forms a closed plant system (suffocating weeds) as early as the end of May. Herbicides with prometrin as the active substance provide efficient weed control if sprayed before sowing in a 4–5 kg/ha dosage.

Instead of the above herbicide, Aresin is preferred in areas infected with cruciferae weeds; this should also be applied before sowing in amounts of 3–3.5 kg/h. Uniformly sprayed Afalon (1.5–2 kg/ha) is efficient if the height of the fennel is over 0.15–0.20 m. In the case of two- or three-year-old fennel, the application of Merkazin (4–5 kg/ha) or Aresin (3–3.5 kg/ha) is efficient if sprayed at the end of the winter (pre-emergent). For these plantations, a post-emergent herbicide treatment can also be done with Afalon (1.5–2 kg/ha) if required.

Fennel is sometimes attacked by the fennel moth, against which the measures described for caraway should be applied. Significant damage has been caused by *Lygus* and *Ortops* species in recent years. This damage can be reduced by spraying methyl parathione, which is applied simultaneously with herbicides.

Harvesting

The seeds of fennel ripen unevenly and the ripe seeds tend to drop out easily. Another feature making harvesting difficult is the fact that the plant is large, the percentage of unused parts is high, and the seeds ripen rather late in the first year.

Fennel may be harvested at the end of September and in October in the first year, and at the end of August and in September in the second and third years.

In order to reduce the harvesting loss, harvesting is performed in two paths. In the first path the plant is cut to windrows with high stubble (0.3–0.5 m). This is done when the seeds, even in the umbels of primary side branches, are in the waxy ripe stage. After post-ripening, which lasts 7–14 days, depending on the weather, the fennel is threshed by a combine harvester provided with a windrow pick-up adapter.

The threshed fennel should be cleaned of green parts and impurities and dried without delay.

The expected yield is 0.4 t/ha in the first year, 1.0–2.0 t/ha in the second year and 0.6–1.5 t/ha in the third year.

Periwinkle (*Catharanthus roseus* (L.) G. Don.)

Periwinkle—Madagascar or Cape or red periwinkle—is an attractive ornamental, widely cultivated and naturalized plant in Madagascar, India, Israel and the USA. It has been used in folk medicine as a diuretic, an antidysenteric, an antihaemorrhagic and for wound healing, and has been considered to be useful in the treatment of diabetes for a long time. During the last 30 years it has been used in modern medical practice as the most important plant in the war against cancer. Noble and Beer, Canadian investigators, noted that a crude extract of this plant drastically reduced white blood cell counts, especially of the granulocytes, and profoundly depressed bone marrow activity in rats (Lewis, 1977). Following this discovery the detailed examination of the plant was started all over the world. An industrial process has been developed in Hungary and in the USA. Thus all parts of the periwinkle have became important raw materials for the pharmaceutical industry.

At present it is cultivated as a medicinal plant in Madagascar, India and other tropical countries, and in all countries of central and southern Europe. As a medicinal plant, periwinkle has been cultivated in Hungary for just over two decades.

The main alkaloids are antimitotic, resulting in tumour cell death during replication. Vinblastine sulphate is used particularly to treat Hodgkin's disease but is also used against lymphosarcoma, choriocarcinoma, neuroblastoma carcinoma of the breast, lungs and other organs, and in acute and chronic leukaemia. Vincristine sulphate is used particularly to treat acute leukaemia in children and lymphocytic leukaemia, but is also used against Hodgkin's disease, Wilms's tumour, neuroblastoma, rhabdosarcoma and reticulum cell sarcoma. Other alkaloids have undergone clinical trials but have not proved satisfactory (ITC UNCTAD/GATT, 1982).

Characteristics

Catharanthus roseus (syn. *Lochnera rosea* Reichb. and *Vinca rosea* L.) is a tropical–subtropical perennial subshrub, but it is an annual herbaceous species in temperate zones. It belongs to the Apocynaceae family. The periwinkle is native to the West Indies, Indonesia, Australia and other tropical areas (Madagascar, Philippines) as well. It is naturalized in Vietnam, Cambodia, Cuba and other subtropical countries.

Fig. 70 Periwinkle

The main root of the periwinkle is white, 0.2–0.4 m long, and scarcely branched. The stem is erect, cylindrical, green or reddish in colour, and branched near the base. It grows to a height of about one metre (in Europe 0.4–0.8 m), and spreads out over an area of about 0.6–0.7 m main diameter (Fig. 70). The leaves are petiolate, round-edged, ovately pointed, oil-green, shiny and glossy. The flowers are located at the branches, generally two-by-two, on short peduncles. The calyx is funnel-like, bright pink or white. The fruit is a follicle which splits early. The seeds are 3 mm long, 1 mm wide, black, and characteristically grooved on the surface. The mass of 1000 seeds is 2–2.4 g.

All parts of the plant contain alkaloids. More than 100 kinds of alkaloid have been isolated from the plant so far. According to Datta (1982), the content of total alkaloids in different parts of the plant is as follows: in the leaf, 0.6–0.67 per cent; in the stem, 0.26–0.31 per cent; in the flowers, 0.005–0.007 per cent; in the roots, 0.78–1.22 per cent; and in the root bark, 4.5–9.0 per cent. The most important alkaloids are vinblastine, vincristine and vinleurosine from leaves, and raubasine (ajmalicine) from roots. In temperate zones, the alkaloid content of the leaves is highest at the end of the summer (middle of September).

Periwinkle (as an annual herb) has a long vegetation period. The first ripe fruits appear within 180–200 days of sowing. The plants develop slowly after sprouting. In Hungary, flowering takes place continuously from the middle of May to the first frosts.

Environmental requirements

Like most plants of tropical and subtropical origin, periwinkle requires lots of warmth, light and water. The optimum temperature for germination is 20–22 °C. An average temperature over 18 °C is also required in the later stages. The development of the plant ceases at 10 °C, and the plant freezes at 0 °C.

It needs 800–1000 mm of precipitation in the vegetation period. For this reason, it can be successfully cultivated only on irrigated fields. The plant grows adequately on those soils which are somewhat loose, have good structure, are easily warmed, have good water-exploitation ability, and are rich in nutrients.

Cultivation

Nutrient supply

The plant requires lots of nutrients; it exploits direct manuring with good efficiency. Most of the time, however, this cannot be done, and for this reason the nutrient supply should be satisfied with fertilizers through the soil and through the leaves. According to domestic and foreign experience, it is necessary to introduce 40–50 kg/ha of N, 60–80 kg/ha of P_2O_5 and 40–70 kg/ha of K_2O into the soil as basic fertilizer. A further addition of 50–70 kg/ha of N in the vegetation period increases the yield economically.

In India, where the plants grow on wastelands under natural conditions, the requirement for fertilizer is not great. However, to obtain a good yield of both leaves and roots, farmyard manure, about 15 t/ha, should be applied.

Soil preparation

The best soil is that which has been ploughed in the autumn; the plant may also be planted in soil ploughed in the spring. The spring soil preparation operations—smoothing, cultivatoring, disc harrowing—are the same as those used for planted vegetables.

Sowing and planting

Periwinkle can be propagated by seed sowing in temperate climatic regions. Owing to the high requirement for warmth and to the long vegetation period, it is propagated through seedlings. The seeds are sown in plastic houses at the end of February. The seedling development requires a temperature around 20 °C. The seeds germinate very slowly. The sprouting time can be reduced by soaking the sowing seeds in water at 22–24 °C for 24–36 hours prior to the sowing. The pretreated sowing seeds are sown shallowly (10–15 mm) in rows.

The previously soaked seeds start to germinate within 7–10 days of sowing in the case of a soil temperature of 20 °C. The sowing seed requirement is 0.4–0.5 kg/ha. The sufficiently strong seedlings are planted into open-air fields after the risk of frost is over. The optimum vegetation area can be maintained with 0.5–0.6 m row spacing and 0.2–0.25 m plant distance.

In India and in other subtropical countries seeds can be sown *directly* in the field or the plants raised in the nursery and transplanted later on.

In the case of direct sowing, about 2.5 kg/ha of seed is required. The seeds are mixed with sand and are sown during the monsoon in rows 0.45 m apart. When the plants have grown in the field in rows, they are thinned out, leaving a distance of 0.3 m between them. The seeds are sown in nursery beds in February–March, and when the plants are about 60–70 mm in height (two months old), they are planted at the onset of the monsoons.

Care of plants

After the initiation of the seddlings the first weeding between the rows should be made; this is repeated three to five times, depending on the development of weeds. Irrigation should also be followed by weeding. Hoeing is generally necessary two to three times. In the vegetation period, four to five irrigations are necessary, depending on the soil conditions and weather. The top-dressing fertilization may also be spread with the irrigation water through the irrigation system.

Harvesting

Most of the active substances accumulate in the leaves and in young shoots; that is why all the parts of the plant above the ground are harvested. The alkaloid content is highest in the second half of September in temperate zones. Harvesting is done in one path, leaving 80–100-mm-high stubble. It is advisable to dry the cut plant at once in artificial driers at 60–70 °C and 12 per cent humidity. The expected yield is 3.5–4.5 t/ha of fresh herb, or 1.0–1.5 t/h of dried herb in Hungary.

In tropical regions, the crop becomes ready for harvesting *for roots* after one year. Two leaf strippings can be taken, the first one after six months and the second after nine months of direct sowing. A third stripping of the leaves can also be taken when the whole plant is harvested after one year.

For the harvesting of roots, the plant is cut about 70 mm above the ground and dried for stems, and leaves, and then the whole field is copiously irrigated and ploughed and the roots are collected. The roots are washed well and dried in the shade.

Under irrigated conditions, 1.5 t/ha of roots, 1.5 t/ha of stems and 3 t/ha of leaves can be obtained.

Valerian (*Valeriana officinalis* L.)

Valerian is one of the oldest medicinal plants. The ancient Greeks knew of its healing effects, and named it Phu. Its present name was derived from the latin word 'valare' ('to be healthy') in the Middle Ages. Its root (Valerianae rhizoma et radix) is an important raw material in modern medical practice. It is regarded as an official drug by almost all of the pharmacopoeias of the world, and it is used in a number of sedatives and tranquilizer tea blends. It is also an ingredient of many Galen preaparations (tinctures, extracts).

From its extracts, many sedative medicines are produced all over the world. The essential oil extracted from the roots (Aetheroleum valerianae) is used as a flavouring in several major food and tobacco products and in alcoholic and non-alcoholic beverages, and is also used in perfumery.

The plant is cultivated in many countries for the pharmaceutical industry. It is cultivated in large areas of the USSR, Belgium, Japan, Germany, Poland, Bulgaria, Romania and France. Its cultivation in Hungary was started early in the 1960s, mainly on small farms.

Characteristics

Valerian is a hardy perennial plant belonging to the Valerianaceae family. Three hundred and sixty different species are known in the thirteen genera of the family. The species belonging to the *Valeriana* genus differ from each other in terms of morphology and chemical composition. Among these the most important is *V. officinalis* L., which is regarded as a group name by Heeger (1956), and he also classifies *V. collina, V. exaltata* and *V. procurrens* here. Three varieties are most frequently distinguished of *V. officinalis,* namely var. *latifolia* Vahl. (= var. *altissima* Koch. = var. *major* Neilz. = *V. exaltata* Miq), which is called Japanese valerian (its root is known as kesso), var. *media* Koch., and var. *tenufolia* Vahl. (var. *minor* Koch. = var. *angustifolia* Tausch). In recent years the species *V. officinalis* has been regarded as a compound species consisting of a number of subspecies according to the classification made by Meyers (1958), including various ecological types and chemotaxons of the European valerian.

The Indian valerian, *V. wallichii* DC., synonymous with *V. jatamansi* Jones, is native to Asia and is found particularly in India and south-western China.

Fig. 71 Flowering valerian

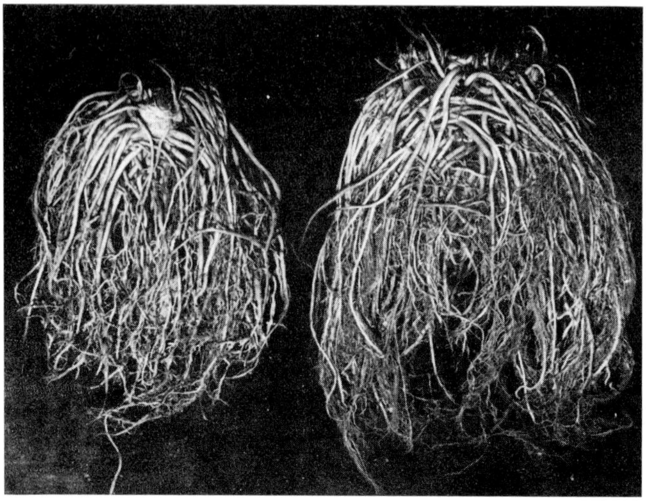

Fig. 72 Root of valerian taxons

The American valerian or ladyslipper, is actually *Cypripedium calceolus* L. of the Orchideaceae family.

The European valerian (Fig. 71) is native to temperate regions of Europe and western Asia (between latitudes 41 °C and 70 °C).

The rhizome is short and cylindrical; from it a number of spindle-like fibrous roots develop that are fleshy, 100–150 mm long, 2–5 mm thick, brownish on the outside and white on the inside (Fig. 72).

The rhizome sometimes develops stolons as well. The stem is erect, 0.4–1.5 m high, cylindrical, grooved, and cavitated inside. The leaves are opposite, alternate, odd-pinnate compound. The base leaves have long petioles, and the stem leaves are sessile. The bottom leaflets are lanceolate and deeply lobed, and the upper ones are almost linear. The inflorescence is a branched dichasium.

The flowers have a characteristic pleasant odour. The fruit is a one-seeded acorn 2–5 mm in length, 1.3–1.5 mm in width, thinning at one end, with an elongated ellipsoid shape, and a yellowish-brown colour. At the tip of the fruit a pappus with 10–15 radii is located, which disseminates the seeds easily with the help of the wind. (This makes the production of sowing seed very troublesome.) The mass of 1000 seeds is 0.5–0.6 g.

A large number of biologically active materials accumulate in the rhizome and in the roots. Among these, the essential oil has been known for the longest time.

European valerian contains 0.5–1.0 per cent essential oil, Indian valerian 0.3–1.0 per cent and Japanese (kesso) valerian up to 8 per cent. The main constituents of the essential oil are valeric acid, isovaleric acid, eugenyl esters, l-pinene, l-camphene, l-borneol and terpineol. The isovaleric acid gives the characteristic unpleasant pungent odour to the root. This odour affects cats strongly.

The most important active compounds of valerian are iridoid compounds called valepotriates (valtrate, acevaltrate, etc.), didrovaltrates (didrovaltrate, deoxyolodidrovaltrate, etc.), and isovaltrates (isovaltrate, 7-epideacetylisovaltrate, etc.). Valtrate and didrovaltrate are the major valepotriates, which amount to 0.5–2 per cent in the root. The valepotriates are mainly responsible for the CNS-depressant and antispasmodic effects.

Valerian also contains the alkaloids actinidine, valerianine, valerine, chatinine, etc.

The seeds of valerian achieve their total germination capacity within 10–20 days of flowering, although this germination capacity is lost relatively quickly. The germination capacity of the seeds falls below 30 per cent by the second year. In advantageous conditions the seeds start to germinate within 7–12 days, but durable sprouting (3–5 weeks) is not rare either. The initial development of valerian is slow; it develops rosettes and a relatively strong root system in the year of sowing. The reproductive organs are formed only in the second year. In the conditions prevailing in Europe, valerian flowers for a long time, from May through June, the fruits ripen unevenly in the second half of July, and the seeds are liable to fall away.

After the ripening of seeds, the parts of the plant in the ground decompose, but this is preceded by bud development on the rhizome, from which shoots grow, which give flowers and fruits in the next year.

The active ingredient content of the root starts to increase at the beginnig of cool weather, and the peak content is reached in the dormant period.

Environmental requirements

With regard to its weather (temperature and precipitation) requirements, valerian can be cultivated successfully in almost any area of central Europe. Its development is especially good in areas with high precipitation (600–700 mm/year). The seed germinates most efficiently in light and at a temperature around 20 °C.

Valerian likes light, thick soil well supplied with moisture and containing proper amounts of nutrients. It thrives in medium loam or sandy loam soils, which have a medium humus content. The root system develops well in chernozem and peat soils with a high humus content, but the drug produced in these soils is very hard to clean (it is contaminated with soil particles).

Cultivation

Valerian can be successfully cultivated after almost any plant; it does not preset special requirements with regard to the previous plant. Plants with perennial roots, like angelica, lovage, liquorice and peppermint, are not regarded as being good previous plants. Valerian itself should not be cultivated again on the same field for 4–5 years.

Nutrient supply

Valerian requires a lot of phosphorus. According to the studies made by Bernáth, Földesi and Lassányi (1973), an increase in nitrogen and phosphorus nutrients enhanced the formation of roots significantly, while an increase in potassium supply did not result significant increase in root formation. Numerous and thin roots are the result of poor nutrient supply, according to their opinion. A balanced nutrient supply helps the development of thick roots. Also, the active ingredient content is highest in this case. In practice, generally 40–60 kg/ha of N, 60–80 kg/ha of P_2O_5 and 40–60 kg/ha of K_2O is introduced into the soil. The phosphorus and the potassium, and half of the nitrogen, are added prior to the planting, and the other half of the nitrogen is introduced into the soil after the winter.

Soil preparation

If planting is done in the autumn after the early harvesting of hoed plants, then shallow or medium to deep ploughing is necessary just after harvesting the previous plant, and this is followed by harrowing or cultivator operation combined with fertilizer introduction. Depending on the development of the weeds, one or two harrowings should be done up to the time of planting. In the case of spring planting, autumn deep ploughing is absolutely necessary, and the planting bed should be prepared early in the spring.

Sowing and planting

Valerian is most frequently propagated using seeds. Propagation by plant division is applicable only for the varieties which develop stolons. This vegetative propagation is not suitable in view of the fact that the plants develop flower-bearing stems frequently in the first year, and this is accompanied by decreased root development.

The seeds of valerian may be sown at various times of the year, and also with various methods (e.g. direct sowing or nursery sowing and transplanting). In Hungary, nursery sowing in open-air seedling beds has proved to be the most reliable and economic way of propagation. Direct sowing (which is widely used in the more humid areas of Poland and the Ukraine, etc.) did not give good enough results in Hungary.

Nursery sowing

The best procedure is to sow the seeds between the end of June and early August into well-prepared open-air seedling beds. The seeds should be sown with 0.15–0.20 m row distance on the surface of the soil, and should be covered with compost of a maximum thickness of 1 mm. The seedling requirement of 1 ha can be grown on a 500–700 m^2 area. An amount of 0.5–0.7 kg of good-quality sowing seed is required for growing the seedlings necessary for 1 ha. One of the conditions for successful seedling growth is regular irrigation. The sown bed should be slightly shadowed in order to reduce the water loss through evaporation until sprouting or until the appearance of the first leaves. The seedling growth period is 2–2.5 months.

Planting

The seedlings reach a height of 0.15–0.17 m by the end of September and can be planted with machines. The optimum planting time for valerian seedlings is the end of September or the beginning of October. If planting is done later than this, then the seedling may freeze during the initial frosts of the winter or in the spring, because they do not have sufficient time to become strong. The spring planting may only be done at the end of April, when the plants are again about 0.15–0.17 m in height. The yield is generally less by 20–25 per cent in the case of spring planting than in the case of autumn planting. In the case of hoeing by hand, the optimum row distance is 0.30–0.35 m and the optimum plant distance is 0.20–0.25 m. In the case of mechanized cultivation, the row distance cannot be reduced below 0.50 m, but the 0.20–0.25 m plant distance should be adhered to.

Care of plants

The first spring work in the valerian field planted in the autumn is rolling. This is especially important if the plants have risen somewhat in the ground owing to the frosts. The valerian plantation requires two or three weedings and hoeings. Chemical weed control can be done by the application of 2.5–3.5 kg/ha of Aresin prior to the spring planting. In the case of autumn planting pre-emergent herbicide treatment should be done with 3.5–4.5 kg/ha of Aresin. According to the results of the latest studies, the plants with heights over 0.20 m may be successfully treated post-emergence with Patoran or Maloran in dosages of 2.5–3.5 kg/ha.

Up to now, valerian plantations have not been attacked by significant fungus or insect pests. With increasing cultivation, however, one can expect damage at the end of the summer from the powdery mildew (*Erysiphe polygoni* DC.) and by peronospora (*Peronospora valerianae* Trail.).

Harvesting

Valerian planted in the autumn should be harvested in the autumn of the next year (after one year of cultivation). That planted in the spring (end of April to beginning of May) can be harvested in the autumn of the same year in the case of exceptionally advantageous cultivating conditions. Only in the autumn of the next year may one expect a suitable yield (two years of cultivation).

Harvesting of the roots should be done in the autumn in every case (second half of October, first half of November). Harvesting just prior to the starting of vegetation at the end of the winter is very seldom done, because the active ingredient content of the roots at this time is already low owing to the early shooting.

Prior to the removal of roots, it is advisable to cut the parts of the plant above ground with the help of mobile chaff-cutter machines. Following this operation, the shallow-laying roots can be removed with a plough with no steering plate attached, or with rotating forked potato-harvesting machines in the case of loose soils. The roots lifted to the surface are gathered by hand and placed in piles, where the green parts are removed and the soil particles between the roots are shaken out.

The roots are washed with liberal amounts of water (with basket immersion or with a water jet), and then they are at once dried at 40–50 °C. The drying and storage of valerian should be done separately from other drugs, owing to the strong smell of the root. It must be stored in closed places, inaccessible to cats, because they would contaminate it. Depending on the conditions prevailing in the cultivation area and also on the mode of cultivation, the expected fresh root yield is 3.5–7.0 t/ha, from which 2.0–4.0 t of dry root may be obtained.

Lavender (*Lavandula* species)

The lavender species cultivated today have been widely known since the Middle Ages. According to the description of Dioskorides and Plinius, only one lavender species *(L. stoechas)* was used in ancient times, and this is rather insignificant today. The true lavender *(Lavandula angustifolia)* was mentioned for the first time in the literature in the thirteenth century. The same period may be regarded as the time of spread of the lavender in central Europe. It is assumed that the hybrid lavender or lavandin *(L. intermedia)* and spike lavender *(L. latifolia)* have been known and also cultivated since the sixteenth century.

The lavender species can be exploited in a number of ways. As a medicinal plant, its dried inflorescense (Lavandulae flos) has traditionally been considered to be an antispasmodic, a carminative, a diuretic, a nervine, a stimulant, and a tonic. It has been used as a folk remedy against colic and headaches. Lavender has also been an important ingredient in herb baths for a long time. The scientific name of the plant was derived from the latin 'lavare' (washing, bathing). The essential oil of lavender (Aetheroleum lavandulae) has antiseptic, carminative, and spasmolytic activity. The flowers are considered to be an insect repellant.

Lavender oil is frequently used in cologne toilet waters, lotions and a wide range of high-quality perfumes, often after extensive blending or 'stretching'. Spike lavender oil scents cheap soaps, polishes, detergents and liquid cleaners, although the more delicate French oil is also used in room sprays, deodorants, disinfectants and insecticides. The major consumer of lavandin oil is the soap industry, but the oil also has many applications in men's colognes and numerous inexpensive perfumes, detergents, cleaning and washing-up liquids, polishes, talcum powders and hair preparations.

Commercial distillation used to take place mainly in France and Spain, but Eastern Europe—principally Bulgaria and the USSR—is now almost certainly the world's largest producing region.

World production of the various lavender oils amounts to around 1000 t/year or perhaps slightly more, of which lavandin oil accounts for some 75 per cent, lavender oil 20 per cent, and spike lavender oil the remaining 5 per cent (ITC UNCTAD/GATT, 1986).

Lavender plants are grown as ornamentals along garden borders, in rock gardens and as potted outdoor plants. They are also grown near highways for beautification and stabilization of soil. Lavender plants are attractive to bees.

Characteristics

Forty-eight species having wide form variability are known in the genus *Lavandula,* which belongs to the Lamiaceae family (Kendz, 1976). As of today, only three species of lavender have significant commercial value.

The best known and the longest cultivated one is true lavender (medical, smelling, thin-leaved), *L. angustifolia* Mill. (= *L. officinalis* Chaix. = *L. vera* DC.). The other species, cultivated in several European countries is spike lavender (male, spikinard), *L. latifolia* (L. f.) Medic. (*L. spica* L.). There is a great practical significance attached to lavandin (hybrid lavender), *L. intermedia* Emeric ap. Lois (= *L. burnati* Brig.), which is the spontaneous hybrid of the former two species, and is widely cultivated.

All three species are native to the Mediterranean region. Lavender is very characteristic of the dry, wild, frequently stony areas between the Alps and the Mediterranean, this is why the symbiotic plant partnership in which lavender lives was named Lavandula-Artemision after lavender by Braun-Blanquet (1961). True lavender lives as high as 1700 m above sea level, while spike lavender grows only in lower regions, and is native in the range 200–700 m above sea level. Lavandin can be found frequently at heights of 700–1000 m above sea level. These areas of growth correspond to their requirements. It is quite understandable, then, that only true lavender and lavandin can be cultivated in eastern Europe; spike lavender freezes in winter.

Both the cultivated species of lavender are shrubs. The roots form a main woody root system, densely branching at the bottom. The water intake roots penetrate the ground to 3–4 m depth. The stem is short, covered with brown cork, and densely branching in a shrub-like manner from its bottom. In natural conditions the lavender shrub has a hemispherical shape. The several-year-old lavender has a height of 0.40–0.60 m, while lavandin frequently reaches 0.80–1.00 m in height.

The diameter of the true lavender shrub is 0.80–1.20 m, and the diameter of the lavandin is over 1.50 m. The flowering shoots of true lavender are 0.20–0.40 m long, and are single-stalked, while that of lavandin may grow to 0.60–0.90 m and generally it is branching. The leaves are linear or thin lanceolate, opposite. The leaves of true lavender are 30–50 mm long, 2–5 mm wide, greyish-green, and felty pubescent. The leaves of lavandin are larger (50–70 mm long, 8–10 mm wide), are not so pubescent and the colour is a brighter green (Fig. 73).

The inflorescence is an indented cylindrical sympodium consisting of verticils. The verticils of the flower of true lavender are more loose than those of the flower of lavandin. The flowers are zygomorphic; the corolla leaves are violet–blue (characteristic lavender blue).

Characteristically for the family, the fruit of true lavender is a calyx enclosing four nutlets, among which only one or two usually develop. A nutlet is 1.8–2.2 mm long, ellipsoid, and shining brownish-black. The mass of 1000 seeds is 0.85–1.1 g. Lavandin (as a hybrid) is sterile; it does not develop germinating seeds.

Both species have a pungent odour which is caused by the essential oil accumulating in the plants. The commercial value of the plants derives from the essential oil extractable from the flower. The essential oil content of true lavender is 0.5–1.5 per cent (depending on the variety) and that of lavandin is 0.9–3.0 per cent.

Fig. 73 Lavender (left) and lavandin (right)

Most of the essential oil accumulates in the oil-holding glands, which are among the ridges of the calyx. A small amount of essential oil with a different composition is also formed in the essential-oil glands found scattered on other parts of the plant (chaff, rachis, young foliage leaves). The most caracteristic, and at the same time the most valuable, components of the lavender essential oils are l-linalyl acetate and l-linalool. The ester content of the essential oil of true lavender is generally 35–60 per cent (expressed in linalyl acetate), while the essential oil of lavandin contains only 7–33 per cent ester. As well as the constituents mentioned above, lavender oil also contains borneol, camphor, geraniol, cineole and several other compounds (more than 40 have been identified).

Lavender lives for 20–30 years in the wild and also in cultivation. The germination of the seed of true lavender is extremely slow. With freezing the germination time can be reduced, and the uniformity of germination can be improved. The seeds generally retain their germination capacity for 3–4 years. The initial development and growth of the seedlings is slow, although flowers are often formed in the first year.

The sprouting of the several-year-old plants begins only after the weather has warmed up (air temperature more than 12 °C). The flowering of true lavender generally starts in the middle of June and lasts to the second half of July. Lavandin begins to flower somewhat later, in the first half of July, and its full flowering takes place around 20–25 July. Full flowering generally lasts for 6–10 days. The flowers contain the most and at the same time the best-quality essential oil.

Environmental requirements

The best conditions for cultivation can be found on warm soil protected from northern winds and on soils sloping towards the southern, south-eastern or south-western directions. The plant does not like the places where frosts remain for a long time, or deep-lying locations. In such places, the frost-damage is frequent. The parts of the lavender above ground bear frosts of −15 °C to −20 °C well in the dormant period. A greater danger is represented by late frosts—especially to lavandin, where the green parts of the plants, which have already started developing in the spring, can be frozen.

Lavender needs a great deal of sunshine. If the duration of the sunshine is not sufficient, then the formation of flowers, the essential oil content of the flowers, and the ester content of the essential oil, are significantly reduced. According to the observations of Hungarian researchers, the essential oil content of the flower growing in an open place is 0.99 per cent, while that of the plant growing in the shade of trees is only 0.45 per cent. The ester content of the essential oil extracted from the plant growing in sunshine is 41 per cent greater than that of essential oil extracted from plants grown in the shade. The generation and quality of essential oil is significantly influenced by the hours of sunshine at the time of flowering. In cloudy, rainy weather the essential oil content might be reduced by 30–50 per cent, and ester accumulation might be reduced by 20–30 per cent.

Lavender is a typical xerophyte. With its cultivation, dry soils may be sufficiently exploited, but it cannot grow in too humid places, or at locations where the ground water is high. In these areas the ability to withstand frosts is also reduced, and the accumulation and quality of essential oil is not sufficient either.

For the cultivation of lavender, shallow soils, wildlands rich in lime, and low-grade agricultural lands established on volcanic soils, are also sufficient. Good conditions for living can also be found in loam soils which have developed on loess. The advantageous conditions of grassland areas result in higher yields.

Cultivation

Nutrient supply

Lavender is not too demanding in terms of nutrients, but sufficient yield can only be achieved with a proper nutrient supply. For the production of 100 kg of inflorescence, lavender extracts 0.8 kg of N, 0.2 kg of P and 0.8 kg of K from the soil (Poludennij, Sotnik and Hlapcev, 1979).

As lavender is left on the same field for a long time, the possibility of manuring is excluded during cultivation. For this reason, the supply of a medium or large amount of manure prior to planting is justified; 35–50 t/ha of manure should be introduced into the top 0.30–0.40 m layer of the soil together with 70–80 kg/ha P_2O_5 and 100–120 kg/ha of K_2O. It is also advisable to apply 70–80 kg/ha of N prior to sprouting in the spring after planting in order to enhance the development of shrubs. For the nutrient supply of plantations, 60–100 kg/ha of N, 50–60 kg/ha P_2O_5 and 80–120 kg/ha of K_2O is necessary yearly, depending on the actual nutrient content of the soil.

The phosphorus and potassium fertilizer and one-third of the nitrogen fertilizer should be applied in the autumn, and two-thirds of the nitrogen fertilizer should be introduced into the soil between the rows in the spring as top-dressing.

Soil preparation

Soil preparation for lavender is started with the manuring and 0.35–0.40-m-deep autumn ploughing one year prior to planting. In the case of shallow soils, the actual depth of ploughing is determined by the stony layer under the soil, but the depth of ploughing should be at least 0.30–0.35 m in any case.

The areas prepared for the cultivation of lavender should be left as ploughed fallow up to planting, or they may be exploited with papilionaceous plants, which leave the field in May at the latest. After spreading the phosphorus and potassium fertilizers, a 0.18–0.20-m-deep ploughing should be done 1.5–2 months before planting. It is especially important that the soil settled, free of weed and even-surfaced at the time of planting.

Propagation

True lavender may be propagated with seed sowing and also with vegetative cuttings. As lavandin ripens no seeds, it can only be propagated with vegetative cuttings.

Nursery sowing

Seed sowing is done in November or early in the spring into well-prepared open-air seedling beds. In the case of spring sowing, the sowing seed should be prepared with freezing.

The seeds are sown with 0.35–0.40 m row spacings in large fields, and with 0.15–0.20 cm row spacings in small fields (the wider row spacing allows the application of machine cultivation). The optimum depth of sowing is 5–10 mm in the autumn and 10–15 mm in the spring. After sowing, the soil should be rolled again.

Sprouting generally starts at the beginning of April and lasts until early May. The care of the seedling consists mainly of weeding and hoeing. If the plants sprout too densely, then it is advisable to thin them out when they have four or five leaves, and the removed plants should be transplanted into open-air beds providing 35–50 mm of growing area. The seedling field is irrigated regularly during the vegetation period. Irrigation after back-cutting and after sprouting is especially important. Back-cutting should be done 80–100 mm from the ground surface at the start of flowering, in order to enhance the development of shrubs. The sowing seed requirement is 7–9 kg/ha, from which 600 000–800 000 seedlings can be developed.

Propagation by cuttings

Both lavandin and lavender can be propagated well with green or woody cuttings. The ripened shoots, which can be applied for developing rooted cuttings, are cut to 150–180 mm lengths between the end of August and the beginning of October, or in the spring (April–May).

The cut and properly prepared shoots are planted into open-air beds, which can be covered during the winter, or into unheated plastic houses. On the surface of the seedling bed, a 0.2–0.25-m-thick layer of lawn soil with a humus content is spread, and on this a 50–60 mm layer of river sand is spread. The cuttings to be rooted are planted in the bed prepared in the above manner to 50–60 mm depth and in a 60 × 40 mm growing area. The cuttings are regularly irrigated (even in the winter) and are covered to protect against strong frosts (but are regularly aerated as well). In the summer, when the first flowers appear, they are cut back to 120–150 mm above ground level. This procedure should be repeated two or three times until the autumn. The rooted cuttings are removed in the autumn prior to planting, and are then bound together and qualified. The number of rooted cutting suitable for planting, which can be grown on a 1 m^2 area, is 600–800.

Planting

The advantageous time for planting lavender lasts from middle September to the end of October. Spring planting should be avoided as much as possible, because the plants planted and abundantly watered until the end of March start to develop sufficiently.

It is very important to use only well-developed rooty and branching cuttings for planting. Roots longer than 150 m should be cut back, as long roots would hinder the planting operations. In order to prevent drying out after planting, it is advisable to prepare pulp around the roots.

Lavender used to be planted in a square, or less frequently in a triangular, arrangement. True lavender was planted into 1 × 1 m or 0.8 × 0.8 m areas, and lavandin was planted in a 1.5 × 1.0 m growing area. These growing areas are only suitable where the slope conditions of the area do not allow the treatment and harvesting of the plants with machines. According to our observations, in the case of cultivating with machines, the optimum pieces of plants are 20 000 of lavender and 14 000–15 000 of lavandin per ha.

Usually, lavender cuttings are planted with 1.0 m row spacing and 0.5 m plant distance, and 1.2–1.5 m row spacing with 0.4–0.6 m plant distance are necessary for lavandin cuttings.

The plantation made in the autumn should be irrigated only in the case of an exceptionally dry autumn, but irrigation is very important after the spring planting made owing to bad weather in the previous autumn, and the irrigation sometimes has to be repeated.

Care of plantation

The young plants should be cut back to 0.10–0.12 m above ground level at the start of flowering (at the beginning of June) in the first year. As a result of this operation, the plant branches and develops dense brushes until the autumn. The back-cutting should be repeated in the spring of the second year just prior to the start of shoot emergence. In this case the cutting heigth is 0.15–0.18 m. This job should be done with the machine which is used for the harvesting of lavender.

In the plantation hoeing with a machine is necessary two or three times yearly. For chemical weed control, in the young plantation (first and second year), a Maloran +

Kerb (propizamid) herbicide combination can be effective in a 6+6 kg/ha dosage, which should be spread pre-emergence. In the older (after two years) plantation, an effective herbicide is Aktikon, which is sprayed late in the autumn in a dosage of 4–5 kg/ha. One of the most dangerous weeds for the plantation is *Agropyron,* which can be destroyed by the application of Geonter (terbacil). It should be spread in a dosage of 4–8 kg/ha in the autumn or winter.

After several years the soil of the plantation gets compacted, contains no air, and cannot take water up sufficiently. In order to avoid this, the row spacings should be loosened to 0.4–0.6 m depth after 3–4 years.

Harvesting

For the purpose of extracting essential oil, lavender is harvested when it is in full flower. This optimum period lasts only for 7–8 days. In practice, harvesting is started at the beginning of flowering and is concluded only at the time of flower shedding, owing to the limited means or harvesting and processing. The lavender flower should be cut together with the stem just under the inflorescence with the first pair of leaves.

It is important to cut the flower axis completely and to uniform height, because the flowers remaining on the shrubs would affect flowering in the next year disadvantageously. Too-deep cutting is also undesirable, as the quality of the product is diminished by the dissimilar composition of the essential oil of the leaves, and the plants get damaged if the woody parts are cut in the vegetation period.

Lavender is harvested by hand even today on sloping fields with slopes of 10–15 per cent or larger. The flowers are cut with well-sharpened sickles and mowing knives,

Fig. 74 Harvesting the lavender by hand

Fig. 75 Harvesting the lavender by machine

gathered in baskets, and then taken to the border of the field. In a ten-hour shift one worker can gather from the 5–8-year-old plantation 0.1–0.5 t in the case of true lavender, and 0.3–0.5 t in the case of lavandin (Fig. 74).

Hand harvesting requires much labour; so countries cultivating lavender in large fields harvest the hedge-like lavender with machines. Special machines were developed in France at first, then in the USSR and in Bulgaria.

The operating principle of the lavender-harvesting machines is that the leaning stalks and flowers are raised with the rotating cone or stable raisers in the direction of the rows, and are pressed together to form a thin hedge, while the spiked reeler leads the stems to the cutting knives working in a horizontal direction. The cut flowers are transported into the gathering tank with the help of inclined conveyor belts or with ventilation (Fig. 75).

The essential oils are extracted from the harvested flowers with steam distillers (or intermittent or containerized equipment).

The harvested flowers are sometimes used to produce flower drug instead of essential oil. In this case the flowers cut with short stems are stored in a dry, ventilated place, or are dried in artificial driers at 40–45 °C, and then the flowers are removed from the flower axis by hand or with threshing machines. After the proper cleaning operations, the flowers are packed into boxes or cases lined with paper. Generally, 8–10 kg of fresh flowers are necessary to produce 1 kg of dry drug.

The average yields in the producing years (from the third year) are as follows: true lavender, 20–40 kg/ha of essential oil or 0.4–0.5 t/ha of dried flowers; lavandin, 50–70 kg/ha of essential oil or 0.5–0.7 t/ha of dried flowers.

Lemon balm (*Melissa officinalis* L.)

Lemon balm has been known and used for a long time as a pleasantly odoured medicinal plant. Its leaves (Melissae folium) and its dried herb (Melissae herba) have traditionally been employed against catarrh, fever, flatulence, headaches, influenza

Fig. 76 Leaves of lemon balm

and toothache. It has also been used as a carminative, diaphoretic and sedative. Recent evidence suggests that lemon balm has a depressant or sedative action on the central nervous system. Essential oil of lemon balm (Aetheroleum melissae) has also been shown to have antiviral, antibacterial and antispasmodic activity. It has been reported to be an insect repellant.

The essential oil is used to only a limited extent in perfumery, because perfumers are able to simulate the odour of lemon balm with less expensive extracts of other aromatic plants (lemongrass, citronella or lemon).

Lemon balm is cultivated in the Mediterranean, central European and Asian areas.

Characteristics

Lemon balm is a perennial, herbaceous plant belonging to the Lamiaceae family. It is native to southern regions of Europe, Asia Minor and North America.

The roots grow from the nodes of the rhizome, and are thin and yellowish-brown. From the axils of the base leaves on the rhizome, stolons grow in a horizontal direction; the tips of these protrude from the soil, giving rise to new plants.

The stem is 0.5–1.0 m high, mainly erect, and sometimes procumbent. The leaves are alternate, petiolate, and ovate (Fig. 76).

The leaf blade is slightly hairy, and its surface is slightly blistered and generally dark green. The flowers are located at the axils of the upper leaves on verticils. The buds are yellow, and later white. The acorn fruits are 1–1.5 mm long, ovate, shining dark, and then black. The mass of 1000 seeds is 0.6–0.7 g.

The odour of the whole plant is similar to that of lemon. The parts of the plant above ground, mainly the leaves, contain 0.1–0.3 per cent essential oil, tanning agent (rosemary acid), oleic acids and ursolic acids. The main components of the essential oil are citronellol (25–50 per cent), geraniol, linalool, nerol and eugenol acetate.

The seeds germinate very slowly and the germination capacity is retained for 3–4 years. Plants sprouting from the seed developed very slowly in the initial period. Plants older than one year start to develop only after the warming of the weather in the spring at the end of March and at the beginning of April. The flowering period is rather long, and lasts from early July to middle August. The flowers generate much nectar, so *Melissa* is a good honey producer. The seeds start to ripen unevenly from the end of August, and are liable to fall away easily.

Environmental requirement

Lemon balm likes warm weather and sunshine. It can bear temperatures of even −20–25 °C if a continuous snow layer is on the ground. In the case of long-lasting cold weather, however, the roots, mainly the older ones, are severely damaged even under snow cover. Its esssential oil content is also reduced, additional to the hampering of development in the case of a long, cool season (in shaded places). Lemon balm can be cultivated on any soil except extremely poor ones. The plant develops fastest in loose, thick, nutritious soils. It is sensitive to excessive or inadequate water levels in the soil.

Cultivation

Lemon balm remains in one location for 6–8 years. It is not advisable to replant with lemon balm at the same location for 4–5 years.

Nutrient supply

Sufficient development of plants and satisfactory yield may only be expected in a soil which contains sufficient nutrients. For this purpose, 25.0–30.0 t/ha of manure should be introduced into the soil to 25–30 cm depth at the time of the autumn deep ploughing. In the years after planting, it is justifiable to spread 50–70 kg/ha of N, 50–60 kg/ha of P_2O_5 and 80–100 kg/ha of K_2O active materials, depending on the nutrient content of the soil.

As well as the base fertilizers spread in the autumn or winter, it is advisable to introduce 20–30 kg/ha of N between the rows as a top dressing.

Propagation

Lemon balm may be propageted by direct sowing, with seedlings or by vegetative root divison. Direct sowing should be done in the autumn, at the end of October and at the beginning of November. Sowing should be done with 0.6 m row spacing and to 5–10 m depth, and using 8–10 kg/ha of sowing seed. The sown seeds start germinating only in the spring. Sprouting is extremely slow, and the initial development of the plants is also slow. For this reason, weeds usually overgrow the lemon balm. Therefore, direct sowing cannot be regarded as a reliable propagation method, and may only be applied in small areas (in gardens).

The most widely used propagation method is nursery sowing. Owing to the slow germination, the sowing seeds are pretreated by soaking in water for 16–20 hours, or by freezing for 1–2 days. Following this procedure, the seeds are sown in a plastic house early in March, or are sown into open-air seedling beds early in April. In both cases the seeds are sown with 0.30 m row spacing and to a depth not exceeding 5 cm.

The amount of seedlings necessary for 1 ha can be developed on a 250–300 m^2 area with the use of 0.5 kg of sowing seed. The seedlings grown in a plastic house are planted in open-air fields at the end of April, and the seedlings grown in open-air seedling beds are planted in the field in late August to early September. The optimum cultivation area can be maintained with 0.60 m row spacing and 0.3–0.4 m plant distance.

For propagation with root division, a method which is seldom used, the healthy 2–3-year-old plants are used. Root division is preferably made in September, as is usually the case with other spreading rooted plants.

Care of plantation

The most important care operation in a lemon balm plantation is the weeding and hoeing of soil. This latter procedure can be done with cultivators between the rows once or twice until the rows close together, then once after the first cutting. Weeding is done by mechanical means in small fields, and can also be performed with herbicides over large areas. Weed control may be accomplished with 3–4 kg/ha of Merkazin or with 2.5–3.5 kg/ha of Patoran herbicides, which should be spread early in the spring in the case of autumn planting, and just after planting if this was done in the spring. The two or more year-old plantation can be protected from weeds up to the first cutting by the application of 4–6 kg/ha of Merkazin, or 3.5–5 kg/ha of Patoran, which should be spread early in the spring prior to sprouting. The herbicide treatment may be repeated with 3 kg/ha of Merkazin just after the first cutting.

Preventive protection can be accomplished against turnip flee beetle and sucking pests (plant louse, cicada) by spraying with Bi-58EC material of 0.1 per cent concentration, or with Phosdrin of 0.1–0.2 per cent concentration. The last spraying should be done at least 25 days before harvesting.

Septoria spottiness (*Septoria melissae* Desm.), which attacks less frequently, can be extinguished by spraying with mankoceb solution at the time when the symptoms appear.

Harvesting

The whole parts of the lemon balm above ground are cut once in the first year and generally twice in the following productive years. The optimum time for the first cutting is when flowering starts. The second cutting is generally made in middle September. Harvesting is done by hand in small fields and with windrowing machines over large areas. Cutting should always be done on sunny days after the dew has dried, and 40–50-mm-high stubble should be left.

In the case of leaf drug production (which is very seldom prepared nowadays), the leaves are removed when still fresh from the cut plant. Immediate transport and fast drying are important. The leaves of the lemon balm go brown rather quickly. For this reason, these are preferably dried at 45–50 °C in an artificial drier. The expected yield is 10–20 t/ha of fresh herb, from which 2–4 t/ha of dry drug can be produced.

Peppermint (*Mentha piperita* L. emend. Huds.)

Mint species have been exploited by man for more than two thousand years. Peppermint itself has only been used for 250 years. In modern medicine there is wide application of its leaves (Mentha piperitae folium), dried herbs (Menthae piperitae herba) and essential oil (Aetheroleum menthae piperitae) as stomachics, carminatives, disinfectants, flavourings, stimulants, etc. The essential oil and the pure menthol extracted from the essential oil have bactericidal effects and a spicy odour. They are used for the preparation of cosmetics, perfumes, toothpastes and mouthwashes. Owing to its spicy flavour and cooling effect, peppermint is popular all over the world, and is used mainly to flavour drinks, chewing gum and confectionery.

Owing to its versatility, peppermit is cultivated in almost every country. According to estimates, *Mentha* oil consumption in the world is over 7000 t/year. The largest peppermint producer is the USA. Among the significant producers are Bulgaria, Brazil, Japan, France, the USSR and Argentina.

In Hungary the cultivation of peppermint was started at the beginning of this century.

Characteristics

The genus *Mentha* L., which belongs to the Lamiaceae family, contains about fifty known species. Among these species, several grow wild in Hungary. Only a few species have significant economic importance. Among these, the most important is the peppermint.

According to present knowledge, the peppermint is a multiple hybrid which has formed spontaneously (Fig. 77). Presumably it developed through the cross-breeding of *M. aquatica* L. and *M. spicata* L. (syn. *M. viridis* L.). *M. spicata* itself has formed from the cross-breeding of *M. longifolia* (L.) Huds. and *M. rotundifolia* (L.) Huds. Among the species taking part in the hybridization process, only *M. aquatica* has essential oil, which contains menthol. Opinions differ with regard to the formation of peppermint; some state that it was formed in the Far East, while others are confident that it developed in England in the seventeenth century. Two forms of peppermint are distinguished in cultivation, i.e. black or red peppermint (*M. piperita* var. *officinalis* Sole f. *rubescens* Camus), and white or green peppermit (*M. piperita* var. *officinalis* Sole f. *pallascens* Camus). In most countries the Mitcham peppermint, which belongs to the rubescens form, is cultivated.

Peppermint is a herbaceous, perennial plant, which survives the winter underground stolons. The roots develop from the nodes of 3–4 mm thick rhizomes or emerge from the underground stolons, and are white, thin and 50–200 mm long. The roots are spread shallowly.

The stolons, emerging from the rhizome, are densely branching, long, have long internodes, are white, and are able to develop turions or roots (propagating material). Stolons above ground (creeping stems) are also developed by the rhizome; these are violet–green, have elliptic leaves, and are also able to form roots and shoots (Fig. 78).

The stem is 0.3–1.0 m high, depending on the cultivation location, four-angled, violet (anthocyanin), and mostly branching at the upper part. The leaves are oblong ovate, and fine serrated. The surface of the leaf is dark green and the veins are conspicuously violet. Both sides of the leaf are convered with glandural hairs which contain essential oil. The inflorescence is a pseudo-spike consisting of six or seven pseudo-whorls. The flowers, which are light violet, are shed within a very short time. The fruit is a minute, reddish-brown, four-parted nutlet. It produces germinating

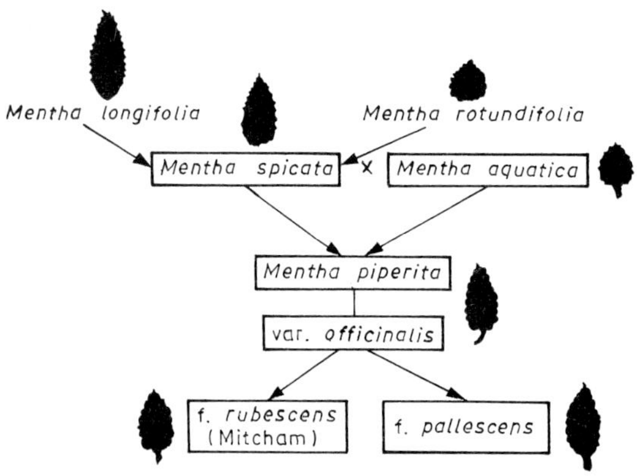

Fig. 77 The establishment of peppermint through hybridization (Heeger, 1956)

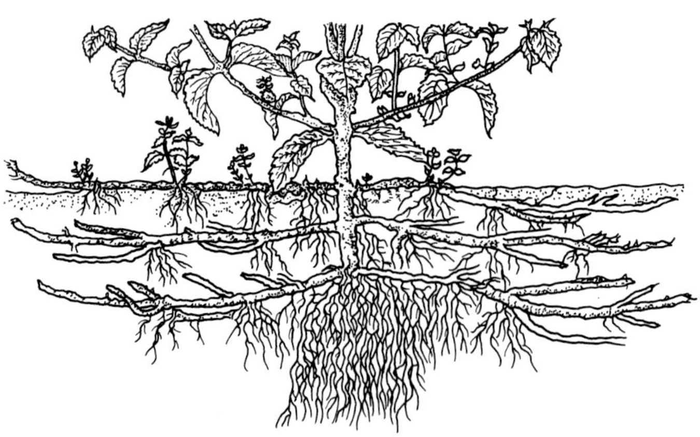

Fig. 78 Underground stolon system of peppermint

Fig. 79 Peppermint

seeds very rarely; however, these are not suitable for propagation owing to the heterogeneous descendants.

The whole plant has a characteristic spicy smell, and a cooling and spicy taste, which originates from the essential oil found in most parts of the plant. The essential oil content accumulated in the plant starts to increase rapidly after the emergence of side branches, and the maximum value is reached at the time of flowering.

The leaves contain 2–2.7 per cent and the inflorescence contain 4–6 per cent essential oil. The stem does not contain essential oil. The essential oil content of the herb is 1–1.5 per cent (relative to the dry material).

According to present knowledge, the essential oil comprises more than 20 ingredients. The most important among these ingredients is menthol (40–60 per cent by weight) (Fig. 79).

The essential oil of young leaves contains the largest amount of menthol. The essential oil extracted from the inflorescence contains mostly the less valuable menthofuran. (This is why it is not worth harvesting at the time of full bloom.) As well as the menthol, the essential oil contains significant amounts of menthone (15–25 per cent), menthyl acetate, piperitone, menthofuran, pinene and sabinene.

The vegetation period of peppermint counted from sprouting to full flowering is 80–100 days. The initial growth after sprouting is slow. After 2–3 weeks, however, growth becomes rapid. The undeground stolons and the stolons on the surface start to develop simultaneously with the side shoots. Flowering starts at the beginning to the middle of July. Growth after the first cutting is similar to the spring growth.

Environmental requirements

Except for very cool areas, peppermint can be cultivated everywhere in central Europe, provided that the nutrient and water requirements can be satisfied.

The underground stolons can survive temperatures of −17 °C, or even −30 °C in the case of snow cover. Sprouting starts at a soil temperature of 2–3 °C, but rapid growth begins only over 10 °C. Repeated cooling of the weather after warmer periods might cause problems, although the young shoots can survive a temperature of −8 °C. For the satisfactory accumulation of vegetative mass and essential oil, a daily mean temperature of 18–22 °C is advantageous in the period of development and especially prior to flowering. According to the studies of Hotyin (1968) and others, the essential oil content will increase at higher (23–25 °C) temperatures, but the menthol content of the essential oil will be less.

Like other members of the mint family, peppermint requires lots of sunshine and long daylight. The yield, essential oil content and quality of essential oil (menthol content) will improve proportionally to the light intensity and duration of light. For this reason the most appropriate locations are those which have lots of sunshine and in which the field slopes in a southerly direction.

Peppermint needs the highest amount of water among the medicinal plants in Hungary. This high water requirement can be attributed to the fact that the shallowly creeping root system has a very low water intake capacity. The transpiration coefficient (312–343) is also an indication of high water consumption (Kerekes, 1960). According to Schröder (1963) and others the peppermint needs at least 700–800 mm of water in the vegetation period, and so it can be cultivated economically only in irrigated lands. Optimum growth and maximum essential oil content can be obtained if the soil contains 80 per cent its full water-holding capacity (Schröder, 1963).

Peppermint is also troublesome in terms of selecting the proper soil for cultivation. The most appropriate is chernozem soil, which has good structure, good water management ability, and is medium-firm. It can also be successfully cultivated in brown sand and peat soils and on the better flood soils, provided that they are rich in nutrients.

Cultivation

The peppermint plantation should be maintained for two, or, rarely, for three years. Peppermint should not be cultivated in the same field for four years.

Nutrient supply

Peppermint requires large amounts of nutrients. According to Hungarian and other observations, the growth and the accumulation of essential oil is mainly influenced by the nitrogen supply. For high yields, however, high amounts of phosphorus and potassium are also required (Table 11). According to Franz (1972) and Scserbakov (1976), peppermint takes up 98.1 kg of N, 34.2 kg of P and 44.2 kg of K for the development of 4.36 t/ha of green mass, until the time of flowering. The nutrient take-up of the plant continuously increases from sprouting to flowering. Regular fertilization in high dosages, and sometimes manuring, is necessary. Peppermint will exploit the 20.0–30.0 t/ha of manure, which is introduced into the soil prior to planting. Depending on the state of the soil, it is advisable to apply 50–90 kg/ha of P_2O_5 and 60–90 kg/ha of K_2O together with the manure.

The required amount of nitrogen is 90–120 kg/ha (or even higher in the case of regular watering) in the producing years followed by the autumn planting. Nitrogen

Table 11 The effect of nutrient supply on peppermint (Hornok, 1980)

Introduced nutrients (kg/ha)			Green mass (t/ha)	Dry mass (t/ha)	Essential oil content as % of dry material	Essential oil yield (kg/ha)	Total menthol content %
N	P_2O_5	K_2O					
—	—	—	2.55	0.83	0.91	7.52	51.6
—	100	200	4.16	1.13	1.04	11.77	54.8
250	—	200	19.36	3.96	1.09	43.34	48.6
250	100	—	17.11	3.88	1.57	61.16	45.8
150	100	200	17.70	3.77	1.33	50.29	51.3
250	100	200	22.29	4.50	1.44	64.99	48.5
350	100	200	24.65	4.85	1.53	74.15	45.7
250	50	200	22.63	4.70	1.34	63.18	49.5
250	150	200	20.98	4.46	1.46	65.39	49.7
250	100	100	19.47	4.44	1.52	67.58	47.4
250	100	300	21.41	4.98	1.40	69.77	51.2
250	100	400	21.52	5.06	1.34	67.78	55.2
Least significant difference	5%		2.47	0.71	0.06	4.58	3.7

fertilizer should be applied in two portions; two-thirds in the early spring prior to sprouting, and one-third just after the first cutting. In the first year the spreading of 50–80 kg/ha of P and 60–80 kg/ha of K, made at the time of the autumn soil cultivation, is unquestionably justified. The trace element content of the peppermint plantation should also be investigated, and replenishment should be made if important trace elements are lacking. The lack of boron, manganese, molybdenum and cobalt will especially hamper the growth of peppermint and the accumulation of essential oil.

Soil preparation

Peppermint is planted mainly in the autumn. The soil should be prepared with stubble stripping (after cereals) and medium-deep (mixing) ploughing, following the introduction of manure. The ploughed soil should be tilled as necessary to keep it free of weeds until planting. Just prior to planting, clod rupturing or smooth rollers should be used, as a precondition of successful planting is a soil which is free of clods and voids. If planting is done in the spring, then the autumn deep ploughing is absolutely necessary, and then the soil should be tilled until free of lumps in the spring.

Propagation (planting)

As a hybrid, the peppermint seldom produces seeds able to germinate. For this reason, propagation is done exclusively with vegetative parts (green shoot, underground stolon and rooty turion).

Fig. 80 Stolon of peppermint ready for planting

Propagation with green cuttings, cannot be done in the case of large-scale cultivation, owing to its high costs and high demand for manual labour. It is appropriate only for the fast propagation of improved clones. Cuttings which are cut just under the lower node, and contain at least two nodes and two pairs of leaves, are put into peaty sand. In the case of regular watering under shade, 2–3 weeks are necessary to develop seedlings, which are suitable for transplantation.

Peppermint plantations are established by the use of underground stolons in every part of the world (Fig. 80). It is very important to take propagating material from peppermint fields which are free of other (wild) mint species (especially *M. Smithiana* Graham) and from the perennial stolonous weeds. The stolons used for propagation are generally gathered from one-year-old mother plantations. The main requirements of propagating material are as follows:

(1) It should be free of other mint species stolons, and decayed, ill parts;
(2) It should include at least 70 per cent white stolon, and the remaining 30 per cent may only be the green stolon of the peppermint;
(3) The humidity content of planted stolons should be around 80 per cent.

In order to increase the ratio of white to green stolons, it is advisable to cover the peppermint rows with 4–5 cm of soil with the help of interrow cultivators, 1–1.5 months prior to the gathering of stolons.

The stolons should be gathered after the second cutting just prior to transplanting. Lifting from the ground can be done with a scarifier plough, or with a potato-gathering machine provided with a vibrating chain. The stolons should be immediately transported to the location of transplanting, or for temporary storage they should be packed into prisms.

The taken-up stolons should be cleaned of old and ill parts, and should be immediately transplanted. From 1 ha the expected amount of good-quality propagating material is 6.0–10.0 t.

One of the advantages of autumn planting is that the water necessary for sprouting and initial growth is available even without irrigation, and the development of plants can start early, as soon as the frosts are over leading to a higher yield. The spring planting can sometimes only be started (owing to the condition of the ground) when the shoots of the autumn-planted plants start to appear. Owing to the late sprouting of spring-planted peppermint, the yield will be significantly poorer than that resulting from autumn planting, and only one cutting can be made. The spring start is regarded as an emergency solution. The optimum time for autumn planting is October, and it should be completed by 20 November at the latest.

The planting of stolons is semi-mechanized. Furrows 13–15 cm deep are prepared with 0.5–0.7 m row distance with the help of an opening plough installed on a cultivator. The stolons, which were previously put there in small heaps, are manually placed in the bottom of the furrows so that they are in contant in one or two string-like rows. The furrow should be opened just prior to planting so as to avoid drying of the soil. The stolons should be covered at once with a soil layer 100–120 mm thick. Stolons planted deeper than this will not be able to sprout, while those planted too shallowly might dry out or freeze. The furrow may be closed with filling ploughs, or with a smoother moved transversely to the rows. Covering and planting should be done in one path, because the stolons wither in the open furrows within 20–30 minutes. After planting and covering, the soil should be compacted with smooth or ringed rollers.

An amount of 1.2–1.5 t/ha of stolons is required for 1 ha plantation.

Peppermint is also propagated with rooty turions at the end of May or early in July (Fig. 81). Turions 8–10 cm high are used; these develop from the underground stolons. These turions can be gathered without damage after irrigation or after rain, and can be planted by any planting machine which is used in vegetable cultivation. For planting 1 ha using 0.6 m row spacing and 0.2–0.3 m plant distance, the amount of necessary rooty turions is 60 000–85 000.

Fig. 81 Rooty turions (from left to right: spearmint, peppermint, japanese mint)

Care of plantation

The care procedure of peppermint plantations consists of nutrient replenishment, irrigation, weed control and plant protection. The time and operations of nutrient supply were discussed above.

Regularly, the rain should be supplemented with 300–500 mm irrigation during the vegetation period. The most valuable irrigation can be made at the following stages (of course, taking the rain into consideration):

(1) After sprouting, when the average height of the plants is 8–10 cm;
(2) At the time of side shoot development;
(3) At the start of budding;
(4) Just after the first cutting;
(5) When the second shoots are 8–10 cm high.

Depending on the soil conditions, 40–60 mm water should be applied at a time with the help of irrigating equipment.

Weeds of high vigour and large mass grow well among peppermint, and for this reason it is very important to control these weeds. After the spring sprouting up to the time of row closing, the weeds can be eliminated from the row spacing with mechanized interrow cultivation. This operation is advisable even if weeds are not present abundantly, as the soil should be aerated and loosened. An interrow operation with a cultivator is applicable until the development of stolons. For this reason, chemical weed control cannot be avoided.

In the first year, the autumn-planted peppermint should be treated early in the spring prior to sprouting, and in the case of spring planting it should be treated just after planting with 3.5–4 kg/ha of Merkazin, which gives good results in weed control. In the second year, Merkazin (4–5 kg/ha) or Aresin (3–4 kg/ha) may be applied early in the spring prior to sprouting. In these plantations, chemical weed control may be done again after the first cutting, and in this case both herbicides can be applied in a dosage of 2–3 kg/ha. Very efficient weed control can be achieved in the plantation in the second and third years with Geonter which has terbacil active substance, applied in a dosage of 4–5 kg/ha prior to sprouting. Ronstar (oxidiazon) can be applied with good effect in the vegetation period in a dosage of 3–4 litres/ha.

The most dangerous and destructive disease of peppermint is rust *Puccinia menthae* Pers.). At first, when the shoots are not over 4-5 cm high, protection is provided by scorching (with chemicals or a flame-thrower). Although scorching destroys the shoots as well as the infecting sources, the peppermint will grow shoots again within several days. If this treatment is made too late, then it might cause serious damage. Following the spring basic treatments, preventive spraying (propiconazol or dichlor-butazol) should be done simultaneously with the protection against the most frequent pathogens (plant louse, red spider mite). Until the first cutting, two or three sprayings should be done with fungicides and acaricides, and this should be repeated once or twice after the first cutting.

Plantations in the first and second year should be ploughed down in the autumn. The purpose of this operation is to prevent the plant from balding in the second and third year, and to lessen the extent of rust infection. Ploughing down should be done in the second half of October, or by the middle of November at the latest. The fertilizers are spread on the ground just prior to the 14–16-cm-deep ploughing. Ringed rollers are used after the ploughing.

Harvesting

Peppermint can be harvested twice in one year. The actucal time and mode of harvest depend on the cultivating goals (i.e. producing essential oil, herb or leaves). In most cases, peppermint is cultivated for the production of essential oil. The optimum time of first harvest is when the main shoots are in full bloom, and the side shoots start to flower. The time of second harvest is in September.

Harvesting into containers can be regarded as the most sophisticated mode of harvest. A self-propelled chaffing machine transmits the cut and chaffed plants into a closed container, which can be attached directly to the steam-generating equipment, and in this way it functions as distillation vessel during the distillation.

The expected yield is 12–20.0 t/ha of green plant, and 30–60 kg/ha of essential oil. At the first cutting the yield of green mass is 8.0–14.0 t/ha, and the essential oil extracted is 2–4 kg/t (related to fresh plant mass), while in the case of the second cutting, the green mass is 6.0–8.0 t/ha and the essential oil amount is 1.0–2.0 kg/t.

For the purpose of dried herb production, peppermint should be harvested at the start of flowering or just prior to flowering, because the ratio of leaves and stems is the best at this time. Harvesting is done with a scythe in small spots and with mower–loader machines on a larger scale. The cut plants are transported at once to driers. The expected yield is 2.5–3.5 t/ha of dry herb.

The production of leaf drug (folium) requires much manual work. For this reason it is applied only in the case of small plot cultivation. For this purpose the peppermint is cut prior to budding in the morning hours by hand, and the leaves are at once removed from the stems. Only the leaves are dried. The expected yield is 1.2–1.8 t/ha of dry leaf drug.

Other mint species

Spearmint (*Mentha spicata* var. *crispa* Bentls./Mausf.). The leaves (Menthae crispae folium), herbs (Menthae crispae herba) and essential oil (Aetheroleum menthae crispae) were used much earlier than those of peppermint. Spearmint is cultivated over large areas of the USA, Brazil and Spain. Its roots and underground shoots are similar to those of peppermint. The stem is 40–80 cm high, and light green. The leaves are crisped. Its smell is similar to that of caraway. The leaves and herb are mainly used today.

The essential oil content is 0.5–1 per cent. Its main component is **L**-carvone (46–60 per cent). Spearmint is also a hybrid. In cultivation, bare and hairy varieties are distinguished, the bare-leaved being more valuable. Its ontogeny, environmental requirements, cultivation and primary processing are the same as those of peppermint.

Japanese mint or Canadian mint (*Mentha arvensis* var. *piperascens* Malinvaud) is mainly cultivated in China, Japan, India, Brazil and Argentina, exclusively for essential oil production. Its essential oil contains 80–85 per cent menthol; thus the pure menthol requirements of medicine are satisfied mainly by Japanese mint. *Mentha arvensis* oil is produced and traded in larger quantities than any of the other mint oils. The total is likely to be between 3000 and 4000 t/year.

Japanese mint is a hybrid; its morphological features differ from those of peppermint mainly in: the large and sightly hairy leaves, and the lower number of turions. It is even more sensitive to dryness than the peppermint. Its cultivation is similar to that of peppermint.

Sakhalin mint (*Mentha arvensis* var. *sachalinensis*) (Briq/Kudo) has been introduced to Hungary in recent years. By selective improvement a variety was established here which is patented as 'mentolcsepp' (menthol drop). Like the Japanese mint, it is applicable only for the production of essential oil and menthol. The essential oil content of the plant is 1.5–1.7 per cent, and the essential oil includes 82–86 per cent menthol. It is presumably a hybrid. It grows more vigorously than the peppermint, to 60–150 cm in height, and develops 25.0–40.0 t/ha of green mass. The whole plant is light green, and tomentose. The leaves are large, ovate, and serrated. It tolerates arid conditions better than peppermint, and is less sensitive to the mint rust. The cultivation is similar to that of peppermint. Owing to its good cultivation properties and high yield, it can be economically cultivated for the production of pure menthol. As well as the described species, other countries (Poland, Portugal, Morocco, Spain), also cultivate the species *Mentha pulegium* L. 'Pennyroyal', which contains pulegone. Its leaf stem (Pulegii herba) is exploited. According to Ruminska (1973) it can be propagated also by seed sowing.

Hyssop (*Hyssopus officinalis* L.)

Hyssop was known in ancient times as a spicy, scented plant. Its dried flowering shoots (Hyssopi herba) are used in tea blends for cough relief, for antiasthmatic effects and to relieve catarrh. This application is made possible by the antispasmodic characteristics of the flowering shoots. It is also a frequent ingredient of stomachic and blood-pressure-increasing preparations. Its brew and extract are used externally, owing to their fungicidal and antisudorific effects. The characteristic bitter taste of the drug allows it to be used as a spice in households and in the food industry for the flavouring of meats and sauces. The essential oil of hyssop (Aetheroleum hyssopi) is used by the canning, liqueur and cosmetics industries. Hyssop can also be used as an ornamental plant on dry hillsides. It is also a good honey-producing plant. Hyssop is produced commercially in France and several other European countries, including Hungary.

Fig. 82 Hyssop

Characteristics

Hyssop is a subshrub belonging to the Lamiaceae family. It is a native of Asia Minor, on the territories ranging from the Altaic mountains and the Caspian Sea to the Black Sea, and on barren rocky hillsides of areas around the Black Sea. The numerous forms and varieties found in the gene centrum of the species are classified into four subspecies by Briquet (Heeger, 1956), among which subsp. *officinalis* and subsp. *canescens* are significant in Central Europe.

The shoots are bare or slightly hairy in the former subspecies, and abundantly haired in the latter subspecies.

The root is a strongly branching, multi-headed tap root. The stem is 0.5–0.7 m in height, straight, or rising, richly branching from the base. It is lignified near the ground, and brownish; then it is soft and green. The leaves are opposite, shiny dark green, entire edged, and lanceolate. The well-developed leaves are 2–4 cm long and 0.5–1 cm wide; both sides of the leaf are densely covered with embedded oil glands, which is a typical xeromorphic feature. The inflorescence is a 20–25-cm-long false spike composed of 7–9 flowered pseudoverticils. The flowers are blue, pink and white. There is no diffference between the various colours as far as the active substances are concerned (Kerekes, 1969). The fruit is dull, dark or blackish-brown, and oblong, with an acute nutlet at one side. Its length is 2–2.5 mm, and its width is 1–1.2 mm. The mass of 1000 seeds is 1–1.2 g (Fig. 82).

The essential oil content of the herb is between 0.3 and 1.0 per cent. The main component of the essential oil is pinocamphene. Other significant components are α pinene, β pinene, camphene and several sesquiterpene alcohols. As well as essential oil, hyssop contains flavonoids, 5–8 per cent tannin, 3–6 per cent bitter substance, diosmin, hyssopin and resins.

The germination of hyssop starts in a wide range of temperature in suitably humid soil. The seeds sown in the early spring will germinate within 2–3 weeks. The seeds which fall to the ground in the summer will germinate in the autumn, and some of the seedlings will even survive the winter. Other seeds remain dormant in the soil and germinate in the spring. The initial development of the plants is moderately vigorous. Hyssop flowers in August in the first year. From the second year on, it flowers in July, and after cutting it will flower for the second time in September in good weather conditions. Its life-span in cultivation is 7–9 years. The fruits ripen continuously in August from the base of the spike upwards. The ripe fruits fall off very easily.

Environmental requirements

The suitable areas for the cultivation of hyssop are dry, warm fields, and hillsides sloping southwards. It is a typical xerophyte. In spite of this fact, however, an increased amount of humidity is very advantageous in certain phases of ontogeny. Such phases are the period of germination and the period after cutting. The second flowering in September takes place only in the case of rainy weather in August and September.

Hyssop is not sensitive to soil conditions. Even soil with a poor nutrient content, and gravelly soils, are suitable for its cultivation. It can also be used for exploitation of wastelands. The best growing soils are those which are limy, warm up quickly, and are medium-firm. Wind-blown sand and areas of high ground-water level are not suitable for the cultivation of hyssop.

Cultivation

Nutrient supply

As hyssop is generally used for the exploitation of wasteland, hillsides and soils with a poor nutrient content, a suitable yield can be ensured only by regular fertilization. Manuring cannot be done most of time, owing to economic considerations. For basic fertilization, it is advisable to introduce 60–80 kg/ha of P_2O_5 and 60–80 kg/ha of K_2O fertilizers into the soil during the autumn deep ploughing, prior to sowing. Generally, it is sufficient to introduce 40–50 kg/ha of nitrogen-containing fertilizer in the spring before sowing.

In the case of older plants, mainly nitrogen fertilization is necessary; the amount of this should be determined according to the vegetative development of hyssop. The production capacity of older plantations can be maintained with higher amounts of nitrogen.

Soil preparation

Autumn deep ploughing should be performed, as hyssop may not be sown into a field ploughed in the spring. Hyssop requires a fine-grained, uniform-surfaced seed-bed, which is free from lumps and furrows.

Sowing

Hyssop should be sown in the early spring, in March, or in April at the latest. If sowing is done too late, then sprouting will be incomplete. The number of seeds capable of germinating should be 120–150/m, which means that 3–5 kg/ha of seed needs to be sown, depending on the germination ability of the seeds. The depth of sowing is 1–2 cm. Deeply sown seeds cannot sprout.

The row spacing for the hyssop should be 0.5–0.7 m, depending on the available soil cultivator and harvesting machines.

Care of plantation

Just after sowing (but before the start of germination) herbicide treatment may be applied with 3–5 kg/ha of Merkazin. This dosage generally ensures sufficient weed control for the first two months of the vegetation season. The plants may be treated, if necessary, with 1–1.5 kg/ha of Afalon or with 1.5–3 kg/ha of Merkazin, at a time when the plants are already 10–15 cm high. The older hyssop may be treated with 5–6 kg/ha of Merkazin early in the spring just prior to the start of vegetation. At the time of sprouting of monocotyledonous weeds, when the hyssop shoots have reached a height of 10–20 cm, the plants may be sprayed with 1.5–2.5 kg/ha of Afalon or with 2–4 kg/ha of Merkazin.

Herbicide treatment should be complemented with mechanical interrow cultivation once a year. With this operation the aeration of the soil is also ensured. In this way the hyssop is properly free of weed at the time of flowering.

Among the insect pests, the blackfly causes the leaves to curl and become yellow. In cases of significant infection, the proper protective measures should be taken prior to flowering.

Harvesting

For the purpose of producing drug, hyssop is harvested from the start of flowering to the main flowering. If the aim is the production of essential oil, then the hyssop is cut when it is in full flower, in July. The time of the second harvest is the end of September. It is preferable to cut the plants just above the lignified stem parts, because the lignified parts would diminish the quality of the drug. Windrowing and mobile mower machines are applicable for harvesting. The cut plant should be put into drying equipment without delay. For the purposes of drying, the most appropriate machines are those which enhance the drying process with cold or slightly warmed air flow. After drying, the drug is packed into bales.

When the purpose is to produce essential oil, hyssop may be harvested with mobile chaffing machines. The harvested amount and the capacity of the distillation vessels should be adjusted to each other accurately, because the harvested hyssop might become mouldy, which would cause loss of essential oil.

The yield of hyssop is 2.0–3.0 t/ha in the first year, and 6.0–10.0 t/ha of fresh plant from the second year on, from which 0.4–0.8 t/ha and 1.5–2.5 t/ha of dry drug can be produced, respectively. The essential oil yield from 1 ha of hyssop is 8–15 kg.

Garden thyme (*Thymus vulgaris* L.)

Garden thyme, or common thyme as a kitchen herb, was a popular plant even in ancient times. Its name is derived from the word 'thymos' (bravery), and it was regarded as the symbol of valour. Greek women embroidered thyme on the garments of their husbands who were bound for war.

The drug of thyme (Thymi vulgaris herba) has been authorized since the sixteenth century in most of the pharmacopoeias, owing to its bactericidal and fungicidal effects. The essential oil (Aetheroleum thymi) is also valuable in modern medicine, owing to its antiseptic character. Its alcoholic extract has antiseptic and expectorant effect and for this reason it is an imporant ingredient of cough medicines and Galen preparations (e.g. Elixirum thymi, Tinctura thymi). Additionaly to its medicinal use, the dried herb of the thyme is also popular. It is used as a flavouring material in the acidifying of paprika, cucumber, etc., by the canning industry. It is also a very popular herb for meat products in some countries. Its essential oil is also widely used by the cosmetics industry.

The plant is grown as an ornamental in rock gardens and along walks, paths and borders. Thyme honey is a well-known and popular variety.

Garden thyme is produced and collected in most European countries, including France, Spain, Portugal and Greece, and in the western USA.

Characteristics

Garden thyme is a perennial subshrub belonging to the Lamiaceae family. Among the numerous species belonging to the genus *Thymus,* wild thyme (*T. serpyllum* L.) is indigenous to Europe, and is used to prepare herb tea blends. The much more significant garden thyme (*T. vulgaris* L.) does not grow wild in all of Europe, but is a native only of southern Europe, in the area of the Mediterranean.

Garden thyme has a tap root, which is lignified, much-branched, and multi-headed. The stem branches from its base, is 20–50 cm tall, is partly lignified, and is mostly erect. The older, not cut-back plants, will form irregular brushes. The shoots and leaves are densely covered with fine hairs. The leaves are opposite, small, and lanceolate. The inflorescence is spike-staggered at its bottom and consists of pseudoverticils (Fig. 83). Two kinds of flowers appear; the smaller ones are pistillate, and the larger

Fig. 83 Thyme

ones are hermaphrodite. The corolla is purplish, or reddish-pink. The fruit is a nutlet, 1 m long, and dark brown. The mass of 1000 seeds is 0.25–0.28 g.

The whole plant has a pleasant, characteristic odour. Its non-lignified herb contains 1.0–2.5 per cent essential oil, 8–10 per cent tannins, bitter substances, flavones and saponins. As many as 36 compounds of the essential oil are known at pressent. Among these the most important is thymol (20–25 per cent), which gives the characteristic odour of the plant. Thymol has bactericidal and fungicidal effects, and inhibits the development of intestinal parasites. For this reason it is used also in pure form in therapeutics. Depending on the actual chemotaxon, the essential oil contains significant amounts of carvacrol, borneol, *p*-cymen (15–15 per cent), linalool, cineole and various esters and acids.

The seed of thyme retains its germination ability for 2–3 years, and it germinates within 14–20 days in suitable conditions. The plants sprouting from seeds have very slow initial development, and they cannot survive in shade. Plants in the second year onwards start growing fast early in the spring (March), and they flower from mid-May to mid-June. If the plants are cut at this time, then the new shoots will produce flowers until the end of August to early September.

Environmental requirements

Garden thyme is a warmth-loving, light-loving plant; it survives arid condition pretty well. Frost may damage it in lowland areas if snow cover is not available. Hillsides sloping southwards are also suitable for its cultivation, but the best soils are those which are moderately firm, contain lots of lime, have good structure, have good water-handling capacity, have proper aeration, and are rich in nutrients. In humid areas the development is slower and the essential oil content is less. Thyme will not grow properly on poor soils with extreme characteristics.

Cultivation

Garden thyme generally remains in the same field for 4–6 years. At least 3–4 years should be allowed before garden thyme is cultivated again at the same location.

Nutrient supply

The nutrient requirement should be ensured by manuring with a dosage of 20–30 t/ha prior to planting. This is especially important in the case of poor soils. Fertilizer should be applied prior to the spring soil preparation operations in dosages of 50–80 kg/ha of K_2O and P_2O_5 and 40–60 kg/ha of N active materials. In the case of older plants, fertilization should be with 30–50 kg/ha of N in the autumn or spring, prior to the herbicide treatment. Nitrogen fertilizers are preferably applied in the form of lime nitrogen.

Soil preparation

Thyme may be planted only in soils deep-ploughed in the autumn. The soil should be smoothed as early as possible in the spring, and should be kept free of weeds until planting. For planting, the soil must be small-granuled, suitably humid, loose and smooth.

Propagation

Thyme can be cultivated either by vegetative parts or by seeds. Vegetative plant division is no longer used, owing to the high demand for manual labour, and to the low propagation ratio. Seeds can be sown into open-air seed-beds either for seedling nursing or directly for growing.

Direct sowing is applicable only in protected, small fields and if proper manual labour is available, owing to the small seeds and the slow initial growth. In the opinion of Rácz (1984), it is preferable to sow directly prior to the winter, because in this way we can be more sure that a suitable humidity content is present for germination in the early spring. According to Hungarian experiments, seeds can also be sown directly in small areas (e.g. in gardens) early in the spring (in the first half of March). The germination of the seeds is reliable; however, protection against weeds in the stage of slow growth (2–3 months) can be ensured only by hoeing and manual weeding. In order to achieve reliable sprouting, the seeds should be sown densely. For this reason, thinning is also necessary. In the case of 0.4–0.5 m row spacing, direct sowing requires the use of 5–6 kg/ha of seed.

Table 12 The influence of sowing time on the development of garden thyme seedlings (Hornok, Földesi, and Szász, 1975)

Time of sowing	Time of sprouting	Amount and quality of grown seedlings				
		Total pcs/plot	From this, Class I		From this, unfit for transplanting	
			pcs	%	pcs	%
15/9/1971	4/10/1971	402	402	100	–	–
29/11/1971	6/4/1972	177	177	100	–	–
2/3/1972	18/4/1972	600	543	89	–	–
22/3/1972	5/5/1972	607	588	97	28	5
17/4/1972	14/5/1972	202	107	53	31	15
5/5/1972	20/5/1972	154	25	16	43	28
L.S.D. 5%		32	103	–	–	–
10/9/1972	25/9/1972	229	229	100	–	–
25/11/1972	5/4/1973	495	495	100	–	–
26/2/1973	17/4/1973	642	583	91	–	–
26/3/1973	24/4/1973	1668	1368	82	–	–
17/4/1973	2/5/1973	907	679	75	38	4
18/5/1973	20/6/1973	742	282	38	87	12
L.S.D. 5%		264	207	–	–	–

In large-scale production, direct sowing is not applied, owing to the high demand for sowing seed and manual labour. The most widespread mode of cultivation is through seedling nursing.

For the purpose of seedling nursing, the seeds should preferably be sown in mid-March into open-air seed-beds.

Although the quality of seedlings is good when the seeds are sown in the autumn, the number of seedlings grown per unit area is small. A poor result is obtained also by sowing prior to the winter, relative to spring sowing (Table 12).

For the purpose of seedling nursing, the seeds are sown with 25–30 cm row spacing and 0.8–1.0 g/m^2 sowing seed is applied. In order to achieve uniform sowing, it is advisable to mix the seeds with compost soil, sand or pearlite. The seeds sown on the surface are covered with maximum 5-mm-thick layer of sand or soil, which is not liable to form cakes. The seeds germinate quickly owing to the thin cover, but may become dry for the same reason. For this reason, regular irrigation and shading up to sprouting is necessary. After sprouting, thyme grows slowly, and is easily overgrown by weeds. For this reason, the seedling bed should be very carefully and regularly weeded.

The seedlings grown from the March sowing can be transplanted to the field from the end of May. The seedlings are suitable for transplanting when their roots are well developed and the aerial parts are 5–7 cm high. It is advisable to transplant the

Table 13 Effect of harvesting time on yield of thyme (Hornok, Földesi and Szász, 1975)

Treatment	Dry drug yield (kg/plot)			Total	yield
	1st cutting	2nd cutting	3rd cutting	kg/plot	t/ha
(1) Cut once (at full flowering)	12.60	—	—	12.60	3.15
(2) Cut twice (at budding + second flowering)	8.56	7.68	—	16.24	4.05
(3) Cut twice (at full flowering + second flowering)	11.55	6.42	—	17.97	4.49
(4) Cut twice (at full flowering + autumn)	12.50	15.07	—	27.57	6.89
(5) Cut three times (at beginning of flowering + second flowering + autumn)	8.72	7.61	4.28	20.64	5.16
L.S.D. 5%	1.52	0.99	—	0.91	0.23

seedlings two-by-two or three-by-three. Small seedlings are transplanted by hand and larger seedlings are transplanted by machines with 0.5 m row spacing and 0.25 m plant distance. After transplanting, abundant irrigation is absolutely necessary. Depending on the mode of transplanting, the seedling requirement is 160 000–240 000/ha and this can be grown by the use of 500–600 g of sowing seed.

Care of plantation

On thyme plantations, weeds have large areas to grow. Just prior to transplanting the seedlings, 7–8 litre/ha Ronstar (oxadiazon) herbicide should be sprayed on the prepared soil. Ronstar keeps the area generally free of weed from May to the end of the growing season. Interrow mechanized hoeing is necessary only if the soil is infected with herbicide resistant perennial weeds.

Herbicide treatment of plants in their second year, and of older plants, can be applied with 2–2.5 kg/ha of Geonter sprayed at the end of the growing season (October–November).

The soil of the plantations should be loosened among the rows at least once a year so that it will be capable of absorbing the winter precipitation.

Harvesting

Garden thyme can be generally cut once int the year of transplanting. From the second year on, it can be cut twice or even three times in a season. According to Hungarian experiments, cutting twice can be regarded as reliable (Table 13).

The optimum time for the first cutting is at full flowering (in May), and the second cutting should be made at the time of second flowering in the first half of September.

Table 14 The change of essential oil content of thyme during the growing season (Hornok, Földesi and Szász 1975)

Time of sampling	Growing stage of the plant	Essential oil content as % of dry material
11 May	Green shoots	0.5786
21 May	Budding	0.6155
31 May	Start of flowering	1.1456
7 June	Full flowering	1.7153
19 June	End of flowering	0.6456
27 June	Green seed growing	0.2975
10 July	Ripening of seeds	0.1877
19 July	Second shoots	0.5820
6 August	Second budding	0.8816
21 Sept.	Second flowering	2.0930
19 Oct.	Third shoots	1.0694

The second cutting should not be made later than the first days of October, because this would hamper the good vernalization of the plants, and also the yield of the following year would be in danger.

According to Hungarian (Table 14) and foreign (Heeger, 1956; Rosengarten, 1969) data, the most valuable drug (which contains the maximum essential oil) can be made from the plant material that is cut at the time of full flowering. Thyme should be cut leaving 10–15 cm stubble so that no lignified parts can get into the drug. Harvesting is done by hand (with a sickle), or, over larger areas, with windrowing machines with strong cutting mechanisms, in sunny periods.

The harvested plants should be at once dried in a shaded place or in an artificial drier (at 35–40 °C). The yield is not too high in the first year, and then it is 1.5–2.5 t/ha of dry herb from the second year on. An amount of 3–4 kg of fresh plant will give 1 kg of drug.

Other species of thyme

Spanish thyme (*Thymus zygis* L.). This is a short, erect shrub found and collected wild in the Iberian peninsula of Spain and used in the production of Spanish thyme oil, which contains 50–60 per cent thymol.

Lemon thyme (Thymus citriodorus). This is a many-branched, lemon-scented shrub cultivated as a culinary herb. It has a semi-erect stem suitable for mechanical harvesting and is the principal thyme of commerce after garden thyme.

Thymus capitatus. This is native to Israel and other Middle Eastern and Mediterranean countries, and is used locally as oregano.

Clary sage (*Salvia sclarea* L.)

Clary sage—also known as clear eye, eyelright, and muscatel sage—has been cultivated for the production of essential oil from the beginning of this century.

The essential oil (Aetheroleum sclareae), which is extracted from the flowers, is used as a fragrance and fixative in the perfume industry. The concrete and absolute, (Concrete means a fatty wax-saturated with an aromatic substance and absolute is an alcohol-soluble liquid or a semiliquid oil saturated with an aromatic substance.) often blended with lavender, jasmine or other scents, are used in soaps, detergents, creams, powders, perfumes and lotions.

The name 'sclarea' is Latin for 'clear' or 'bright' with reference to the colour of the flowers, and the name 'clear eye' refers to the traditional use of its mucilaginous seed for clearing the eyes.

The principal production centres include France, the USSR, Hungary and Bulgaria; it is widely cultivated throughout the temperate regions of the world.

Characteristics

Clary sage is a biennial or sometimes three-year-old herbaceous plant belonging to the Lamiaceae family. Two taxons of the species are distinguished by Hegi (Heeger, 1956); i.e. *piramidalis* (Petagna), which grows vigorously, and var. *hirsuta,* which is very hairy. The species is indigenous to the coastal area of the Mediterranean in Europe and in Africa up to the Atlantic Ocean. It can generally be found in dry, rocky regions.

The deeply penetrating taproot of clary is densely covered with secondary rootlets (penetrating to 100–130 cm depth). The stem is erect, up to 1–1.5 m tall including the 40–60-cm-long inflorescence, and is covered with glandules. The rosette leaves (Fig. 84) are 10–20 cm wide, ovate, oblong, clasping, silvery-grey from the glandule cover,

Fig. 84 Rosette of clary sage in its first year

Fig. 85 Clary sage in flower

later greyish-green, and rough to the touch. The stem leaves continuously get smaller downwards, their petioles are shortened, and the upper ones are sheathed.

The inflorescence (Fig. 85) is 40–60-cm-long pseudoraceme, and is composed of pseudoverticils with 3–6 flowers each. The flowers are pink, purplish-pink, or sometimes white. It is a good honey producer. The fruit is an ovate, three-edged nutlet, which is 2–3 mm long, medium or dark brown, with dark brown veins. The mass of 1000 seeds is 4–5 g.

The economic significance of clary sage results from the essential oil accumulating in its inflorescence. The leaves and stem contain essential oil only in traces. The essential oil content of the inflorescence varies over a relatively wide range, generally between 0.04 and 0.3 per cent, depending on the ecological conditions and on the developmental stage of the plant. The main compound of the essential oil is linalyl acetate, which amounts to 45–87 per cent of the total volume, depending on the time of harvesting. Additional compounds of the essential oil are: sclareol, linalool, nerol, β-pinene, α- and β-thujone, borneol and a very small amount of mircene and camphor (Verzár, Petri and Then, 1974). The sclareol content of essential oil extracted with solvent may be as much as 45 per cent (Ivancsenko, 1976).

The seeds retain their germination capacity for 4–5 years. The germination will start within 5–6 days under optimal conditions, but the sprouting time is generally 15–20 days. The initial development of the plant is medium-slow. In the year of sowing, only the rosettes will develop; the peduncles emerge only after vernalization. Flowering starts at the end of June of the year following the sowing and lasts for 25–27 days.

Under favourable conditions a second flowering might occur (in September), but the essential oil content of the flowers is generally very low.

In the case of sowing early in the spring, some plants might flower in the first year. Using this capability, annual varieties are sometimes selected and cultivated, as for instance in the USSR and Bulgaria. It may also happen that clary sage remains alive for three or even for four years. The fruits ripen non-uniformly and for a long period of time; they are liable to fall off.

Environmental requirements

Clary is a xerophyte, and requires warmth and sunshine. The essential oil content at the time of flowering is 0.1–0.2 per cent in dry weather. In the case of rainy, cool, June weather, vegetative development is rapid, but the essential oil content is less than the above value. At the time of sprouting and the development of rosette, it requires more water, like most plants.

The seeds start to germinate at a temperature of 8–10 °C, although the optimum germination temperature is 25–28 °C. It is moderately resistant to winter frost when it is in the rosette stage. The well-developed plant with 5–7 rosette leaves will survive the average winter in central Europe without damage. The weaker plants, however, are frequently killed by the frost. The same applies to the loose-tissued, too rapidly developed plants.

Clary sage can be successfully cultivated on almost any soil. It also grows sufficiently on dry, gravelly areas. For this reason, it can be used for the exploitation of wastelands and slopes with southern exposure.

Cultivation

Nutrient supply

According to the investigations of Ivancsenko (1976), clary sage takes up 1.78 kg of N, 0.29 kg of P_2O_5 and 2.37 kg of K_2O from the soil the development of 100 kg of inflorescence. In the initial period of growth (at the time of developing the rosette) it requires mainly nitrogen and phosphorus, while the potassium take-up is highest after the emergence of the stem.

As a basic fertilization, 30–60 kg/ha of N, 45–60 kg/ha of P_2O_5 and 50–90 kg/ha of K_2O active material should be applied at the time of soil preparation. According to the cultivation practice of Moldavia, 10–15 kg/ha of superphosphate granules is applied to the rows simultaneously with sowing. This fertilization increases the yield. It is advisable to perform top-dressing with 30–50 kg/ha of N early in the spring of the second year. It is also efficient to apply 20–30 kg/ha of phosphorus after sprouting in the spring of the second year, simultaneously with the interrow cultivation.

Soil preparation

As clary sage has small seeds, the seed-bed should be prepared so as to provide sufficiently clod-free soil. It is a general rule that autumn deep ploughing is necessary for the spring sowing and that the lump-free state of the seed-bed should be obtained by treating the top layer of the soil as shallowly as possible. In dry weather, uniform sprouting can be facilitated by rolling the soil.

Sowing

Clary sage can be cultivated as a two-year plant by sowing in the spring, or as a one-year winter plant by sowing in the summer. In Hungary, spring sowing is regarded as more reliable. In this case it can be cultivated in association with other one-year plants. This mode of cultivation is more economical than the shorter but unreliable summersowing method.

In the spring, sowing can be done as early as the second half of March in rows 0.5–0.6 m apart, with 90–110 pieces/m of seeds able to germinate, which means a seed requirement of 5–7 kg/ha. The optimum depth of sowing is 2–3 cm.

The best associate plant for spring sowing is dill cultivated for the production of vegetable oil. In this case the dill should be sown first with a 0.12 m row distance. Clary sage should be sown right after the sowing of dill (prior to the sprouting of dill at the latest).

The summer sowing is done in a similar way to the spring sowing, in July or in the first days of August at the latest. This will allow enough time for clary to develop rosettes, and ensures overwintering without damage to clary sage.

Care of plants

The most important operation is weeding, which can be done either with herbicides or by interrow cultivation.

In the case of sowing in association with dill, the herbicide treatment can be applied just after sowing with 3–3.5 kg/ha of Maloran or with 1.5–2 kg/ha of Afalon. Post-emergent spraying should be done with 1–1.5 kg/ha of Afalon when the plants are 10–15 cm tall.

After harvesting the associate dill, it is advisable to perform a mechanical interrow cultivation.

In the case of summer sowing, the herbicide treatment should also be applied immediately after sowing with 2.5–3.5 kg/ha of Afalon.

In the second year, 3.5–5 kg/ha of Afalon can be used before the start of vegetation.

Harvesting

According to observations, the essential oil content continuously increases from the start of flowering and reaches its maximum at the milky ripening of seeds. Following this period, when the seeds begin to turn brown, the essential oil content decreases continuously. Practical experience indicates that the maximum essential oil yield can be reached if harvesting is started 8–10 days after full flowering. The optimum duration of harvesting is 10–15 days, depending on the weather.

The essential oil content varies significantly, also in accordance with the time of day. The essential oil content is highest within the period from 21.00 hours to 03.00, and is lowest around noon 12.00 to 15.00 hours). For this reason the cutting should always be done in the night and dawn hours, or in the early morning hours.

The inflorescences are best harvested by mobile chaffing machines. In this case the cut and chaffed material should be transported to distillation vessels without delay, as significant loss of essential oil might occur if the cut material is stored for several hours. The cutting mechanism of the machine should be adjusted to such a height that it will cut the inflorescence with the top pair of leaves, and not the plant parts containing essential oil.

The expected yield is 6–12 t/ha of fresh inflorescence, from which 7–15 kg/ha of essential oil can be distilled.

Sage (*Salvia officinalis* L.)

Sage has been known and used as a medicinal plant for 1000 years. Its dried leaf (Salviae folium) is an authorized drug in most pharmacopoeias. It is widely used in therapeutics, mainly as an ingredient of mouth-washing and throat-flushing tea blends. Its decoction has disinfectant and astringent effects. In folk medicine it is also used as a haemostat and diuretic, and against sweating. The essential oil (Aetheroleum salviae) is applied in the food industry (as a seasoning), in the perfume industry and also in the pharmaceutical industry. The name of *Salvia* comes from the Latin 'salvare', meaning 'to heal'.

Sage is cultivated in the countries of the Balkan peninsula, over large areas of the USSR, and in Italy and the USA. It has also been cultivated for a few decades in central Europe in small fields.

Characteristics

Sage is a perennial subshrub belonging to the Lamiaceae family. Its height is 0.5–0.8 m. It is a native of the northern shore region of the Mediterranean, and grows wild in large masses in the calcareous, karstic mountains of Dalmatia, Montenegro and Hercegovina. In the Iberian peninsula, subsp. *lavandulifolia* is native; the composition of its essential oil is different from that of Dalmatian ('true') sage. In the Crimean Peninsula and in the mountains of Asia Minor, the native ecotype is the so-called large-leaved (subsp. *major*) sage (Fig. 86).

The root is deeply penetrating and abundantly branching. The stem branches when older, becoming greyish-brown and lignified. The young shoots are densely covered with greyish-white hairs, which usually have a purple hue. The leaves are oblong lanceolate or elongated obovate; the bottom leaves are petioled and the upper ones

Fig. 86 Sage

are sessile. The upper side, but mostly the underside, of the leaves is tomentose. The inflorescence is a pseudospike composed of 5–8 pseudowhorls, including two or three flowers. The corolla is violet-blue, pink or white. The fruit is a nutlet, which is 2–3 mm long, 2 mm wide, and chestnut reddish-brown is colour. The mass of 1000 seeds is 7.6–7.8 g.

The herbs of the plant, especially the leaves, contain 1–2.5 per cent essential oil, the main compounds of which are thujone (30–60 per cent) cineol (15 per cent), camphor (8 per cent), borneol (6 per cent) and sabinyl acetate. Additionally, the plant contains 3–8 per cent tannin, bitter substances, glycosides and resinous materials.

Sage lives for 5–7 years, but the yield is rather poor in the fifth year of cultivation. The seeds retain their germination ability for 3–4 years. Germination starts at 12–15 °C. The plants grow very slowly in the first year. They produce flowers from the second year onwards. Flowering is rather prolonged; in Hungary it generally lasts from the end of May to the first half of July.

Environmental requirements

Sage is known as a warmth-loving plant tolerating arid conditions; however, the young plants sprouted from seeds require lots of water in the initial period. The sage will be killed in the winter, if the temperature reaches about −15 °C over 5–6 days.

The soil quality is of no particular importance in the cultivation of sage; it can be cultivated successfully in most soil types. However, it likes medium–firm soils best, which get warm easily, are suitably aerated, and are limy. Sandy, deeply laying clayey, cool fields are not suitable for its cultivation.

Cultivation

Nutrient supply

Sage exhausts the nutrients of the soil. It exploits manure and also fertilizers. The manure should be applied in amounts of 20–30 t/ha at the time of soil preparation, prior to the establishment of the plantation. If manure is not available, then 60–80 kg/ha of N, 40–60 kg/ha of P_2O_5 and 40–60 kg/ha of K_2O should be introduced into the soil prior to planting. In the producing years, a uniform nutrient supply is necessary, with medium dosages of all the three main nutrition elements.

Propagation

The propagation of sage is made by seed sowing or plant divisions, or sometimes by cutting. Generally, seed sowing is used. The seeds can be sown directly in the field just prior to the winter (by using 15–20 kg/ha of sowing seed) or into an open-air seed-bed for the purpose of seedling production.

As long as no reliable mode of chemical weed control is available, direct sowing in the field is only advisable if adequate manual labour is available for eliminating the weeds from among the slowly growing sage in the initial period.

Today, the most cost-efficient method of propagation is seedling nursing in open-air seeds-beds. For this purpose, the seeds are sown in April or May into well-prepared soil in rows 15–20 cm apart. The seedling amount necessary for a 1 ha plantation can be produced by using 1.5–2 kg of sowing seed on a 250–300 m^2 surface area. The proper development of seedlings should be facilitated by frequent irrigation and careful weeding during the summer. By the autumn (October) the seedlings reach a height of 150–200 mm, and can be transplanted with machines to rows 0.7 m apart, and with plants 0.4–0.5 m apart.

Care of plantation

In order to provide an abundant and good-quality yield, it is advisable to cut the plants back to 80–100 mm from the ground in the spring of the second year. In this way the number of young shoots will be higher, and the leaves developing on these shoots will contain more active substances.

In the year following planting, weed control can be achieved by mechanized interrow cultivation and hoeing the rows. From the second year onwards, in the early winter, chemical weed control can also be applied, using 3.5–4.5 kg/ha of Aktikon herbicide.

Harvesting

In order to produce drug, the leaves and the herbs of sage are used. The plants transplanted in the autumn can be cut in the following year. The first cutting takes place prior to flowering. The cutting is made just above the lignified part, with a sickle or sickle-knife. The hedge-like plantations are cut with mobile mowers. The second cutting can be made in August or September. At this time the cutting height should be selected very carefully, because plants cut too low will freeze in the winter. The leaves are removed from the stem immediately after cutting, similarly to those of peppermint. The expected yield in the case of two cuttings is 2.5–3.0 t/ha of fresh leaves, or 5.0–8.0 t/ha of fresh herbs. The drying ratio is 5 : 1.

Sage is cut at the time of full bloom if the purpose is the production of essential oil. As opposed to the case of clary sage, the essential oil content of sage is highest around noon, and for this reason cutting should be done in the noon hours in full sunshine, by cutting the plants just above the lignified parts. The expected essential oil yield is 8–10 kg/ha.

Marjoram (*Majorana hortensis* Mönch)

Marjoram is a widely known, old spice. It was cultivated as the secret plant of Osiris in ancient Egypt. It was the symbol of happiness in the Greek and Roman empires. In the kitchen book of Apicius (in A.D. 30), marjoram is mentioned among the versatile spices. It has been regularly cultivated in Europe since the fourteenth century. The dried herbs of marjoram *(Majoranae herba)*, which have a characteristic odour and taste, are usually sold in the ground (cleaned from stem parts) state. It is a pleasant tasting stomachic spice used in the food industry and in households. It is an important ingredient of almost any spice used for meat. Sometimes it is used in herb teas, in the case of stomach and bowel troubles, and as a sedative. The essential oil (Aetheroleum majoranae) is used mainly by the meat industry and more rarely by the perfume industry.

Owing to its popularity, marjoram is frequently cultivated in ploughed fields and in gardens around houses. The major countries of marjoram cultivation are: France, Hungary, Spain, the USA, Egypt, Greece, Mexico and Turkey.

Hungary is a traditional cultivator and exporter of marjoram. Marjoram has been cultivated in Hungary since the sixteenth century.

Characteristics

Marjoram is an annual plant belonging to the Lamiaceae family. It is an indigenous species in North Africa, eastern India and the eastern coastal area of the Mediterranean. It is perennial in its original area and in the warm regions of southern Europe.

The root is thin, and is generally very liable to develop adventitious roots. The stem is erect, 25–50 cm high, much-branched, and four-edged, which is characteristic of the family. Young shoots are greyish-green, and hairy; older parts of the stem are balding, and reddish-brown. The leaves are opposite, short-petioled, ovate, entire-edged, and densely covered by whitish-grey fluffy hairs (Fig. 87). The inflorescence is a dense pseudospike composed of 7–9 pseudoverticils. The flowers are small, white or green-

Fig. 87 Marjoram

ish-pink, and originate from the axils of greyish-green bracts. The round or ovate bracts cling to one another in four rows (like in a cone). The fruit is a four-part nutlet, very small, ovate, slightly irregular in shape, and yellowish-brown. The mass of 1000 seeds is 0.2–0.4 g.

The whole plant has a pleasant spicy odour owing to the essential oil accumulated in the glandular bracts of the epidermis. The average essential oil content of herbs of marjoram is 0.5–1.3 per cent. The essential oil contains terpen-4-ol, gamma-terpinene, alpha-terpineol, alpha-terpinene, cis-sabinenen hydrate, linalool and several other compounds. As well as the essential oil, 9–10 per cent tannin, bitter substances and resins accumulate in the plant.

The seeds of marjoram retain their germination capacity for 2–3 years. In optimal circumstances the germination of the seed starts within 5–6 days. Sprouting occurs within 8–12 days, or about 20–22 days in an open seed-bed. The initial development of the seedlings is very slow. The plants start to develop branches 35–45 days after sprouting, and growth is more rapid from here on. In Hungary, budding generally starts at the end of June, and the first flowers appear early in July. Marjoram flowers for a long time (25–30 days). If the plants are cut at the beginning of flowering, then new shoots will develop on the plants, and in good conditions the plants will flower again at the end of August.

The fruits of marjoram ripen unevenly at the end of August; the seeds are not particularly liable to fall away.

Environmental requirements

Marjoram needs warmth and lots of light. Although its seeds start to germinate at 12–15 °C, the optimum temperature for germination is 20–22 °C. Small plants can survive a frost of −1 to −2 °C temporarily, but their growth stops at a temperature around 10 °C. Cool, cloudy weather just prior to budding results in a smaller yield and lower essential oil content. The water requirement is high owing to the shallow root system. According to the investigations of Schröder (1963), the drug yield of marjoram is highest when the water content is 60 per cent of the water-holding capacity of the soil. The proper water supply is especially important during germination, at the initial period of shoot emergence and at the time of budding. It requires 600–650 mm of precipitation during the vegetation season.

Majoram, can be effectively cultivated only on soils which are easily warmed, are rich in humus, have good water managemenet ability, and are medium-firm.

Cultivation

Soil preparation

The precondition for successful cultivation is the autumn deep ploughing. Excellent soil preparation is very important, especially if the marjoram is directly sown.

Just prior to sowing or planting, the soil must be prepared with harrowing, smoothing and rolling so as to give a very smooth surface.

Nutrient supply

Marjoram needs lots of nutrients. In the initial growing stage (from sprouting to the start of branching) marjoram takes up lots of potassium and a small amount of nitrogen from the soil. In the second stage (from branching to budding) the take-up of the three main nutrients increases and the difference between the built-in amounts is much less significant. In the third stage (from budding to ripening of seed) the take-up of potassium decreases until it becomes the least important of the three nutrients. In the case of the usual density of 500 000–600 000 plants/ha, marjoram takes up 51.93 kg of N, 13.81 kg of P and 43.44 kg of K from the soil of 1 ha. There is an additional consumption of 90–100 kg/ha of CaO and 35–48 kg/ha of MgO (Schröder, 1959). In the opinion of Schröder (1964) 80–100 kg/ha of N, 36–45 kg/ha of P_2O_5 and 80 kg/ha of K_2O are required for the fertilization of marjoram. The following nutrient supply system is used in Hungary with very good effect:

(1) Basic fertilization. This is applied at the time of autumn soil preparation; 40–60 kg/ha of N, 60–80 kg/ha of P_2O_5 and 120–140 kg/ha of K_2O are applied.
(2) Starting fertilization. This is applied at the time of spring soil preparation; 40–60 kg/ha of N and 18–20 kg/ha of P_2O_5 are applied.
(3) Foliage fertilization. This is applied after the first cutting together with irrigation in three stages, when 60–70 kg/ha of additional N are spread altogether.

Sowing

Marjoram can be cultivated by direct sowing in fields and also by seedling nursing. Both methods are successful.

In Hungary, propagation with seedling nursing was traditional practice. For a long time, marjoram seddlings had been grown in warm seed-beds with large manure bottoms, which were used for the nursing of paprika seedlings and tobacco seedlings, and these beds were covered with rush matting or with plain cloth. In later years, however, it has become more economical, in terms of manual work and cost, to nurse marjoram in plastic houses without heating.

In recent years, seeds have been sown directly in fields in the first half of March, into well-prepared soil, with a row spacing of 0.4–0.5 m.

The maximum depth of sowing is 5 mm. For this reason the seeds are spread on the surface, and covered with the help of a smooth roller just after sowing. The optimum plant density is 120–130/m, which can be established with the use of 8–10 kg/ha of seed. Uniform and reliable sprouting can be expected only if rolling is performed after sowing. The sprouting time of marjoram is 20–22 days when sown into open seed-beds. This is a relatively long time, during which fast-sprouting weeds might cause problems. Weeds growing prior to the sprouting of marjoram can be destroyed by spraying with Gramoxone of 0.6–0.8 per cent concentration.

Care of plants

One of the most important nursing operations is regular soil cultivation. The soil-loosening operations give rise to a number of effects, which, together with watering, are the most effective factors in increasing the yield. For this reason the first soil loosening in the row spacing should be made 10–12 days after transplanting the seedlings, or several days after sprouting in the case of direct sowing. The interrow cultivation must be done after every irrigation or significant rain, and after the first cutting. During the breeding season, at least four to five interrow cultivations are necessary.

Another basic requirement for successful marjoram cultivation is an adequate water supply. In Hungary, marjoram requires 200–300 mm watering supplementary to the rain in the vegetation period. The efficiency of the watering is highest at the time of initial branching, at the time of budding, just after the first cutting and, of course, if the humidity content of the soil is low. It is advisable to combine irrigation with foliage fertilization (with nitrogen).

Among the diseases harming marjoram, *Alternaria* on the leaves, stems and fruit is the most dangerous. Owing to the infection, necrotic spots appear (brownish-black) on the leaves, stems and the bracts of inflorescences, then the plants become bald and die. In unfavourable conditions, the fungus (*Alternaria* sp.) might cause the complete destruction of marjoram fields. For this reason, regular preventive operations should be performed. The critical times for infection are the beginning of budding, and the shooting after the first cutting. Preventive spraying (with fungicides containing copper) should be started in the stage of branching, and should be repeated after 8–10 days, depending on the weather.

Harvesting

The first cutting should be made when the marjoram is still in flower, because the good-quality, greyish-green drug can be prepared only from plant material cut during the flowering period. The optimum level of active substance is reached at flowering.

The first cutting is generally made in mid-July, leaving 5-cm-high stubble. The time of the second cutting is September or October.

The cutting is preferably made in sunny weather after the dew has dried up. A sickle-knife or sickle is used for cutting. In large fields of well-developed plants, windrowing is also applied successfully.

In order to ensure good-quality drug, the cut plants are dried without delay. In small farms drying is done in shady, airy places (lofts, sheds), where the plants are spread in a 10–20-cm-thick layer. Large-scale production requires the use of artificial driers, which dry the plant at 35–40 °C. Artificial drying results in better quality drug.

After drying, the leaves and inflorescence parts of the marjoram should be removed from the stem. In earlier times this operation was performed by hand, with beating or by treading on sacks full of marjoram. However, high-capacity shelling machines are generally used today. The shelling machine ensures the sand-free state of drug as well as removing the stems.

The material resulting from the shelling operation is screened on 5 mm mesh screens, then 3.5 mm mesh screens, in order to separate it from the stem parts and

Fig. 88 Harvesting marjoram

sand. Special care should be taken during the operations of shelling and cleaning to remove soil (sand) particles, because materials with more than 2.5 per cent sand content cannot be used for drug or for spice.

The end product is the sand-free plant which gets through the screen of 3.5 mm mesh size (drug or spice). This material consists of leaves, bracts and other stem parts of marjoram. The expected yield is 1.5–1.8 t/ha of dry, shelled drug (Fig. 88).

Summer savory (*Satureja hortensis* L.)

Summer savory is an old spice plant. It is mentioned by Virgil, who states that one of the most fragrant spice plants is 'summer savory, which grows to provide food for Gods'. The Romans used summer savory for spicing foods and drinks prior to the discovery of pepper. It is one of the most important spices of the kitchens of the English-speaking world. Its aerial parts (Saturejae herba), were also used in folk medicine. A number of researchers share the opinion that the scientific name of 'Satureja' was derived from the word 'satyr', as it was mostly used as an aphrodisiac. It has no particular significance in modern medicine; however, it is an ingredient of several tea blends which are used to increase blood pressure and to remedy coughing. Owing to its pleasant spice odour, and a taste that is similar to that of pepper, summer savory is used for flavouring foods, drinks and pickles. The essential oil (Aetheroleum saturejae) is used by the canning and liqueur industries. It is cultivated in fields and in gardens around houses. The main cultivating countries are Yugoslavia, France, Spain and the USA. It has been cultivated for a long time in Hungary, mainly in the areas of marjoram cultivation (Fig. 89).

Fig. 89 Summer savory

Characteristics

Summer savory is an annual herbaceous plant belonging to the Lamiaceae family. Among the species belonging to the genus, *S. montana* winter savory finds limited use as a spice plant, similarly to the summer savory, in several countries. *S. montana* is a perennial shrub plant.

The root of summer savory is abundantly branched. The stem is erect, richly branching from the base and 0.3–0.6 m high. The shoots are lignified at the base, slightly hairy, dark green, and purplish- or brownish-green when the plant is in flower. The leaves are opposite, subsessile, fibrous lanceolate.

Both sides of the leaf blade are densely covered with glands containing essential oil. The inflorescence is a leafy false spike composed of 1–5 pseudoverticils. The flowers are bisexual, tiny, and purplish-pink or white. The fruit is a nutlet which splits into four mericarps; it is small, globose, and blackish-brown. The mass of 1000 seeds is 0.5–0.6 g.

The characteristic spicy odour and taste of summer savory result from the essential oil accumulated in the leaves and flowers. The essential oil content of the herb is 1–2 per cent. The plant contains relatively high amounts of iron compounds, tannins and sugar. The essential oil contains 30–40 per cent carvacrol, 20–30 per cent cimol, and other phenol-type compounds.

The vegetation period, counted from sprouting to the ripening of seeds, is 140–160 days. The seeds retain their germination capacity for 1–2 years. The sprouting time seeds sown into fields is 14–21 days; it may be 25–30 days under disadvantageous conditions. The initial development and growth of plants is fast. Flowering starts within 75–80 days of sprouting, at the end of June. Flowering is continuous and lasts for 25–30 days. Fruits ripen continuously, and are liable to fall away.

Environmental requirements

The environmental requirements (soil, climate) of summer savory are almost identical to those of marjoram, except for the water requirement. Summer savory does not require as much water as marjoram, and for this reason it can be cultivated without watering.

Cultivation

Nutrient supply

Summer savory exploits the after-effects of manuring sufficiently. Complementary fertilization, however, is absolutely necessary. A high potassium level in the soil is especially important. The uniform development of plants can be ensured by introducing medium amounts of fertilizers (50–60 kg/ha of P_2O_5 and 60–80 kg/ha of K_2O) into the soil at the time of autumn preparation operations. In order to help vegetative growth, and to increase the yield, it is advisable to apply 50–80 kg/ha of N at the time of soil preparation in the spring.

Soil preparation

This is identical in every respect to that of sown marjoram.

Sowing

The seed of summer savory germinates at 14–16 °C. For this reason it is not advisable to sow it too early. The best sowing time is the end of March to the first half of April. At the end of April it can be sown on soils where irrigation is available, leading to reliable yields.

The sowing depth should be only 0.5–1.5 cm, depending on the humidity content of the soil. In order to obtain uniform sprouting, the soil should be rolled before and after sowing. The optimum distance between rows is 25–30 cm, which is increased to 45–50 cm in the case of mechanical cultivation of row spacings. The required amount of sowing seed is 4–8 kg/ha, if 120–140 seeds capable of germination are to be obtained on 1 m.

Care of plants

Thinning may be necessary if sprouting is too dense. In this case, thinning should be done at a time when the plants have 4–6 leaves. Thinning may be done with a light harrow pulled perpendicularly to the rows. Weed control can be achieved with interrow hoeing, or with herbicides. Proper chemical weed control can be achieved by spraying 3–4 kg/ha of Aresin prior to sowing without harming the summer savory. In the case of prolonged sprouting, fast-growing weeds can be eliminated by spraying a Gramoxone solution of 1 per cent concentration prior to the sprouting of summer savory. Yet another spraying can be made, if necessary, when the plants are 10–20 cm high, with a 1–2 kg/ha dosage of Aresin.

Harvesting

The summer savory should be harvested when it is in flower. In good weather it can be cut twice. The total aerial part of the plant is cut just above the bottom branches with the help of mobile mowers. The harvested parts of the plants are dried, shelled and cleaned with the methods described for marjoram. The expected drug yield is 1.0–1.5 t/ha of cleaned drug.

Sweet basil (*Ocimum basilicum* L.)

Sweet basil, or common basil, is a fragrant medicinal plant which has been used for a very long time. Its aerial parts (Basilici herba) were used in folk medicine to treat a number of diseases. In modern medicine the common basil is not particularly significant; it is used as an ingredient of cough-relieving, stomachic, diuretic and carminative tea blends. The drug is used by the food industry for spicing canned foods and drinks. Its essential oil (Aetheroleum basilici) is used in perfumes, soaps and shampoos. It also has antibacterial properties. Sweet basil is cultivated in Europe, India, Africa and the USA.

Characteristics

Sweet basil is an annual plant of the genus *Ocimum,* which belongs to the Lamiaceae family. There are a number of species in this genus which can be used for producing essential oil. Among these species, the perennial *O. gratissimum* is rather significant, as it is cultivated in many countries, as well as the annual *O. basilicum,* which is also cultivated in Hungary. In the opinion of Hegi (Heeger, 1956), sweet basil is a species of extremely variable shape, and appears in many varieties and forms. Kandelaki (1962) describes nine varieties of common basil. The cultivated basils are classified into three intraspecific taxa by Moschini as follows: maximum (large leaves, high), minimum (small leaves, abundantly branched) and bullatum (blistery, crisped leaves, late flowering).

The sweet basil cultivated in Hungary is a native of Iran and India. It has a well-branched tap root. The stem is erect, 40–60 cm high, and densely branching from the base.

The leaves are petiolate, ovate, and wide lanceolate. The surface of the leaves is smooth, green or anthocyanin. The inflorescence is a terminal, loose, false spike

Fig. 90 Sweet basil

consisting of 17–18 pseudoverticils. The flowers are small, and white or light pink. The fruit is a black or dark brown nutlet. The weight of 1000 seeds is 1.2–1.8 g.

In the aerial parts of the common basil essential oil acummulates in an amount of 0.5–1.1 per cent. The main compounds of the essential oil are methyl chavicol (55 per cent), and linalool. As well as the main compounds, the essential oil also contains small amounts of cineole, camphene, geraniol, ocimene and pinene. Besides the essential oil, the plant contains vitamin B_1 (0.1–0.2 mg/100 g), vitamin C (150–250 mg/100 g), carotene (1.2–2.8 mg/100 g) and bitter substances (Fig. 90).

The seeds retain their germination capacity for 4–5 years. The germination time of the seeds is 14–21 days under average conditions. Flowering starts at the end of June or early in July, and lasts for up to two months. If the plants are cut when flowering starts, then the new shoots will flower again in September. Ripening begins in the second half of August. Ripening is continuous, and the fruits are highly liable to drop away.

Environmental requirements

By and large, the requirements of sweet basil with regard to the environment are the same as those of marjoram, except that the common basil needs more warmth. The optimum germination temperature of common basil is 18–20 °C. Young and old plants die at a temperature of 0 °C. For the cultivation of common basil, suitable soils are those which have the capability to warm up quickly, have good water management, and are rich in nutrients.

Cultivation

Nutrition supply

Common basil has an ability to exploit the after-effects of organic fertilizer. For this reason, direct manuring is not necessary. Fertilization and foliage fertilization should be done as described for marjoram.

Sowing

Reliable cultivation in Hungary can be established by seedling nursing, and direct sowing the field. Seeds are generally sown in the second half of March in unheated foil-covered sowing beds. Seeds are placed in rows (10–15 cm row spacing) or scattered, and the depth of sowing is 0.5 cm. Depending on the quality of the sowing seed, the required amount of seed is 10–20 g for 1 m^2 of surface. The amount of seedlings needed for 1 ha can be grown on a surface of 350–400 m^2 area by the use of 0.6–1.2 kg of sowing seed. The duration of seedling nursing is 5–6 weeks. The seedlings are transplanted to the field only after the risk of frost is over, i.e. in mid-May. Seedlings should preferably be planted with row spacings of 40–50 cm, and with a plant distance of 25–30 cm, when the seedlings are 6–10 cm tall. Planting can be done manually or with machines. Direct sowing is only done later. In this case the soil is compacted by rolling prior to sowing the seeds, and then the seeds are sown to 0.5 cm depth with 40–50 cm row spacing in the second half of April. Yet another rolling is necessary after sowing is completed. The required amount of seeds is 2–3 kg for 1 ha.

Care of plants

The care procedures for common basil after planting or sowing are as follows: mechanical weed control regularly, regular irrigation (as with marjoram) and top dressing.

Harvesting

The flowering, leafy stem can be cut twice in the breeding season. The first cutting should be made at the start of flowering. If the cutting is performed later than that, then the produced drug will be of poor quality, as it will contain developed seeds as well. The second cutting is generally made around the end of September when the risk of frost is not significant. The cutting is made by hand, or by windrowing or

Fig. 91 Cutting of sweet basil

mower–loader machines, leaving 10–12 cm stubble. All these methods are capable of producing good-quality drug (Fig. 91).

The harvested plants are dried, ground and cleaned with the operations described for marjoram. The expected yield is 1.2–1.5 t/ha of dry, cleaned drug.

Henbane (*Hyoscyamus niger* L.)

Henbane (henblain, Jusquaime) was described by Dioscorides. It was used by ancient peoples and was regarded as a deadly poison. It was employed as an ingredient of witch ointments fashionable in the Middle Ages. It was restored to pharmacopoeias at the beginning of the eighteenth century, thanks to the studies of Störck. The dried leaves (Hyoscyami folium) are official in almost all European pharmacopoeias.

In earlier times the seeds of henbane (Hyoscyami semen) were used in folk medicine. Both drugs are strongly poisonous. The extracts of drugs are used in Galen preparations. The dried leaves are used in the treatment of alcohol and morphine habits, as mydriatics, sedatives and pain-killers, and in motion sickness. In Europe, henbane is administered for earache and rheumatism. The Chinese apply the plant externally for neuralgia and odontalgia. Henbane is cultivated in a number of European countries.

Characteristics

Henbane is one of the species of the genus *Hyoscyamus,* belonging to the Solanaceae family. According to Dános, two separate species are called henbane in Hungary. One is *H. niger* L., which is a biennial herbaceous plant, and a ruderal weed. The other is *H. agrestis* Kit. (syn. *H. niger* var. *agrestis* K. = var. *annuus* Sims), which is an annual weed found mainly in fields among cultivated plants. This latter has become almost extinct owing to the extensive use of herbicides. The biennial henbane is preferred in cultivation, because it develops larger leaves. Both plants are natives of Eurasia, but have been introduced to southern Asia, North America and Australia as well.

The root of cultivated henbane is a taproot, 15–25 cm long, and somewhat branched. The stem may attain a height of 1 m, and is thick as a finger at the base, cylindrical or edged, and branching at the top. The leaves are sessile, and deeply lobed. The whole plant is viscid and hairy. The corolla of the flower is funnel-shaped, and the flower is short-stalked, and dirty white with violet veins and a purple-violet throat. The fruit is a capsule with a lid which opens on maturity. The seeds are small, brown, verrucose. The mass of 1000 seeds is 0.5–0.9 g (Fig. 92).

All parts of the plant are poisonous. The organs of henbane contain tropine-based alkaloids. The alkaloid content is 0.08–0.15 per cent in the roots, 0.06–0.17 per cent in the leaves, 0.02 per cent in the stem and 0.06–0.10 per cent in the seeds. The most important alkaloids are **L**-hyoscyamine, atropine and **L**-scopolamine. The alkaloid content of the leaves is highest at the time of flowering. Besides the alkaloids, the leaves also contain flavone glycosides (hyoscypicrin, hyoscerin) and coumarins. The seed also accumulates 25–35 per cent fatty acids.

Fig. 92 Henbane in its first year

The seed of henbane germinates rather early in the spring when the soil temperature reaches 6–8 °C. The seed of biennial henbane germinates only after vernalization. The initial development of both species is very slow. The biennial henbane develops rosettes in the first year, the stems develop early in the spring of the second year, and the first flowers appear in the middle of May. The annual henbane flowers in June–July. The flowering of both species is rather long; it lasts for 1.5–2 months. The fruits of the biennial henbane mature in the middle of June, while those of the annual henbane mature at the end of the summer. The seeds easily fall off the opening capsule. The seeds retain their germination ability for 4–6 years (Fig. 93).

Environmental requirements

Henbane needs lots of sunshine, and develops at the time of long daylight. It can be regarded as a nitrogen-indicating plant, as it grows vigorously mainly on soils rich in humus and nitrogen. It requires no special properties of the soil; however, for successful cultivation, rich, loose soils are the best.

Cultivation

Nutrient supply

Henbane needs fairly large amounts of nutrients. It also exploits direct manuring efficiently. An extensive nutrient supply increases the amount of active substances as well as the yield. If the soil is medium–rich in nutrients, then as basic fertilization

Fig. 93 Henbane in flower

80–100 kg/ha of N, 70–80 kg/ha of P_2O_5 and 100–120 kg/ha of K_2O should be applied in the autumn. In order to facilitate the initial development of the plants, 20–30 kg/ha of N should be introduced into the top 10–15 cm layer of the soil at the time of spring soil preparation. Top-dressing is also efficient, and should be done simultaneously with the first interrow cultivation.

Soil preparation

The precondition for successful cultivation is autumn deep ploughing. The operations of sowing bed preparation should also be performed in the autumn, if sowing is done before the winter. In order to achieve uniformly shallow sowing, the sowing bed should be of smooth and well-compacted soil.

Sowing

The seeds of the henbane can be sown either in the autumn (prior to the winter) or in the spring. In the case of sowing prior to the winter, the seeds should be sown in the beginning to the middle of November, so that they will germinate only in the spring. In the case of spring sowing, it should be kept in mind that the seeds are to be sown as soon as the frost in the soil thaws. In this case, the seeds should be vernalized in order to avoid poor germination or even complete failure of germination.

The vernalization should be started three weeks prior to sowing, treating the seeds at −8 to −10 °C. Spring sowing (with previously vernalized seeds) is better than autumn sowing. The seeds of annual henbane germinate properly even without vernalization. The seeds should be sown in rows 50–60 cm apart, and to a depth of 1–1.5 cm. Rolling should be done after sowing. The seed requirement is 4–6 kg/ha.

Care of plants

Even if the preparation is sufficient and in average weather, the germination of henbane is rather slow; it sprouts after 20–30 days. In the meantime the field might become very weedy. In order to avoid this situation, the whole field should be treated with 2.5–3 kg/ha of Gramoxone, when some of the weeds have sprouted, but the henbane has not yet come to the surface. The too-densely sprouted henbane can be thinned by spike harrowing perpendicularly to the rows when the plants have two or three leaves. In the first year, an additional two or three interrow cultivations and one or two row hoeings are necessary to keep the field free of weeds.

The most frequent pest of henbane is the potato beetle (*Leptinotarsa decemlineata* Say). Protection against this potato beetle is achieved by regular spraying.

Harvesting

The purpose of cultivation is most frequently the production of leaf drug. For this, the biennial henbane should be cut in the first year when the rosettes are already well developed. Generally, the cutting is made by hand in early to middle August. If cutting is going to be made only once a year, then all the leaves are cut simultaneously at ground level. Continuous harvesting is also performed fairly frequently. In this case, the first cutting is made when the external two or three circles of leaves in the rosettes have reached their final size. At this time the inside cordiform leaves are not cut, so that the harvesting can be done again after 2–3 weeks. In this way henbane can be harvested twice or three times in one year. At the time of the last cutting, the cordiform leaves should not be left on the plant unless the purpose is to gather seed in the next year.

The gathered leaves must be dried immediately. Special care should be taken to prevent the leaves becoming mouldy during transport. The thick, juicy leaves of henbane dry slowly. For this reason they should be dried in an artificial drier at 50–60 °C.

Sometimes the pharmaceutical industry requires herb as well. For this purpose the annual henbane is cut down with a mower–loader or mobile cutter at the beginning of flowering.

The expected yield is 0.8–1.0 t/ha of dry leaves, which can be prepared from 5.0–7.0 t/ha of fresh leaves, or 1.7–2.0 t/ha of herb (from 11.0–14.0 t/ha of leafy stems).

Grecian foxglove (*Digitalis lanata* Ehrh.)

Grecian foxglove (woolly foxglove) has been used from the middle of the nineteenth century, and today it is still an indispensable medicinal plant. Formerly, from the eigth century, only the leaves of common foxglove *(D. purpurea)* were used. Around the end of the nineteenth century it was discovered that the active substances of Grecian foxglove are also effective, and could be extracted from the plant on an industrial scale more easily than those of the common foxglove. Cardenolides derivable from the leaves of Grecian foxglove are basic materials for heart medicines (Digoxin, Isolanid, Neoadigan).

The main cultivators of Grecian foxglove are Germany, Poland, the Netherlands, Hungary, England, the USSR and India.

Characteristics

Grecian foxglove is a biennial, herbaceous plant belonging to the Scrophulariaceae family. It is indigenous to south-eastern Europe, in the Balkan peninsula. The northern boundary of its area is the line between Budapest and Vienna.

The root is 10–20 cm long, has at most the thickness of a finger, and is abundantly branching. The stem, developed from the second year, is erect, 80–150 cm high, not branching, and wooly–hairy on the upper part. The rosette leaves are 15–25 cm long and lanceolate, and the blade of the leaf, thinning towards the petiole, is glanduar and hairy. The stem leaves are sessile, 10–12 cm long, elongated lanceolate, and pointed. The inflorescence is a terminal, dense raceme. The peduncle and the sepals are woolly-hairy. The petals are white with rust-brown veins. The fruit is a capsule splitting into two parts, and contains plenty of seeds. The seed is tiny, and reddish-brown. The mass of 1000 seeds is 0.4–0.5 g.

Fig. 94 Grecian foxglove ready for harvesting in the first year

Fig. 95 Grecian foxglove in full bloom

The medicinal significance of Grecian foxglove is provided by cardenolide steroid glycosides (cardiac glycosides), which accumulate in almost all parts of the plant. Although these substances are present in flowers, stem and root only in traces, the rosette leaves contain 0.4–1.0 per cent. The most important among these are the cardiac glycosides, especially digitoxin, digoxin, and gitoxin. As well as the cardenolide glycosides, the leaves contain important saponins, including digitonin, tigonin and gitonin.

In the first year, Grecian foxglove develops rosette (Fig. 94). The generative organs develop only in the second year (Fig. 95). The seed retains its germination capacity for 3–4 years. The sprouting time is 25–30 days under average conditions. The initial growth is extremely slow. The first rosette leaves appear 20–30 days after sprouting. The rosette develops completely by the end of July, or the beginning of August. In the second year, development starts early; the peduncle develops at the end of April, and flowering lasts for 20–25 days from early June. The fruits mature continuously according to the rate of flowering. The first mature fruits appear at the end of July. The mature fruits open by splitting, and spread the seeds easily.

Environmental requirements

The plant likes a warm and relatively dry climate; however, it survives harsh winters without damage. The seeds start to germinate at 7 °C, but the optimum germination temperature is 18–20 °C. For proper development, the plant requires lots of sunshine and warmth in the stage of vegetation. It will not grow sufficiently if the average mean daily temperature is below 15 °C. It tolerates arid conditions, and if the annual precipitation is over 650 mm, then this will hamper the growth and lower the active substance content. For cultivation, medium–firm or loose, brown sand is adequate, provided that its location allows easy warming up, and it is rich in lime and nutrients. Firm, dry clay, light drift sand and too-humid soils are absolutely unusable for the cultivation of Grecian foxglove.

Cultivation

Nutrient supply

In order to achieve high yield, the proper nutrient supply is absolutely necessary. According to experience in Hungary and other countries, fertilization with nitrogen and potassium is especially advantageous. As basic fertilization, 40–50 kg/ha of N, 40–50 kg/ha of P_2O_5 and 70–90 kg/ha of K_2O are introduced into the soil at the time of the autumn soil preparation. As a top-dressing, 60–70 kg/ha of N is added, introduced into the spaces between the rows or applied together with irrigation, when the plants have 6–8 rosette leaves each in the spring.

Soil preparation

Seeds should be sown in the early spring, and extremely shallowly, owing to their small size. This is also the reason why the mode of soil preparation is crucial. The autumn deep or medium-deep ploughing should be done by the autumn, and if sowing is done prior to the winter, then the sowing bed should also be prepared. In the case of spring sowing, it must be done as soon as the machines can move on the ground. Sowing is allowed only into a smooth and compacted (as a table) sowing bed, and for this reason smoothing and rolling should be done prior to the sowing.

Sowing

The seeds are sown in the field in November or early in the spring. In the case of loose soils (brown sand), it is preferable to sow the seed in the second half of November or early in December. In the case of soils liable to form cakes, sowing should be done early in the spring, at the end of February, or early in March at the latest.

Seeds should be sown rather shallowly, to a maximum depth of 0.5 cm, in rows 45–50 cm apart, using 5–6 kg/ha of seeds, so that 70–80 embryos will be in 1 m.

Care of plants

The most dangerous enemies of the slow-growing Grecian foxglove are weeds. For protection against weeds, the herbicide Dachtal (chlortalmethyl) should be sprayed on the field prior to sowing, in a dosage of 8–10 kg/ha. The relatively long sprouting time allows herbicide treatment with a Gramoxone solution of 0.8 per cent concentration, which is performed after sowing, but prior to the sprouting of Grecian foxglove. In this way the weeds germinating faster than the Grecian foxglove can be eliminated. For the prevention of weed development, Asulox may also be applied. This herbicide can be sprayed when the plants are more developed than the four or five basal leaves stage, in an amount of 6–8 litre/ha, or prior to sowing in a dosage of 8–10 litre/ha. As well as herbicide treatment, interrow cultivation is necessary; this should be performed twice or three times until the rows close together. It is especially important to loosen the soil after substantial rain or irrigation if it becomes compacted. Interrow cultivation, which is done when the plants have 6–8 leaves, can be done simultaneously with the nitrogen top-dressing. Although the plant requires no particular irrigation, the yield can be economically increased by irrigating with a medium amount of water (30 mm) in the dry period.

Septoria digitalis often attacks the Grecian foxglove. This can be prevented by treating the seeds and by regular protective spraying. For this purpose, the fungicides containing copper can be applied. The preventive spraying is started at the end of May or early in June, and is repeated every 10–12 days, as required.

Harvesting

In earlier times, the rosette leaves were cut twice or even three times in the first year, paying no attention to the active substance content. The investigations completed around the end of the 1960s confirmed that cutting more than once is unjustified in terms of active substance content and economic considerations. The glycoside content of the rosette leaves is highest at the end of September, and the amount of leaves harvested at this time is almost the same as the sum of two cuttings.

Harvesting should be started after 4–5 days of sunny, warm weather. The rosette leaves are cut at ground level by sickles or sickle-knives.

It is advisable to leave the cut leaves to wither in a well-ventilated, shady location for 5–7 days, and to dry them in an artificial drier at a temperature of 50–60 °C. The pre-withered drug contains much more lanatoside C than the one which is dried immediately in an artificial drier. According to recent investigations, the extraction of digoxin can be improved by sweating the leaves for a while prior to drying. The fresh leaves are sweated for 4–5 days at a temperature around 30 °C prior to drying. The drug, with a final humidity content of 12 per cent is generally baled by pressing. The expected yield is 2.5–3.5 t/ha dry drug.

Common foxglove (*Digitalis purpurea* L.)

Common foxglove (purple foxglove) is often planted as an ornamental. The dried leaves are used medicinally for their effects on the central nervous system, and the heart and blood vessels. It contains glucosides which act as heart tonics, stimulants, and diuretics. Its most important use is in congestive heart failure, because it increases the force of systolic contractions. It is native to western Europe, and is a biennial plant. (Sometimes it remains alive for 3–4 years.) Its root is 20–30 cm long. The rosette leaves are large, wide, oblong, and short-petioled.

The leaves and the stem are downy. The stem is 1–1.5 m high. The inflorescence is a one-sided terminal raceme. The corolla is bell-shaped, and dark red, whitish or pink. The fruit is a capsule containing lots of seeds. The seed is tiny, and reddish-brown. The mass of 1000 seeds is 0.05–0.06 g. The leaves contain several cardenolic glycosides, e.g. digitalin, digitin, digitonin and digitoxin. The ontogeny and environmental requirements of common foxglove are the same as those of Grecian foxglove. With the exception of propagation, the cultivation procedure for foxglove is the same as that for Grecian foxglove. The seeds are sown on the surface of the soil in a heated foil tent (at the end of February) or in an unheated foil tent (first half of March). The seeds may be covered with a layer of maximum thickness 0.2–0.3 cm. The seedlings are transplanted to the field at the end of April to early in May. As the foxglove grows rapidly, it should be planted in rows 60 cm apart, and the plants should be 30–40 cm apart.

The expected yield is 1.5–2.5 t/ha of dry leaves.

Mustard (*Sinapis alba* L., *Brassica* species)

Mustard was known and consumed in ancient times. In the times of the Roman emperor Diocletian, young mustard plants were consumed like spinach and the mustard seeds were used as spice. In the Middle Ages, it was used as a flavouring material for the otherwise dull, salted meats in the winter. Its cultivation was extended up to Spain by the Arabs. It was introduced into Germany and England in the seventeenth century.

The mature seeds of mustard species (*Sinapis albae* and *Sinapis nigrae semen*) are used today as spices by the canning industry. Table mustard paste is prepared from finely ground seeds, flavoured with salt, vinegar and spice mixture. The fatty oil of the mustard is an important foodstuff, especially in eastern countries. It is also used as a lubricant and an illuminant. In recent years its therapeutic usage has declined (relieving rheumatic and neuralgic pains). Most of the mustard species are also cultivated for green fodder and for green fertilizer. It is regarded as a good honey producer, giving honey up to 40 kg/ha.

Owing to its excellent spicy taste and good effects in the diet, mustard is very popular all over the world. The mustard requirement is 200 000 t annually in the world. It is cultivated over large areas of Canada, Denmark, France and the USSR. In Hungary mustard is cultivated over an area of 15 000–20 000 ha.

Characteristics

The cultivated mustard species are members of the genera *Sinapis* and *Brassica,* which belong to the Cruciferae family. It is indigenous to the Mediterranean region and western Asia.

White (or English) mustard—*Sinapis alba* L.—is medium—large, up to 0.6–1.0 m high. The root is a thin taproot, sparsely branching. The stem is cylindrical. The leaves are alternate, deeply lobed pinnate, and petioled. The stem and the leaves are covered with stiff, widely spreading hairs. The inflorescence is a corymbose raceme, which elongates when older to form a loose raceme. At this latter stage the inflorescence contains mature seeds at its bottom and has open flowers at its tip. The fruit is a swollen, 3–4-mm-thick silique, which stands almost perpendicularly to the peduncle. The siliquae are hairy, and 20–80 mm long. The mature fruit is yellow, has rough hairs, and is swollen around the seeds. One silique contains 3–7 seeds. These seeds have a 2–2.5 mm diameter. The mass of 1000 seeds is 4–7 g.

Brown (or Indian) mustard is the species *Brassica juncea* (Cosson) Czern. It grows vigorously. Its root is thin and spindle-like. The stem is smooth, downy, and sparsely hairy. The side shoots stand at a 30–60° angle relative to the main axis. The leaves are runcinate, and lobate, divided on the bottom two-thirds of the stem, and elongated and wavy-edged on the upper part of the stem. The inflorescence is a terminal, dense, umbellate raceme located on the peduncle and on the side shoots. The petals are sulphur-yellow, and obovate. The fruit is a 2–3 mm wide, long-beaked silique, often contains anthocyanin, and stands at a 30–45° angle relative to the peduncle. One silique contains 10–12 seeds. The seed is 1.4–1.8 mm in diameter. It is globose, and reddish-brown or yellow (eastern or oriental mustard, var. *lutea*). The mass of 1000 seeds is 2.5–3.5 g.

Black (or French) mustard—*Brassica nigra* L.—has been cultivated widely, and is almost the same as brown mustard with regard to its morphological characteristics. The seed of black mustard is smaller than that of the two mustards described above; the mass of 1000 seeds is 1.5–2.9 g.

Mustard have short vegetation periods, of 110–130 days. The seeds retain their germination capacity for 3–4 years; they germinate properly even at a soil temperature of 4–5 °C. The sprouting time is short, generally 4–6 days. In the case of sowing in March, flowering starts early in June. The brown and black mustards generally flower 12–16 days later than white mustard. The seeds of brown mustard are less liable to spread than those of white and black mustards.

The aroma and flavour of mustard come from the essential oils contained as glycosides inside the seeds. The enzymic action of myrosin on the glycoside sinigrin in black and brown mustards or on sinalbin in white mustard releases mustard oil, which consists principally of allyl isothiocyanate in black and brown mustards and of *p*-hydroxybenzyl isothiocyanate in white mustard; these are the compounds responsible for the pungency.

Environmental requirements

Mustard grows in arid conditions, and is not sensitive to frost. For this reason it can be sown early in the spring. It is not particularly sensitive to the quality of soil; however, large yields can only be expected on soils which contain proper amounts of nutrients. The best are the calciferous soils, but it tolerates sour (up to pH 5.5) soil as well. Brown mustard grows better on light soils than does white mustard. It may also grow on gravelly sand, but it cannot tolerate drift sand. In the initial stage of the growing season, warm conditions cause damage to the plants. A dry season after the emergence of fruits (carpogenesis) will hamper the development of seeds, thus causing severe loss of yield.

Cultivation

On the same field, mustard should not be sown within two years. It should not be sown after rape (*Brassica* napus.) as they would mix owing to the sprouting of fallen seeds, and they have common pests and diseases.

Nutrient supply

It is not advisable to sow after manuring the soil directly, because in this case the vegetative parts of the plant will grow at the expense of seed yield. The best way to provide the necessary nutrient is by fertilizing at the time of autumn deep ploughing. Although mustard likes nitrogen, overfertilization should be avoided. The fast development of the vegetative mass might cause the plants to lean on the ground in this case.

Mustard takes up 5 kg of N, 2.5 kg of P_2O_5 and 4 kg of K_2O from the soil for the production of 100 kg of seeds. Fertilization should be so applied that the available phosphorus content of the soil will be 10–14 mg/100 g, and the potassium content easy to take up will be 15 mg/100 g. At least 2 per cent humus content is necessary in the soil for successful cultivation.

Soil preparation

Mustard may be sown only in soil deep ploughed in the autumn. It should be sown early in the spring, and for this reason the soil should be treated with a smoother as soon as possible. After sowing, the soil should be rolled in order to facilitate fast sprouting.

Sowing

Sowing should be done as early as possible in the spring. The plants sown early will fertilize better, have better aridity resistance, and resist the damage caused by *Phyllotreta* sp. better. A 7–8 days delay in sowing may cause a loss of yield of up to 100–200 kg/ha.

The seeds should be sown by sowing machines designed for small seeds to 1–2 cm depth and in rows 0.20–0.24 m apart, in the case of white mustard and brown mustard. The sowing seed requirement of white mustard is 8–10 kg/ha (= 30–40 seeds/m) and that of brown mustard is 7–8 kg/ha (= 25–30 seeds/m).

Care of plants

Chemical weed control can be achieved by the application of 4–5 litre/ha of herbicide containing trifluoraline just prior to sowing. The herbicide should be mixed into the soil to 70–150 mm depth with a single operation together with spraying.

Regular pests of the mustard are turnip flee beetle (*Phyllotreta* sp.), flower beetle *(Meligethes aeneus)* and the pseudo-caterpillar of the turnipsaw-fly *(Athalia rosae)*. Against the turnip flea beetle protection should be applied when they appear (when mustard is in the seminal leafy stage) with methylparathion. In order to protect the plants against the two other pest and the cabbage aphid *(Brevicoryne brassicae)*, spraying shall be done with the insecticide methylparathion, if the pest appears to the flowering, and with Rovlinka (dioxacarb) or Zolone (phosalone) in 1.3–1.5 kg/ha dosage if the pest appears during flowering. In the latter case, spraying should be done after sunset in order to protect the bees. At the time of flowering, the mustard should be sprayed from the air.

Harvesting

The selection of the harvesting time is one of the most difficult operations of mustard cultivation. If harvesting is too early, the seeds get mouldy easily owing to the high humidity content; thus mildew and poor resistance to mechanical damage will cause deterioration of both the quality and the germination ability. (The drying of high-humidity mustard seed is very costly.) If harvesting is too late, then the yield will be significantly reduced, owing to the fallen seeds.

Mustard is preferably harvested with a cereal combine harvester in a single path. Harvesting can be started when most of the siliqua are yellow or yellowish-brown. The stem part just below the inflorescence is also yellow, snapping when bent, while the base of the stem is still green. The seeds in the siliqua should be uniformly coloured, characteristic of the given variety, be hard to chew, and have an average humidity content of 11–12 per cent.

The humidity content of the seeds should not be over 15–16 per cent in any case.

After threshing, the mustard seeds should be cleaned, and if the humidity content is over 12 per cent, then drying is also to be performed. The properly cleaned seeds containing 11–12 per cent humidity can be placed into sacks. The expected yield is 0.5–1.2 t/ha in the case of white mustard and 0.8–1.5 t/ha in the case of brown and oriental mustard.

Safflower (*Carthamus tinctorius* L.)

Safflower is a medicinal and oil-producing plant which has been cultivated for a long time. It was even used by the ancient Egyptians, and was known in Palestine. Its dried petals (Carthami flos) have been used in folk medicine as an ingredient of expectorant tea blends. Today, its pigments are used all over the world for colouring foods, drinks and medicines, and sometimes for flavouring as well. The drug is also used by the

Fig. 96 Safflower

cosmetics industry. It can be used as a substitute for the much more expensive drug (Croci stigma) of saffron (*Crocus sativus* L.)

In a number of countries safflower is cultivated for the fine fatty oil, which can be extracted from the fruit (seed), and which is used either as a foodstuff or for industrial purposes. The valuable unsaturated fatty acids of safflowers are also used in therapeutics for preventing and curing arteriosclerosis. The fruit is an excellent poultry feed. The biggest producers are Mexico, the USA, Italy and Spain.

Characteristics

Safflower is an annual plant belonging to the Asteraceae family. In the opinion of several investigators (Kerekes, 1969; Rácz, 1984), it originates from tropical Africa, while others are confident (Kárpáti and Terpó, 1968) that it is a native of Asia Minor and eastern India.

The root is a taproot, 20–30 cm long, branching, and whitish. The stem is straight erect, shiny yellowish-white, branching at the upper part, and 60–170 cm high depending on the location of growth.

The leaves are alternate, 5–10 cm long, ovate, pointed, thorny, hard and shiny green. The inflorescence is a corymb composed of several terminal capitula. The capitulum is globose and 2–4 cm thick. It is surrounded by hard, leathery, thorny bracts. The flowers are tubular, bisexual, and orange-coloured, then falme-red at a later period. The fruit is a 5–8-mm-long white achene, similar to that of sunflower. The mass of 1000 seeds is 35–45 g (Fig. 96).

In the flowers of safflower, two pigments accumulate, namely safflor yellow and carthamin. The yellow pigment, safflor yellow ($C_{24}H_{30}O_{15}$), amounts to 24–30 per cent of the flower. It dissolves in water and alcohol easily. Carthamin, or safflor red

($C_{21}H_{22}O_{11}$), accumulates to a lesser extent (0.3–0.6 per cent). It cannot be dissolved in water or alcohol, but it dissolves in fat. The fruits contain 20–22 per cent fatty oils and 20–25 per cent raw protein.

The breeding time of safflower is 140–160 days from sprouting to maturing of fruits. The initial growth is fast; the first flowers appear at the beginning of July. Flowering is prolonged (35–40 days). The fruits ripen continuously, and spread in the case of over-ripening, owing to the catapulting hairs. Under advantageous conditions (warm, protected area), the fallen seeds sprout spontaneously. The seed retains its germination capacity for 4–5 years.

Environmental requirements

Safflower requires warmth and sunshine. The seeds start to germinate at 8–10 °C, but the optimum germination temperature is 20 °C. In the rosette leafy state it survives slight frosts, but the foliage leaves are destroyed at -2 °C. The frost-bitten safflower does not generally die, but growth will be very slow, and lots of branches will develop.

The well-branched roots exploit the water content of the soil very well; for this reason, safflower survives relatively arid conditions. It can be cultivated with good results in warm, deep-layered soils with good water-management capacity, but it can grow almost everywhere, except in soils with extreme characterisitcs.

Cultivation

Nutrient supply

Safflower exploits the nutrients of the soil very well, and takes up even the raw mineral nutrients. Fertilizers are economically exploited by the plant, especially on poorer soils. As basic fertilization, generally 35–40 kg/ha of N, 40–60 kg/ha of P_2O_5 and 50–70 kg/ha of K_2O should be introduced into the soil simultaneously with soil preparation in the autumn.

Soil preparation

The operations of soil preparation are the same as those usually applied for plants sown in the spring. Spring-ploughed soils are not suitable for the cultivation of safflower.

Sowing

Safflower is sown directly into the field. It is preferable to sow early in April if the soil gets warm easily. If sowing is done too early, then sprouting might be prolonged and non-uniform. Sowing is generally done in rows 40–50 cm apart. The sowing machine should be so adjusted that 8–10 germinating seeds will be sown in 1 m. The required amount of sowing seed is 18–20 kg/ha. The depth of sowing is 40–60 mm.

Care of plants

In suitable weather, safflower will form rows 6–8 days after sowing.

The first interrow hoeing should be done when the rows are properly established, and the rows should be hoed when the plants have four or five leaves; the necessary thinning is also to be done at this time. Too much thinning should be avoided, as the plants will develop abundant branches if space is available, and the number of good-quality flowers will be less. The optimum number of plants is 150 000–160 000 on 1 ha. Depending on the weed development and the compaction of the soil, an additional 2–4 interrow cultivations are necessary during the breeding season.

Harvesting

The active ingredient content of the flower petals, which give the drug, is highest after full flowering. For this reason, the tubular corollas should be gathered (plucked from the calyx) when their colour is already vivid red, and the flowers in the outside ring start to lean on the bracts. The continuously flowering inflorescences (1.5–2 months flowering time) should be gathered continuously every three or four days. The gathered flowers should be dried quickly. Owing to the high labour requirement of gathering, safflower is mainly cultivated on plots around houses, and dried in shady warm locations (lofts), spread in a thin layer. It is advisable to dry the flowers with an artificial drier at 50–60 °C, in the case of a large number of flowers.

The dry drug is highly hygroscopic, and for this reason should be packed into cases lined with plastic or paper, or into sacks. The expected yield is 600–700 kg/ha of fresh petals, giving 120–150 kg/h of dry drug.

The flowers are plucked from the calyx, leaving the ovary in place; this means that the fruits will develop and can be gathered after maturing. In the case of small fields, the plants are cut with the stems when three-quarters of the fruits are ripe, and then the seeds are threshed and cleaned. In the case of large fields, the plants are cut in a single path with combine harvesters. In this case, harvesting should be done at the time of completely dry fruits, and the combine harvester should be so adjusted that the seeds will not be damaged. The expected seed yield is 1.1–1.7 t/ha.

Blessed thistle (*Cnicus benedictus* L.)

As a medicinal plant, blessed thistle (or St. Benedict thistle) has been known from the thirteenth century. The dried aerial parts of the plant (Cardui benedicti herba) are on official drug in the 7th edition of the Hungarian Pharmacopoeia, as well as in most other pharmacopoeias. It has been used for a long time in therapeutics, owing to its digestive and stomachic effects. It was used in the Middle Ages against tumours of the stomach and bowel. Halmai-Novák (1963) mentions that it was used in the sixteenth century for preventing and treating plague. As of today it is an ingredient of tea blends and tinctures, and heals certain diseases of the respiratory system and digestive

Fig. 97 Blessed thistle

system. The drug is also used by the liqueur industry for preparing aperitif drinks and for spicing other drink.

Blessed thistle is cultivated in several European countries: in Germany, Czechoslovakia, the USSR and in the Balkan countries.

Characteristics

Blessed thistle is an annual plant belonging to the Asteraceae family (Fig. 97). It has a taproot which is 10–15 cm long. The stem branches from ground level, and is erect, 20–80 cm high, five-edged, bristly on the lower part, and viscous glandular and cobwebby on the upper part. The leaves are 5–30 cm long, lanceolate, sinuate, lobed; they have sharp, pointed teeth, and are cobwebby lanate. The capitulum is terminal, ovate, and surrounded by acicular, involucral bracts. The flowers are light yellow, bisexual, and tubular. The fruit is an achene, 1 cm long, cylindrical, slightly curved, and with a 0.5–1-cm-long prickly pappus at its top.

The mass of 1000 seeds is 28–34 g. The whole plant has an unpleasant spiniferous character. Its name, which indicates this feature, was presumably derived from the Greek 'chnisein' (to wound).

All parts of the plant are irritatingly bitter, either dried or fresh. This property comes from the sesquiterpene bitter substance, cnicin, which accumulates in all parts of the plant, especially in the leaves. The cnicin content of the herbs is 0.2 per cent on an average. Additionally, it contains 5–10 per cent mucilage, about 8 per cent tannins and about 0.3 per cent essential oil.

The seed of blessed thistle germinates within 8–10 days, under advantageous conditions. After sprouting, the first part to develop is the rosette, and then the plants start to grow quickly and produce flowers very soon. Only 60–70 days are necessary from sprouting to the appearance of flowers; blessed thistle can therefore be sown as a second plant if the purpose is the production of herbs. The plants sown in the spring flower continuously from June to September. The fruits also ripen continuously according to the flowering, from the middle of August.

Environmental requirements

The climatic conditions are suitable for the cultivation of blessed thistle in every part of central and southern Europe. It does not require particularly rich soil, but grows in soils with moderate nutrient content. For economic cultivation, however, the best choices are medium–firm or brown, sandy soils rich in nutrients.

Cultivation

Nutrient supply

Not too many nutrients are required. In order to get high yields, a basic fertilization should be made at the time of autumn ploughing by introducing 40–60 kg/ha of N, 40–45 kg/ha of P_2O_5 and 40–50 kg/ha of K_2O into the soil. If the soil is poor in nutrients, mainly if nitrogen is lacking, then 20–25 kg/ha of N should be applied prior to sowing in the spring.

Soil preparation

The soil preparation operations are the same as described for safflower.

Sowing

Blessed thistle can be sown in almost any season. In several countries sowing is done in the autumn or just prior to the winter. It is regularly sown in the second part of March or early in April at the latest. It can be sown with any available sowing machine in rows 40–50 cm apart. The seeds can be sown after abrading the pappi from them. The seeds should be sown so as to provide 7–10 plants in 1 m. The optimum depth of sowing is 3–4 cm. The required amount of sowing seed is 17–20 kg/ha.

Care of plants

Depending on the weather, the rows will be established within 8–14 days after sowing. If sprouting is too dense, then thinning can be performed when the first foliage leaves appear, by spike harrowing perpendicularly to the rows. For weed control, the effectiveness of Merkazin is well proven, if sprayed uniformly prior to sowing in a dosage of 4.5–5 kg/ha. This herbicide treatment is generally sufficient, as the plants cover the ground completely until May. At most a supplementary interrow cultivation is needed.

Harvesting

The drug is produced from the leafy stem of the blessed thistle. The first cutting takes place in the middle of June, when the first flowers appear. In the case of small plots, cutting is done by hand, using protective gloves. In the case of large-scale cultivation, it is harvested with mower-loaders, and transported immediately to artificial driers. The plants cut to 8–10 cm stubbles develop shoots after the cutting and can be cut again in the same year. The second cutting is done in the same way as the first, in the second part of August. The yield is unquestionably increased by interrow cultivation after the first cutting. In warm, rainy weather, the plant can even be cut three times. The cut plants should be dried immediately so that the drug remains light green.

The expected yield is 1.8–2.2 t/ha of dry herb. The fresh amount of the plant is five times as much (9.0–11.0 t/ha).

Pot Marigold (Calendula) (*Calendula officinalis* L.)

Pot marigold has been cultivated in Europe since the seventeenth century, mostly as an ornamental plant and sometimes as a medicinal plant. Its inflorescence (Calendulae flos) is used in tea blends as a remedy for stomach and bowel diseases. Its extract is mixed in creams for treating wounds, and its pigments are used for colouring fats and foods. An infusion of the flowers can be used for ulcers of the doudenum.

It is cultivated as a medicinal plant in Germany, Czechoslovakia, the USSR, Poland, Austria, Switzerland, and recently also in Syria and Egypt. In Hungary it is an old medicinal plant.

Characteristics

Pot marigold is an annual or rarely biennial herbaceous plant belonging to the Asteraceae family. It is native along the Mediterranean coast and in the western areas of Asia.

Its taproot penetrates deeply. The stem is 40–70 cm high, has short internodes, and densely branches from the base. The leaves are alternate, elongated, spatulate, light green, and tomentose hairy. The inflorescence is a terminal capitulum of 4–7 cm diameter. The pistillate ray flowers are arranged in two or three or more rows. The inflorescence is vivid orange. The fruit is a brown, curved achene, and its surface is prickly. The mass of 1000 seeds is 5–10 g (Fig. 98).

The active substances of the plant are the flavonoids, carotenoids (3 per cent), essential oils (0.02 per cent), and resinous and bitter substances, which accumulate in the inflorescence. According to Duke (1985), the flowers contain calendulin, analogous to bassorin, mucilage, oleanic acid, a gum, resin, a saponin, a sterol, cholesterol, and esters of lauric acid.

Fig. 98 Pot marigold

Pot marigold has a long breeding season; its vegetation is only suspended at the onset of the autumn frosts. Its seed retains its germination capacity for 5–6 years. The sown seed will germinate within 4–5 days under average conditions. The plants grow rapidly, and produce flowers after 40–50 days. If the flowers are picked continuously, then new flowers will develop. Flowering is continuous from early June to the autumn frosts.

Environmental requirements

Pot marigold grows best in full sunshine in rather dry locations. It can be cultivated in areas where the average temperature in June, July and August is between 17 and 20 °C and the field is open and sunny. The seed germinates properly at 8–10 °C.

Although it grows on soils with a poor nutrient content, for cultivation at least medium nutrient amounts are necessary, and easily warmed, loose soils should be selected. It should not be cultivated in humid, deep-lying areas, because the risk of powdery mildew infection is higher here.

Cultivation

Nutrient supply

It should not be manured or given too much nitrogen, because the number of developed flowers would be less. For basic fertilization, 40–60 kg/ha of N, 60–80 kg/ha of P_2O_5 and 80–100 kg/ha of K_2O should be applied at the time of autumn soil preparation.

Soil preparation

The operations of soil preparation are the same as those for safflower.

Sowing

The seeds of the pot marigold should be sown directly into the field early in the spring, in the first half of March. The row spacing should be 40–50 cm. The sowing depth is 2–3 cm. The amount of sowing seed applied should be 5–6 kg/ha. During sowing, the hopper of the sowing machine should be continuously checked, because the incurved seeds might stick.

Care of plants

After sprouting, the too-dense plants should be thinned when the plants have 3–5 leaves. Thinning is preferably done with a light spike harrow moved perpendicularly to the rows so that the remaining plants will be 5–8 cm from each other. During the vegetation period, the plants should be kept free of weeds by regular interrow cultivation and 1–2 row hoeing. The plants are cut with sickles 15 cm above ground level when the lower leaves are turning yellow and the inflorescence is small. The cut plant material is removed from the field, and 30–40 kg/ha of N is deeply hoed into the soil as top-dressing. Within three weeks the plants produce flowers again, and the petals can be gathered up to the onset of the frosts.

Harvesting

The gathering of opened flowers can be generally started by the end of May. The gathering can be repeated every two or three days, in accordance with the rate of flowering, until the frosts.

The whole heads of the flowers are gathered, and then the petals are plucked from the fresh inflorescence. Sometimes the inflorescence is dried together with the sepals.

The gathered inflorescences retain their vivid colour and quality only with immediate and fast drying. Small amounts are dried in the shade spread in thin layers, but high amounts should be dried in artificial driers at 35–40 °C. The dried drug is strongly hygroscopic, and for this reason it should be packed in foil bags, or lined paper bags. The expected yield is 200–300 kg/ha of dry petals of 500–800 kg/ha of complete inflorescence.

Variegated thistle (*Silybum marianum* (L.) Gaertn.)

Variegated thistle has been used in medicine since ancient times, but has been cultivated as a medicinal plant only in recent decades. In earlier times, mostly the leaves (Cardui mariani folium) and ripe fruit (Cardui mariani fructus) were used for curing spleen, liver and gallbladder diseases. Modern therapeutics uses the active substances extracted from the fruit mainly for the treatment of liver diseases.

Variegated thistle is cultivated over significant areas, of several hundred hectares, in European countries (Germany, the USSR, Romania). It is also widespread in South Africa. In Hungary it has been cultivated for about 15 years.

Characteristics

Variegated thistle is an annual overwintering plant belonging to the Asteraceae family. In Hungary it can be cultivated as a one-summer herbaceous plant.

The sunny, stony slopes of the mediterranean region are the growing locations of the warmth-loving milk thistle, which forms thorny forests. It has spread to the whole of Europe, and is common in countries of the Mediterranean region. It has been introduced into Australia and South America as well.

It has a taproot. Its stem is erect, branching, and 2–2.5 m high. Its rosette consists of stiff, basal leaves lying close to the ground. The leaves have no petioles and are elongated elliptic, lobed, with thorny edges. The veins of the shiny green leaves are

Fig. 99 Variegated thistle

surrounded by leaf tissue with no chlorophyll, giving a marble-like appearance to the leaves.

The capitula are large, 5–8 cm in diameter, and ovate; the flowers are purple, sometimes white, and solitary at the tips of the branches. The bracts of the multi-flowered capitulum terminate in hard thorns. The fruit is an incurved achene with a brush-like pappus, elongated ovate, and 8 × 4 mm large; the colour is brown when ripe, with light spots, and a yellow ring at the tip. The mass of 1000 seeds is 22–31 g (Fig. 99).

Flavonoids accumulate in the fruit walls of the variegated thistle up to an amount of 2–5 per cent. The varieties with purple flowers contain silybin (silymarine) and silydianin as the main flavonoids. Bitter substances, resins and fatty acid can also be found in the fruits. The fatty acid content of the fruit is 20–25 per cent. The main ingredients are linoleic acid (50–60 per cent) and oleic acid (20–35 per cent).

The breeding season of variegated thistle is 110–140 days long. The seeds sprout within 8–10 days after sowing. The first flowers appear 80–90 days after sprouting.

Environmental requirements

Variegated thistle grows well in sunny, warm weather. Rainy, cool weather hinders development. It is not particularly sensitive to the soil type, and can be cultivated on very different types. Poor sandy soils are not suitable for the cultivation of variegated thistle.

Cultivation

Nutrient supply

For the satisfaction of nutritional requirements, it is sufficient to apply 40–50 kg/ha of P_2O_5 and 80–100 kg/ha of K_2O at the time of autumn soil preparation. A moderate nitrogen supply of 25–30 kg/ha is advisable as too much nitrogen will cause the plant to lean.

Soil preparation

Autumn deep ploughing is also a must for variegated thistle. In the spring, the usual methods should be used for preparing the sowing bed.

Sowing

The sowing of variegated thistle should be started as soon as possible. It should be sown not later than early in April, because otherwise the vegetation season will be shortened, and the yield will be smaller. In warm locations, where no risk of freezing is faced, the seeds are sown at the end of the summer of the previous year (in August or September). A 0.40–0.50 m row spacing is preferred. The sowing depth is 2–3 cm. For proper cultivation (25–30 pieces/m), 12–14 kg/ha of sowing seed is necessary. After sowing, the application of a ringed roller is advisable.

Care of plants

After sprouting, when the plants have four leaves, mechanized interrow hoeing and manual row hoeing should be performed. The mechanized interrow hoeing should be repeated after three or four weeks. Additional hoeing is not necessary, because the plants soon cover the ground completely, and this prevents the development of weeds.

Variegated thistle is tolerant to herbicides containing trifluraline, but otherwise it is very sensitive to herbicides, especially when post-emergent. In the case of chemical weed control, cultivation should be done using smaller amounts of sowing seed and without hoeing or thinning.

Harvesting

Harvesting can be started when most of the brachts of the side primary shoots have dried, the middle of the inflorescences start to become white, and the shiny pappus is visible. In the brown capitula most of the seeds become dark. Manual harvesting can only be advised in small plots. The arms and feet should be provided with proper protection, because the plant is rather prickly, and often causes wounds which get easily inflamed.

In large fields, harvesing is done in a single path with a cereal harvesting combine. The cutting table of the harvester should be raised as much as possible, and the clearance of the drum should be increased so as to avoid the large green mass sticking in the machine. The seeds mixed with green plant parts coming from the harvester should be dried for a few days, and then cleaning is done with suitable machines. The expected yield is 1.0–1.6 t/ha, or sometimes up to 2–2.5 t/ha. The dried and cleaned seeds can be stored in sacks in a ventilated place.

German camomile (*Matricaria recutita* L., *M. chamomilla* L., *Chamomilla recutita* L., Rauschert)

German camomile (also known as matricaria, Hungarian camomile and wild camomile) is one of the medicinal plants known for the longest time. Even ancient peoples, the Egyptians and Greeks, knew the healing effects of the plant. In the works of Pliny and Dioscorides it is mentioned as the most important medicinal plant, by the name of chamaemelon. Its inflorencence (Chamomillae flos) and essential oil (Aetheroleum chamomillae) are also indispensable in modern medical practice; it is an official drug in all of the pharmacopoeias of the world. The tea made from the inflorescence (camomile tea) is an anti-inflammatory, antispasmodic and antiseptic, as is well known. Its essential oil is used by the pharmaceutical industry to prepare anti-inflammatory medicines, and by the cosmetics industry to prepare skin-care products. Extracts of German camomile have antibacterial and antifungal properties.

The application of this plant is significant all over the world. The amount of wild-growing plant has declined in recent years. For this reason, cultivation was started 20–30 years ago in a number of European countries. Finland started cultivating then Czechoslovakia, Germany and Poland followed suit. The method of cultivation has been developed by Kerekes (1960) in Hungary. The countries with the most extended production of German camomile are Argentina, Egypt, Hungary, Germany, the USSR and Czechoslovakia. Most of the drug appearing in the world market originates from Argentina and Egypt.

Owing to its excellent quality, the camomile grown in Hungary is marketed all over the world. Its economic advantages are enhanced by the fact that poor soils (calcareous) can be exploited by its cultivation.

Characteristics

German camomile is an annual overwintering plant belonging to the Asteraceae family. Its gene centre is the area around the eastern basin of the Mediterranean. As a weed seed of wheat, it was introduced and naturalized in every region of temperate climate. The northern border of its area is along the 63°–64° north latitude. It is also indigenous to Asia Minor and can be found in North and South America and in Australia. It grows wild in all parts of Europe, especially on saline soils.

The root is a spindle root, which branches when older, and does not penetrate deeply. The stem is cylindrical, erect or lying, depending on the taxon and the growing location, and 50–80 cm high. The leaves are alternate, sessile, bare, elongated lanceolate, fringed incised, and the fringes are fibrous. The inflorescence is a terminal capitulum. The receptacle is hemispherical initially, then elongated during the flowering, becoming conic and empty inside. This characteristic is very good for distinguishing it from the mayweed *(Anthemis)* species, which have no value in therapeutics. The inflorescence consists of 12–18 pistillate, white, ligulate or ray flowers, and a number of hermaphrodite tubular flowers, the latter being located on the receptacle cone (Kerekes, 1969).

The fruit is a 1–1.5-mm-long greyish-white achene. The mass of 1000 seeds is 0.02–0.03 g (Fig. 100).

Camomile is a species with a variety of forms. The minute camomile growing on saline soil was described earlier as an individual variety, namely var. *salina*. This variety, however, will grow normally on better soils. The small size is only the result of ecological effects. According to Máthé and Tyihák (1962), Tétényi (1970) and others, a number of intraspecific taxa exist within the species.

The main active substances of the flower are the essential oil, accumulating to 0.2–1.5 per cent, and the antispasmodic apigenin and kolin (2.5–3 per cent). The inflorescence contains significant amount of flavonoids, glycosides, polysaccharides (Szendrei, 1986), resins and vitamin C. The essential-oil-holding glands are mainly located on the bottom one-third of the tubular flowers, like beads (Fig. 101). Signifi-

Fig. 100 German camomile

cant amounts of essential oil can also be found in the intercellular fluid of collateral bundles of receptacles. The colour of camomile oil is dark blue, greenish or brownish. The colour of the essential oil depends on the amount of otherwise colourless prochamazulene (matricine contained in the flowers). Prochamazulene transforms into chamazulene during distillation, and this has a blue colour. The taxa containing essential oils rich in prochamazulene are ink-blue. One of the anti-inflammatory substances of German camomile is chamazulene. Other very significant compounds of the essential oil are /–/-α-bisabolol, pharnesene, bisabolol oxides and spathulenol.

Although German camomile is an overwintering annual plant, frost effects are not necessary for flowering. It will produce flowers even if it is sown in the spring. However, the number of flowers will be less than after overwintering. The stored seeds lose their germination capacity within 2–3 years. The seeds that are dormant in the soil retain their germination ability for up to 10–15 years. It germinates relatively quickly (in 6–8 days). The initial growth is slow, and small rosettes are formed in the first half of the vegetation period. It starts to develop early in the spring, develops many branches, and then flowers appear. The start of flowering depends on the variety and on the growth location. The duration of the main flowering is 10–16 days. The earliest-flowering type is the wild–growing camomile on saline soils (at the end of April to the first days of May). The cultivated varieties produce flowers somewhat later, in the middle of May. In the case of loose, sandy soils, flowering takes place earlier, while in the case of firm soils, flowering takes place 5–8 days later. If the flowering plant is cut, then it will produce flowers again (but in lower amounts). The fruit matures continuously, according to the rate of flowering. The receptacle will become conic during ripening, and the ligulate flowers press close to the stem. The ripening of fruits starts at the bottom part of the receptacle. The mature fruits are liable to fall off.

Cultivars

Two cultivars are cultivated in Hungary, which have been improved and authorized in Hungary.

The cultivar 'Budakalászi 2' has large flowers, and is tetraploid. Its breeding period is long; it grows rapidly and produces a high yield. Its essential oil content is average (0.7–0.9 per cent) and the essential oil contains 12–20 per cent chamazulene and about 10 per cent /–/-α-bisabolol.

The cultivar 'Soroksári 40' has medium-large flowers, and is diploid. The breeding time is medium-long; it grows thick rapidly, and produces a high yield. Its essential oil content is 0.8–1.1 per cent, and the essential oil contains 16–19 per cent chamazulene and 6–10 per cent /–/-α-bisabolol.

Environmental requirements

Being a plant of temperature region, it requires a medium amount of warmth. Germination starts at 6–7 °C, but the optimum germination temperature is 20–25 °C. In the stage of growth, a mean daily temperature of 19–20 °C is advantageous. For

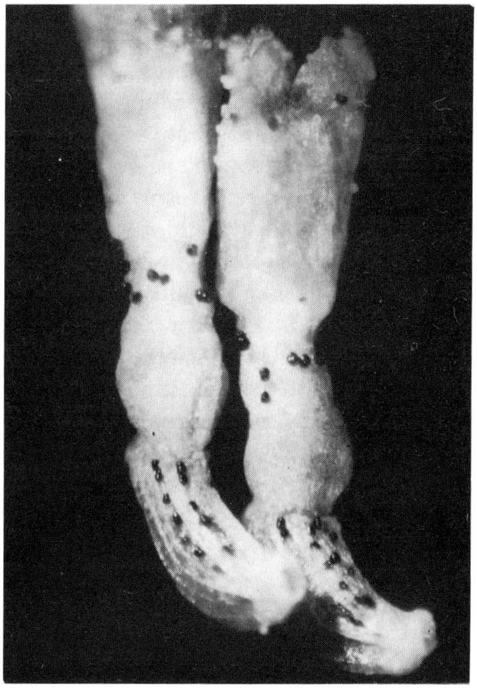

Fig. 101 Tubular flowers of German camomile with essential-oil-holding glands

the acceptable accumulation of essential oil, 20–25 °C is necessary during flowering. It is not sensitive to winter frosts in the rosette stage, but will be damaged by spring frosts in the period of thickening; the development will stop and flower formation will be poor. The plant requires much light. It requires light even at sprouting, and the light requirement is highest in the period from budding to full flowering. If the proper amount of sunlight is not available, then the accumulated amount of prochamazulene will be small. It tolerates arid conditons well, but high amounts of water are necessary for germination and stem emergence.

German camomile is a characteristic plant of saline soils, and for this reason it was regarded as a salt-loving plant for a long time. The results of experiments done by Kerekes (1960) confirm that the plant is only tolerating salinity, and cannot be regarded as a salt-loving plant, because it develops well even in the absence of sodium salts. The plant can accumulate sodium salts in its root cells in amounts up to 10 mg/g. This could be the reason why German camomile is still in good condition when other plants have already died owing to lack of water, i.e. it exploits the water content of saline soils which is no longer available for other plants. Except for too-loose and strongly calcareous soils, German camomile can be cultivated on any soil. In terms of economic considerations, however, German camomile should be cultivated on soils which cannot be used economically for the cultivation of other plants.

Cultivation

Cultivation of monoculture

German camomile can be cultivated in monoculture. In its original growth location, on the saline grazing field, it is self-sowing, and flowers in the same places. This is why cultivation in monoculture can be considered.

Soviet, Czechoslovakian and German investigators had already dealt with the possibility of cultivating the German camomile for a number of years in the 1950s. In their opinion German camomile could have been cultivated for 2–3 years in the same field thorugh the fallen-off seeds, but they did not prevent it from weeds.

Similar results were obtained by Kerekes (1969) from experiments performed in Hungary. According to recent experience in Hungary, it can be cultivated for 4–5 years, or even for a longer time, with proper agronomical techniques (not only with self-sowing), in the same field. The field should be changed when a weed flora resistant to the herbicides has developed. There is another advantage of cultivation in monoculture; namely the German camomile will not appear as a weed among other cultivated plants, as the fallen-off seeds will perpetuate the appearance of German camomile for years.

Nutrient supply

German camomile is an eurytropic plant, and grows on poor soils. According to the studies of Schröder (1964), it takes up 85 kg of K_2O, 53 kg of N and 21 kg of P_2O_5 from the soil for the production of 1.0 t of inflorescence.

Table 15 Influence of fertilization on German camomile in its fourth year in monoculture, on medium–firm soil

Applied active substance with basic fertilization (kg/ha)			Average value at full flowering (t/ha raw flower)	
N	P_2O_5	K_2O	Basic fertilization	Basic fertilization 50 kg/ha N, in spring
—	—	—	0.37	1.25
—	34	—	0.37	1.12
—	68	—	0.50	0.87
—	34	34	0.37	1.00
—	68	51	1.00	1.25
—	34	51	0.83	1.16
12	71	71	1.75	2.25
60	119	119	2.12	2.37

According to Hungarian cultivation and fertilization experiments, it gives good yields on medium-firm or firm soils even without fertilization. A uniform, small amount of fertilizer increases the flower yield, but high amounts of fertilizer will cause the plant to lean on the ground, resulting in less yield. In exceptionally poor soils, the following fertilizer dosages can be applied; 20–30 kg/ha of N, and 20–30 kg/ha of P_2O_5. The introduction of potassium will increase the yield only if the soil lacks potassium.

In loose sandy soil, even in the first year, fertilization is necessary for the production of a good yield. On eroded sands, with neutral chemical effects, and which are poor in nutrients, 40–60 kg/ha of P_2O_5 should be applied to achieve a proper yield. The yield-increasing effect of the basic fertilization is enhanced by nitrogen top-dressing in a dosage of 30–40 kg/ha.

The precondition for cultivation in monoculture for years is regular fertilization. In this kind of cultivation, the nutrient stock of the soil is exhausted by the third year to such an extent that the German camomile disappears from the field. The consumed nutrients can only be replaced by large amounts of phosphorus and potassium fertilization. In this case the addition of a small dosage of nitrogen in the autumn is also advantageous for the development of the rosette. The application of too much fertilizer is not harmful, only uneconomical. In the case of cultivating German camomile in monoculture, an annual 60–70 kg/ha of P_2O_5, 50–70 kg/ha of K_2O and 10–20 kg/ha of N fertilization from the second year on (applied prior to sowing) proved to be sufficient to keep the plants in proper condition. Nitrogen top-dressing, 40–60 kg/ha, is necessary in the spring (Table 15).

Late frosts (in March and April) harm German camomile. The development stops, the leaves become yellow, the plants have only one stem, and the number of flowers is low. In this case, nitrogen top-dressing in a dosage of 30–40 kg/ha will take the plants through the critical period. Foliage fertilization will not particularly help the poorly growing German camomile.

Table 16 Influence of sowing time on the yield of German camomile (Kerekes, 1960)

Sowing time	Yield (dry flower) (kg/ha)
Sowing at the end of summer or in autumn:	
13–16 July	688
3–6 August	683
24–27 August	737
14–17 September	730
5–8 October	600
26–29 October	528
Sowing in the spring:	
29 February–15 March	447
14 March–1 April	311
28 March–15 April	170
25 April–12 May	137
9 May–27 May	59

Soil preparation

Another precondition of successful German camomile production is the proper preparation of the sowing bed. In the case of one-year-long production, the stubbles of the predecessor plant should be immediately turned over, preferably with a unidirectional disc. Ploughs should not be used for this operation, except when the stubbles cannot be turned over by disc owing to the rainy weather. The field should be worked over with disc tillers and crushing rollers until the soil is smooth, free of lumps, well compacted, hard, and is not liable to settle.

In the case of cultivation in monoculture, the stem residues after harvesting should be removed by mobile chaffing machines. This is followed by stubble stripping and rolling.

Sowing

The optimum time for the sowing of German camomile is the end of August or the beginning of September. Spring sowing is unreliable, independently of how soon the sowing is done, and brings much less yield than in the plants sown at the end of the summer (Table 16). In the case of one-year-long cultivation, sowing should be done camomile flour in an amount corresponding to 3–4.5 kg/ha of clean (standard quality) sowing seed. The camomile flour consisting of 20–25 per cent seed and 60–75 per cent dry disc floret allows uniform sowing. The seed germinates in daylight, and for this reason it should be spread on the soil surface in rows 12–15 cm apart. In order to facilitate uniform sowing on the surface, the coulter of the sowing machine should be adjusted according to the usual row distance for cereals, and should be locked in the raised position. After sowing, a Cambridge roller should be used; this prevents the blowing away of small seeds and compaction of the soil. A seed-covering harrow should not be used after sowing German camomile.

In the case of cultivation in monoculture, no sowing is necessary of the seeds germinate well after the stubble turing and rolling. Generally, sprouting is poor, or sprouting takes place only in spots, and for this reason sowing should be done with a reduced amount of seed.

Care of plants

Chemical weed control among German camomile can be achieved with Maloran (3–4 kg/ha). The herbicide should be applied early in April at the time of weed sprouting in order to achieve optimum results. The herbicide will not harm German camomile with two or three rosette leaves. German camomile is also tolerant of Afalon treatment of 3–4 kg/ha dosage; however, this will hamper flowering in disadvantageous weather. In the case of cultivation in monoculture, the establishment of herbicide-resistant weed flora is expected. In Hungary the most dangerous monocotydelonous weed is *Bromus tectorum*. It can be controlled by spraying 2 kg/ha of Kerb (propizamid) herbicide late in the autumn. Against the dictydelonous weeds, the application of Sys 67 Prop proved to be effective.

Harvesting

German camomile should be harvested when most of the flowers have already opened. The essential oil content of the inflorescence increases continuously from budding, and reaches its maximum value when the ligulate flowers stand in a horizontal direction. As ripening proceeds, the essential oil content, and the chamazulene content, decrease continuously. The daily cycle of essential oil accumulation also depends on the weather. The highest level is around noon, on sunny days. In cool, cloudy weather, the usual daily cycle cannot happen, and so the essential oil content is also lower.

The flowers should be gathered when they are in full bloom. If gathering is done earlier, then the total amount of drug is smaller, while late gathering causes disintegration of the drug. The inflorescences are mainly gathered with a camomile comb. With this method a very good quality drug can be prepared, not only from the wild-growing plants, but also from the cultivated plants. An experienced worker can gather 100–150 kg of flowers from cultivated plants in a ten-hour shift. In the case of cultivated plants, the stem parts should be carefully removed, as the amount of green parts torn by the tool is much more than in the case of plants growing wild on saline soil.

One of the conditions for economical production is a mechanized harvest. A well-known camomile-gathering machine is the Ebert–Schubert equipment. These kinds of machine produce excellent quality drug in the case of weed-free German camomile, but their capacity is low. The Argentine camomile combine harvester operates in a similar manner. This machine has a higher capacity. By the modification of a cereal combine harvester, a high-capacity gathering machine has been constructed in Hungary, which is provided with a special screening system. This is excellent for harvesting large camomile fields.

The gathered flowers should be transported at once to the drying location. The fresh flower will heat and go brown within a few hours if it is stored in sacks, thus rendering the flowers worthless.

The continuous transportation of the gathered camomile is a basic organizational task. If the delivered flowers cannot be put into the drier immediately, then they should be spread in the shade for temporary storage for several hours.

In order to reduce the costs of drying the gathered flowers should be screened on motorized camomile screens of 7–12 mm mesh size, thus removing the coarse impurities at the field or at the drying premises. German camomile can be preserved by either natural or artificial drying. Artificial drying results in more durable and more aesthetic drug, which is not too liable to disintegrate to dust. According to Sváb *et al.* (1967), the drug dried in counterflow driers (belt conveyor and tunnel types) at 40–70 °C has a better active ingredient content than that dried in Schilde driers.

For the purpose of essential oil production or for other industrial usage, German camomile can also be gathered with mobile mower or mobile ensilage machines. The cutting height of the machines should be so adjusted that the cutting takes place below the average height of the flowers, thus avoiding cutting a high number of stems. The green parts impose an extra load on the essential oil distillation and drying equipment. The gathered flowery stems, especially of gathered with mobile ensilage machines, should be immediately taken to the distillation vessel.

The expected yield is 0.5–2.0 t/ha of fresh flowers, from which 0.1–0.5 t/ha of dry drug can be produced. With mechanized harvesting, 4.0–8.0 t/ha of flowery stems can be gathered for the production of essential oil. The yield varies widley as is usual for other plants, because the plant produces flowers once or twice depending on the weather. After the harvesting of the main flowers, now flowering shoots emerge from the bottom parts of the plants, thus allowing the second or sometimes the third harvest. In bad conditions, the second harvest cannot be made.

Tarragon (*Artemisia dracunculus* L.)

Tarragon has been a popular spice plant in Europe since the sixteenth century. With its dried herbs (Dracunculi herba), soups, sauces, fried meats, salads, pickles and acidified juices are flavoured. Its essential oil, known as estragon oil (Aetheroleum dracunculi), is a base material for spices, mainly, and to a lesser extent it is an ingredient of perfumes. Digestive and stomachic effects are also important. In recent years the canning industry has used increasing amounts of these materials. The therapeutic significance of the drug has declined. It is a household preparation for treating catarrhal and metabolic diseases, and hypertension.

Tarragon is cultivated over large areas of Europe, in France, Italy and Germany. There are also large plantations in California. In Hungary, tarragon has been cultivated on a large scale for three decades.

Fig. 102 Tarragon

Characteristics

Tarragon is a herbaceous, perennial plant of the genus *Artemisia,* which belongs to the Astraceae family. A number of other members of the genus (e.g. wormwood and mugwort) are used in medicine and for spicing. According to Heeger (1956), two varieties of tarragon are cultivated, namely French or German tarragon (*A. dracunculus* L.) and Russian tarragon (*A. dracunculus* L. var. *redowskyi* Turcs). In the opinion of Rosengarten (1969), two individual species are cultivated, and Russian tarragon is identical to *A. dracunculoides* Pursh. In Hungary, as well as everywhere else in the world, French tarragon is cultivated; it is richer in flavour and scent materials than other species.

It is indigenous to central and northern Asia, the Caucasus and the Far East. According to Heeger (1956), it is also a native of the western regions of North America. It grows on the banks of rivers (Fig. 102).

It has spreading roots; a dense root system develops from the short rhizome. Most of the roots penetrate to 20–30 cm depth. The stolon, which protrudes from the surface to form a shoot, also develops from the rhizome. It is 5–15 cm long, white or whitish-brown, and densely covered with adventitious roots. About 10–15 such stolons develop from the rhizome. The shoots are green, brownish-red at the base, lignified and 80–150 cm high, depending on the conditions of the growth area. The leaves are 3–8 cm long, 0.5–1.2 cm wide, entire-edged, fibrous lanceolate, and bare, and both sides are shiny. The capitulum inflorescences form a densely branching, leafy, green panicle. The flowers are yellow or brownish-red. French tarragon will not produce seeds. Russian tarragon does produce seed.

The value of tarragon comes from the essential oil, which accumulates in the herbs. The essential oil has an odour somewhat similar to that of anise; however, it is not sweet, but spicy and hot-tasting. French tarragon contains the maximum amount of essential oil (0.5–2.8 per cent) at the time of flowering. The most important compound of the essential oil is methyl chavicol (estragole), which is the isomer of anethole, and amounts to 40–70 per cent of the total volume. The minor components of the essential oil are camphene, ocimene, sabinene, myrcene, anisole and eugenol.

Tarragon lives for 8–10 years, but the plantations are maintained only for 4–6 years, because the number and vigour of the shoots decline significantly in old plants.

Shoot growth starts early, at the beginning of March, and lasts till the late autumn, until the onset of frosts. It flowers from the middle of July for 10–15 days in Europe.

Environmental requirements

The most appropriate locations for cultivation are sunny, warm places, where the essential oil content of the plant is the highest, and the quality of the essential oil is also better. Frost of −15 °C will not damage tarragon even in winters without snow, and the late spring frost will not damage the shooting plants either.

Tarragon requires a relatively large amount of water. For this reason, humid areas with good water management capacity should be selected for cultivation. It will, however, not tolerate long-term water cover, especially in the vegetation period. In these places the plants die, or grow slowly in the soil. The water need is highest at the time of peduncle development and after the main cutting (Wali, Hornok and Héthelyi, 1980).

The best soils for cultivation are deep-layered, medium-firm, and rich in nutrients. Loose sands with no structure, and airless firm soils, or calcareous soils, are not appropriate for cultivation.

Cultivation

Nutrient supply

Tarragon requires lots of nutrients; it is preferable to plant in freshly manured fields, and the regular replenishment of nutrients is also very important.

Prior to planting, 40–50 t/ha of manure, 80–100 kg/ha of P_2O_5 and 120–130 kg/ha of K_2O fertilizer should be introduced into the soil to 35–40 cm depth. At the time of shoot emergence in the spring, and after the first cutting, 140–150 kg/ha of N active material should be applied in two portions as top-dressing. It is advisable to provide the plants with foliage fertilizers 6–8 times. Foliage fertilization should be started when the plants are 30–40 cm high. Foliage fertilization may be combined with weed-controlling spraying after the first cutting.

Soil preparation

Depending on the previous plant, the soil preparation consists of stubble stripping, and subsequent soil cultivation and autumn deep ploughing. In the spring the soil is prepared with smoothers, spike harrows and rollers. In order to achieve successful mechanized planting, the soil must be properly settled, free of lumps, and smooth.

Propagation

French tarragon can be propagated with green cuttings, with turions (rooty shoots) and by root division.

The preparation of green cuttings and the root initiation is expensive; it requires lots of manual labour and careful treatment. The root development of cuttings is very slow, and the rooty seedlings will not develop well.

In large-scale production, propagation with turions (rooty shoots) is applied generally. From the short stolons of 3–4-year-old tarragon, a number of strong shoots, i.e. turions, develop in the spring. Adventitious roots develop on the turions in the ground. This development may be enhanced by filling up the rows with at 10–15 cm layer late in the autumn or early in the spring. The rooty turions are torn away from the mother plants of 30–40 cm height in May, after opening the soil around the plants. The turions should be transplanted immediately to the final location. From a 3–4-year-old plant, up to 20–40 turions can be gathered. If all the turions are torn, then the mother plant may get weak or may even die. It is advisable to gather only 10–15 turions from one plant. This mode of propagation is reliable only in rainy weather or with proper irrigation.

The oldest and most reliable propagation mode of tarragon is root division. The plants allocated for transplanting are taken from the ground by shovels or ploughs, and then the plants are cut into a number of pieces so that each part will include at least two overwintering buds and adventitious roots. The well-developed 3–4-year-old plants can be cut into 5–10 parts.

The turions can be planted by hand or with planting machines. Tarragon is planted in rows 0.6 m apart, with the plants 0.4–0.5 m apart. Irrigation is provided two or three times with abundant water after planting, even in rainy weather.

The parts prepared by root division are planted into nests prepared by hoe in the autumn or spring. In recent years the root parts have been planted into 150–200-mm-deep furrows made by a tool installed on a cultivator. The backfilling of furrows is done with the same tool or with hoes. The growing area is also 0.6×0.4 m in this case. The required amount of propagation material is 35 000–40 000/ha.

Care of plantation

The plantation can be treated with herbicides. From the second year on, spraying with 8 kg/ha of Merkazin prior to sprouting and after the first cutting is efficient. It is also efficient to apply 6 kg/ha of Afalon + 2 kg/ha of Aresin prior to sprouting and after the first cutting. In terms of effect, coverage and duration, Sencor (metribuzin) herbicide is best; this should be applied in dosage of 1–1.2 kg/ha on newly planted areas, and in a dosage of 2–3.5 kg/ha on older plants.

Tarragon develops well in aerated soil, and for this reason it is preferable to perform mechanized interrow cultivation in the spring and after the first cutting.

Regular protection procedures should be carried out against the often-occurring rust fungus (*Puccinia dracunculina* Fahr.) and against the recently appearing *Diplodina dracunculi* Vör. The damage caused by these fungi is most frequent after the first cutting. For the purpose of spraying, it is adequate to use either Vigil (diklorbutrazol) or Tilt (propiconazol). The spraying should be repeated every 20–21 days.

Harvesting

Tarragon may be cut once in the year of planting and twice in later years. For the purpose of essential oil production, the first cutting is generally made at the second half of July, at the start of flowering. The second cutting is made in the period between mid-September and the onset of frosts. The harvest is gathered with mower–loaders. The cutting height is 70–80 mm. If the cutting is made lower than this, then the repeated shoot emergence will be delayed or the plant might die.

For the preparation of herbal drug, tarragon should be harvested prior to the appearance of flowers, in the second half of June. In this case even three cuttings are possible. The fresh plants should be dried in artificial driers at 40–50 °C, or spread in a shady place in a thin layer.

The expected yield in the first year is 3.5–5.0 t/ha of green plants, and 5–8 kg/ha of essential oil. In the following years, the annual yield is 15.0–20.0 t/ha of green plants, and 25–35 kg/ha of essential oil. In the case of dried drug, the expected yield is 2.0–2.5 t/ha of herbal drug.

Pyrethrum (*Tanacetum (Chrysanthemum) cinerariaefolium* (Trev./Vis))

The pyrethrum that has been applied since the middle of the nineteenth century (Dalmatian pyrethrum, pyrethrum) contains twice as much material with insecticidal effects, than was contained by the Persian insect powder (*P. carneum* Bieb) and by the Armenian insect powder (*P. roseum* Weber at Mohr) which were used earlier.

Its cultivation was reduced in the whole world in the initial years of the 1940s after the invention of less expensive synthetic insecticides (HCH, DDT). When the use of chlorinated hydrocarbon derivatives was banned at the end of the 1960s, then the cultivation of pyrethrum was started again over larger areas, and the cultivated area is still increasing.

In earlier years, the finely ground powder (Pulvis insectorum) of the inflorescence of the plant (Chrysanthemi flos, or Pyrethri flos) was used for killing insects. Today, what we find in commercial distribution is the preparation made from the ground particles of the flower and active ingredients including the powder Pyrovir, Piretrin 89-PHYL, the fluid Solutox, Pyrotox ULV and the aerosols Chemotox and Pyrotox.

Major advantages of these insecticides is that they are not toxic to human beings and other warm-blooded animals, their effect is quick and wide-ranging, and the insects do not become resistant. They are mainly used in households to keep cereal and food stores and animal breeding places free from insects, and in recent years they have also been used against insect pests of house plants.

The demand for pyrethrin has also been increased by the realization that sezamin, which has no insect-killing effect itself, increases the insect-killing effect of the pyrethrin when mixed with it, and the duration of the active stage of pyrethrin is also lengthened in this way.

Pyrethrum is purchased in large quantities all over the world, owing to its advantageous properties. The world production of pyrethrum drug and extract was over 22 000 t in 1972. The countries producing the largest amounts are Kenya (14 000 t/year), Tanzania, Uganda, the countries of Central and South America, Japan, Yugoslavia, and recently, India and Pakistan.

Characteristics

Pyrethrum belongs to the Asteraceae family. It is an annual herbaceous plant. It is native to the shores of the Adriatic Sea on the Balkan peninsula. It can be mainly found on stony, warm hillsides.

It has a short rhizome with the thickness of a finger, from which a large number of 15–20-cm-long dark brown cylindrical roots develop. The leaves are alternate, and the leaf-blades are divided two or three times, fibrous, and obtuse. The leaves are

Fig. 103 Pyretrum

silvery on the back and more or less bald on the front. From the multi-head rhizome a number of 30–50-cm-high grey, pubescent flower stems develop, which have solitary composite flowers at their ends (Fig. 103).

The well-developed plants have 100–200 inflorescences, but the number is sometimes over 400. The opened composite flower has a 2.5–5 cm diameter. It has white fruit-bearing flowers, and yellow, tubular hermaphrodite disc flowers. The 3–4-cm-long achene fruit is yellowish-brown, its thickness is 1–1.5 mm, it has five edges, and it is glandular pubescent in the longitudinal groove. The mass of 1000 seeds is 1–1.2 g.

All the parts of the plant above ground contain pyrethrins. The leaves contain pyrethrin only in traces, while the dried stem has 0.15 per cent, and the flowers have 0.5–2.5 per cent. Most of the pyrethrin accumulates in the testa, 2.2–4.5 per cent, while the petals contain 0.2–0.4 per cent and the involucral bracts contain 0.15 per cent.

Six pyrethrin-type compounds have been isolated from the plant so far. To the group named Pyrethroid I (pyrethrine I, cinerine I, jazmolin I) belong the monocarbonic acid esters, and the group Pyrethroid II (pyrethrine II, cinerine II, jazmolin II) includes the dicarbonic acid esters. Of the total pyrethrine content, the percentage of monocarbonic esters, which have ten times as much effect, is 30–70 per cent, of which 40–65 per cent is pyrethrine I.

Pyrethrum lives for 8–10 years, but after 4–5 years of cultivation its yield diminishes. The seeds retain their germination capacity for 3–4 years. Under average conditions, the seeds begin to germinate slowly within 15–20 days. The initial development of the plant is very slow; the 8–10 cm height is reached in 2.5–3 months, even under optimal conditions. The flower-bearing stem develops only in the year following the sprouting year. The complete flowering of plants two or more years old takes place at the end of May to the beginning of June in Hungary. Some flowers may develop up to the end of the summer, especially in the year of planting. The fruits ripen relatively uniformly in the second half of July.

Environmental requirements

The climatic conditions are satisfactory for cultivation in most of the areas of Europe. In spite of its warmth requirement, the plant bears the winter frosts relatively well. Except in freezing spots and too-humid areas, freezing out is rare, and for proper flowering a cold winter is even a condition.

The plant requires a lot of sunshine. Shading reduces the size of the flowers, and the yield and the pyrethrin content are also reduced. The plant developing in shade may have an active ingredient content 50 per cent less than the normal value.

The plant bears arid conditions well. For this reason, it can also be used for the exploitation of shallow soils which have a poor humus content, wastelands rich in lime, stony hillsides with a southern slope, and also sandy soils. It can also be successfully cultivated in areas with abundant precipitation if the soil is permeable and loose. According to Hungarian experience, the salt resistance of the plant is also good; for this reason, alkaline soils, on which water would not accumulate, may also be considered for cultivation.

Cultivation

Nutrient supply

The plant does not need much fertilization, but sufficient yield can only be attained if the nutrient supply is adequate. It is advisable to add, as basic fertilization prior to planting, 70–80 kg/ha of N, 100–120 kg/ha of P_2O_5 and 90–100 kg/ha of K_2O.

As a top-dressing, it is advisable to apply 60–70 kg/ha of N in the productive years just after sprouting. In the autumn, 60–70 kg/ha of P active material should be introduced between the rows.

Soil preparation

The depth of ploughing depends on the thickness of the soil; however, at least medium–deep ploughing is necessary, even in stony soil. The larger stone boulders are removed from the field in the spring, and then the soil is kept free from weeds by repeated mechanical processes until the time of planting. Prior to planting, the soil should be well worked over, because the seedlings have short, fibrous root systems, and for this reason their planting is only successful in soils free from lumps.

Sowing

Pyrethrum is propagated by seed sowing, or by root division in a number of countries. For the purpose of seedling cultivation it is sown in March–April, but it can also be sown in August–September if required. In the latter case, the young plants should be covered with greenhouse windows, foil or a thin layer of straw, in order to protect them from the winter frosts. The seeds are sown with 30–40 cm row spacing and to 0.5–1 cm depth. After sowing, the field is repeatedly rolled and watered as necessary. The seeds sprout after 2–3 weeks, and the seedlings usable for planting develop by September. When they are sown in plastichouses, the time for development of seedlings can be reduced to 2.5–3 months. For planting 1 ha of field, 60 000–80 000 seedlings are necessary; these develop from 1–1.5 kg of sowing seeds.

Root division can be done in September–October, or in the spring (March–April). The 3–4 year-old plants are removed from the soil and separated into 5–10 parts, which all have roots, and these parts are planted.

In some places pyrethrum is sown directly with the use of 6–7 kg of sowing seeds, but this is uneconomical owing to the high requirement for manual weeding.

Depending on the development, the seedlings are planted their final location at the end of August to early October or in April–May. The seedlings are planted in the soil with the help of dibbles, providing a cultivation area of 60 × 20–30 cm.

Care of plants

In earlier years, weed-killing was performed with machines between the rows and by hoeing the rows. In recent years, protection against annual weeds has been achieved by the application of 4–5 kg/ha of Merkazin, 4–5 kg/ha of Afalon, or 4–5 kg/ha of Maloran. The herbicides must be applied prior to the start of vegetation.

Harvesting

The flowers can be gathered in the first year, but sufficient yield can only be expected from the second year on. Harvesting should be started when the tongued flowers are in the horizontal position, and three-quarters of the tubular flowers have already opened.

This phase of flowering lasts only for 2–3 days, which is not enough for harvesting. Consequently, harvesting is started when the tubular flowers open, and it is continued until the wilting of the tongued flowers. In this way the harvesting period lasts for 9–10 days.

Owing to the high labour requirement, the flowers are seldom gathered by hand. For harvesting pyrethrum, windrowing machines and mower–loaders are generally used, and also combine harvesters have been applied in recent years. The flowers are cut with raised cutters, so that a relatively short stem is cut with the flowers.

The cut flowers are generally dried in artificial driers at 50–60 °C, or in cool-air hay driers. The flowers are threshed after the completion of drying, and then the unnecessary stem and leaf parts are removed by repeated screening. The material thus obtained consists of tongued and tubular flowers, fruits, involucral bracts and leaf particles and its active ingredient content is frequently 20–50 per cent, more than that of the flower drug. The expected yield in the first year is 0.1–0.2 t/ha, and 0.7–1.2 t/ha of dry drug in the productive years. The drug should be stored in a cool place in well-sealed paper or plastic bags, so as to avoid humidity absorption and fermentation.

In order to avoid oxidation loss, it is advisable to put oxidation retarder into the drug prior to closing the bag; otherwise, a significant amount of active ingredient will be damaged within several months.

Roman camomile (*Chamaemelum nobile* (L.) All.) (syn. *Anthemis nobilis* L.)

Roman camomile has been used for a long time, but its cultivation was started only a few decades ago. It is a medicinal plant containing very valuable essential oil. The brew of its inflorescence (Chamomillae romanae flos) is used for treating digestive troubles, and for the treatment of hair and head skin. In folk medicine it is used as a sedative, pain killer, antispasmodic, and sudatory agent. The essential oil (Aetheroleum chamomillae romanae) is used mainly by the cosmetics and perfume industry. It is also planted for ornamental purposes.

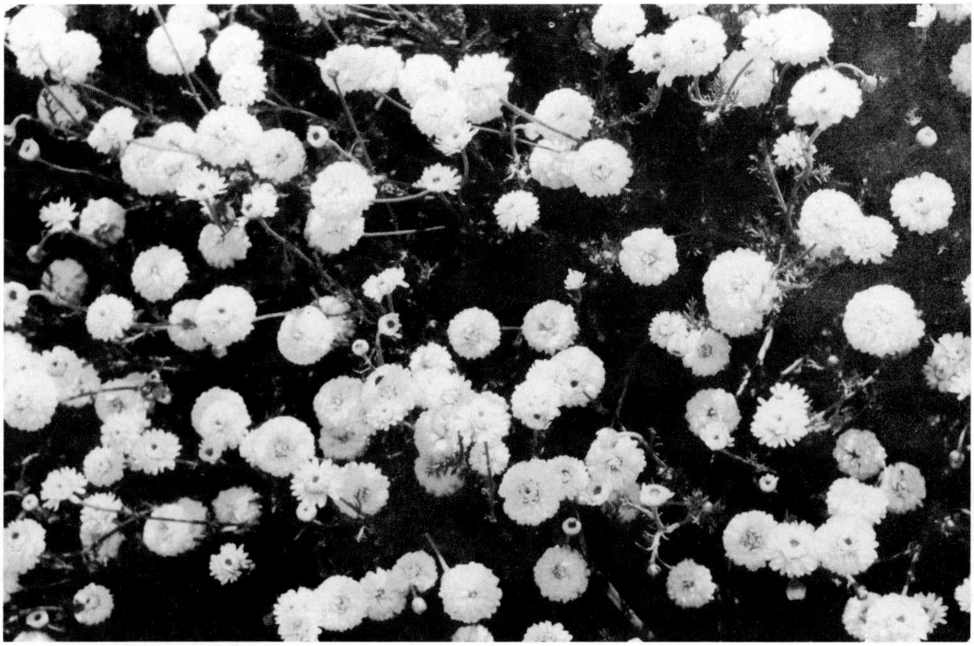

Fig. 104 Roman camomile

The most significant producers are in western Europe: Belgium, the Netherlands, England, France and Italy. Smaller areas are cultivated in almost all other European countries.

Characteristics

The Roman camomile is a perennial, herbaceous plant belonging to the Asteraceae family. It is of Mediterranean origin, native mainly to Portugal, France and Algeria. Only the variety with full flowers and containing large amounts of essential oil is cultivated (var. *flora plena* = var. *ligulosa*); this does not grow in the wild.

From the multi-headed, spreading rhizome, a number of creeping shoots develop, which frequently develop auxiliary rhizomes at the nodes. The stems are simple or branched, tomatose, and their tops are 20–40 cm high. The leaves are alternate, bifoliate, fibrously divided, and greyish-green. The long, white capitula are terminal standing and 1–3 cm in diameter (Fig. 104). The fruit is a three-edged silvery-grey achene. The full-flowered variety seldom develops seeds which are capable of germination, but the descendants are never full-flowered. Pubescent glands containing essential oil can be found on any part of the inflorescence. The essential oil extracted from the fresh plant is light blue, but it becomes greenish or brownish-yellow in a few days. The essential oil extracted from the dried plant is yellowish-brown. The essential oil has a strong odour and spicy taste. The main ingredients are: several alcohols, angelic acid, angelic acid butyl ester, isobutyric acid ester, anthemol, methacrylic acid and its ester, and chamazulen. The dry inflorescence contains 0.6–1.0 per cent (sometimes 1.5 per cent) essential oil, and the flowering plant contains 0.2–0.3 per cent essential oil. As well as the essential oil, the plant contains bitter materials, choline, apiine, inositol and quercitin, and sugar.

Enviromental requirements

Camomile requires warmth and sunshine. It grows on warm soils. It is not particularly sensitive to frosts; freezing out seldom occurs even in very cold winters. In warm weather the plant starts to develop early, even at the end of winter. In this case the late frosts may cause damage to the plant.

It grows on any soil; however, the best soils are medium-firm humid soils which are rich in humus. Excessively firm, water-bearing areas and loose sands are not good for cultivation.

The water requirement is high, especially at the time of root development and in May-June in the initial period of thick growth and stem emergence. A significant increase in yield can be achieved by abundant irrigation two or three times in this period. It can be economically cultivated even without watering in areas which have abundant precipitation and where the soil is good in terms of water exploitation.

Cultivation

The Roman camomile plantation remains in one place for 3–4 years.

Nutrient supply

Organic fertilizer is introduced into the soil when the previous plant is still in cultivation. As basic fertilizer, 100–140 kg/ha of P_2O_5 and the same amount of potassium active material, and 60–70 kg/ha of N should be introduced into the soil at the time of autumn deep ploughing.

After planting and at the time of shoot emergence in every year, 80–100 kg/ha, and after harvesting, 40–50 kg/ha of N should be applied as top-dressing.

Soil preparation

Roman camomile requires weed-free soil. For this reason, proper soil conditions should be obtained with operations started in the previous year. The 25–30-cm-deep autumn ploughing should be smoothed in the spring, and a completely lump-free state should be prepared with repeated spike harrowing and rolling, because the seedlings have rather short roots and may only be planted shallowly.

Propagation

The full-flowered Roman camomile may only be propagated by root division. The plants to be replanted are loosened from the soil with a plough, from which the steering plate has been removed, or the plants are lifted out by cultivators or with a shovel; then the plants are divided into pieces so that every plant part includes at least two or three rosettes.

Planting should be started when the developing new shoots are 2–4 cm long.

From one matured old plant, 10–15 seedlings may be obtained. The seedling requirement is 70 000–80 000 for 1 ha.

Planting should be done after abundant rain, or the field should be irrigated, prior to planting. Planting is done with 50–60 cm row spacing and 20–25 cm plant distance. Care should be taken to avoid putting the seedlings with no collars too deep. The plantation should be repeatedly irrigated, because the seedlings with short roots may dry out quickly.

Care of plantation

As long as the seedlings are small, the field may only be kept free of weeds by very careful hoeing, as the shallow-rooted seedlings might be easily loosened. From the second year on, the dry parts of the plant are cut with machines in the early spring, and then the field is repeatedly spike harrowed, the places between the rows are machine-hoed, and the rows are hoed by hand.

In recent years the application of herbicides has increased. It has proved to be successful to introduce trifluraline prior to planting.

Prior to the start of vegetation, Aresin, having a longer effective period and a wider weed-killing range, may also be used (4.5–5 kg/ha), but overdosing will prevent the formation of flowers.

Harvesting

Roman camomile flowers continuously from the middle of June to August. For the purpose of producing flower drug, the flowers are gathered at the time of full flowering in dry weather with the help of a device similar to the camomile-gathering comb, or are picked by hand with short stems. As flowering takes place over a longer period, the gathering is repeated several times.

The gathered flowers are spread as soon as possible in a thin layer in a shaded place, or are dried in an artificial drier at 35–40 °C. The fresh flowers might get mouldy within 1–2 hours, becoming brown and unusable.

For the purpose of extracting essential oil, the plants in full flower are cut in clear weather with windrowing machines or by hand mowing in small fields, leaving 5–6 cm high stubble. The cut plant is generally dried in the field. In this way the extraction period can be extended, and the extraction itself is more simple and cheaper.

According to experiments, the essential oil content increases during drying.

The expected yield in the first year are 200–250 kg/ha of flower drug, or 4–6 t/ha of fresh plant, or 3–5 kg/ha of essential oil, while in the productive years they are 400–600 kg/ha of flower drug, or 10–12 t/ha of green plant, or 6–10 kg/ha of essential oil.

References

Békésy, M. (1938). 'Investigations of ergot cultivation' (in Hungarian), *Kísérletügyi Közl.*, 1–14.

Békésy, M., and Garay, A. (1960). 'The ergot' (in Hungarian), *Culturflora of Hungary,* Vol. I, No. 10.

Bernáth, J., Földesi, D., and Lassányi, Zs. (1973). 'Influence of nutrient supply and type of soil on the catnip (*Valeriana off.* L. ssp. *collina* Wallr.) I.' (in Hungarian), *Herba Hungarica,* **12,** 2–3, 46–53.

Boros, Á., and Szujkó, I. (1970). 'The dill' (in Hungarian), *Culturflora of Hungary,* Vol. IV, No. 10, 1–69.

Braun-Blanquet, I. (1961). *Die inneralpine Trockenvegetation,* G. Fisher, Stuttgart.

Datta, S. C. (1982). 'Cultivation of *Catharanthus roseus* in West Bengal.' In C. K. Atal and B. M. Kapur (eds) *Cultivation and Utilization of Medicinal Plants,* Jammu-Tawi, Reg. Res. Lab., 297–283.

Duke, J. A. (1985). *Handbook of Medicinal Herbs,* CRC Publishers, Boca Raton.

Földesi, D. (1972). 'Production of poppy without thinning by sowing irradiated seeds' (in Hungarian), *Magyar Mezőgazdaság,* **6,** 8–9.

Földesi, D. (1975). 'Chemical weed control in poppy cultivation with Dicurán' (in Hungarian), *Magyar Mezőgazdaság,* **12,** 10–11.

Földesi, D., Bernáth, J., and Pál, J. (1980). 'Effective method for control of tolerant weed in poppy cultivation (*Papaver somniferum* L.)'. *Acta Horticulturae,* **96,** 261–8.

Francz, Ch. (1972). 'Einfluss der Nahrstoffe Kalium und Stickstoff auf die Bildung des ätherischen Öls von Mentha piperita', Planta Medica, 160–83.

Glushchenko, N. N. (1976). 'Caraway' (in Russian), In A. M. Szmoljanov (ed.) *Efiromaslichnye kultury,* Kolos, Moscow, 74–114.

Halmai, J., and Novák, I. (1963). *Acology* (in Hungarian), Medicina, Budapest.

Heeger, E. F. (1956). *Handbuch des Arznei- und Gewürzpflanzenbaues,* Deursch. Bauer Verlag, Leipzig.

Hornok, L. (1980). 'Effect of nutrition supply on yield of dill (*Anethum graveolens* L.) and the essential oil content', *Acta Horticulturae,* **96,** 337–342.

Hornok, L., Földesi, D., and Szász, K. (1975). 'Experiments on the development of cultivating procedures of thyme (*Thymus vulgaris* L.)' (in Hungarian), *Herba Hungarica,* 47–64.

Hotin, A. A. (1968) 'Role of environmental factors in essential oil accumulation'. In *Essential Oil Plants and their Processing* (Smolyanov, A. M. ed) (in Russian), Pishchevaya Promisnlenoszt, Moscow, 35–44.

ITC UNCTAD/GATT (1982). Markets for Selected Medicinal Plants and their Derivatives, ITC UNCTAD/GATT, Geneva.

ITC UNCTAD/GATT (1986). Essential Oils and Oleoresins. A Study of Selected Producers and Major Markets. ITC UNCTAD/GATT, Geneva.

Ivanchenko, N. Yu. (1976). 'Clary sage'. In A. M. Smolyanov (ed.) *Essential Oil Plants Protection* (in Russian), Pishchevaya Promislenost, Moscow, 199–229.

Kandelaki, G. V. (1962). *Studies on Ocimum basilicum* L. (in Russian), Trudi Tbiliskava Bot. Inst., Tbiliszi.

Kárpáti, Z., and Terpó, A. (1968). *Horticultural Botany* I–II. (in Hungarian), Mezőgazdasági Kiadó, Budapest.

Kerekes, J. (1960). 'Water management of peppermint (*Mentha piperita* L.)' (in Hungarian), *MTA Agrártudományok Osztálya Közleményei* **1,** 105–16.

Kerekes, J. (1969). *Cultivating Medicinal Plants* (in Hungarian), Mezőgazdasági Kiadó, Budapest.

Ksendz, A. T. (1976). 'Lavandel' In A. M. Smolyanov (ed.) Essential oil-plant production (in Russian), Kolos, Moscow, 159–99.

Lewis, W. H. (1977). *Medical Botany,* John Wiley and Sons, New York, London Sydney, Toronto.

Lukyanov, I. A. (1976). 'Anise'. In A. M. Smolyanov (ed.) *Essential oil-plant production* (in Russian), Kolos, Moscow, 57–74.

Máthé, I., and Tyihák, E. (1962). 'Further data on the L-alphabisabolene content of camomile, in relation to the prokamazulene content' (in Hungarian), *Herba Hungarica,* **1,** 29–38.

Meyers, T. (1958). *Een Onderzoek van Het Linneon* Valeriana officinalis *in Nederland,* Gravenhagen, Wageningen.

Petrov, A. Sz. (1980). 'Diseases of coriander', *Trudi Vniiemk, Szimferopol,* **13,** 113–122.

Pillai, P. K. T., and Nambiar, M. C. (1982). 'Condiments'. In C. K. Atal and B. M. Kapur (eds) *Cultivation and Utilisation of Aromatic Plants,* Jammu-Tawi, Reg. Res. Lab., 167–177.

Poludennii, L. V., Shotnik, V. F., and Hlaptsev, E. E. (1979). *Essential Oils and Medicinal Plants* (in Russian), Kolos, Moscow.

Rácz, G. (1984). Medical plants (in Hungarian), Ceres, Bucharest.

Rosengarten F. (1969). *The Book of Spices,* Living Publishing Company, Wynnewood, Pennsylvania.

Ruminska, A. (1973). *Medicinal Plants* (in Polish), Panst. Wydaw, Nauk., Warsaw.

Schröder, H. (1959). 'Der Einfluss von Stickstoff Phosphorsäure und Kali auf Ertrag und Gehalt and ätherischem Öl bei Majoran (*Majorana hortensis* Moench)', Pharmazie, **5,** 329–46.

Schröder, H. (1963). 'Untersuchungen über den Einfluss unterschiedlicher Wasserversorgung auf Erträge, Gehalte an ätherischem Öl.' *Pharmazie,* **1–3,** 150–8, 241–5.

Schröder, H. (1964). *Arznei- und Gewürzpflanzen,* Bernburg-Saale.

Shcherbakov, S. E. (1976). 'Peppermint' In A. M. Smolyanov (ed) *Essential oil-plant production,* (in Russian) Kolos, Moscow, 229–54.

Szendrei, K. (1986). *Die Suche nach Quinta Essentia Heute.* Abstracts of the 7th Hungarian Medicinal Plants Symposium.

Tétényi, P. (1970). *Intraspecific chemical Taxa of Medicinal Plants,* Akadémiai Kiadó, Budapest.

Verzár-Petri, G., and Then, M. (1974). 'Comparative thin layer and gas chromatographic investigation of *Salvia officinalis* and *Salvia sclarea* essential oil compositions' (in Hungarian), *Herba Hungarica,* **1,** 51–56.

Wali, A. K., Hornok, L., and Héthelyi, I. (1980). 'Effect of water supply on yield and volatile oil content in tarragon (*Artemisia dracunculus* L.)' (in Hungarian), *Herba Hungarica,* **19,** (3) 55–63.

Sváb, I., El Din-Awad, C., and Fahmy, T. (1967). The influence of highly different oecological effects on the volatile oil content and composition in the camomile. *Herba Hungarica,* **6,** 2, 177–188.

PART 3 MISCELLANEOUS

In this part those medicinal plant species are introduced which are cultivated on a small scale in temperate zones, and those tropical and subtropical plant species which are not suitable for cultivation in temperate zone but whose consumption is significant in Europe. Also, this part also includes summary tables of the important data on cultivated and collected medicinal plants.

Belladonna (*Atropa belladonna* L.)

The scientific name of belladonna is derived from the Greek Atropos, that one of the three mythological Fates who cuts the thread of life with scissors. Thus even its name indicates its strongly poisonous character. In Italian, belladonna means beautiful woman, which is an allusion to the fact that in earlier times Italian women used the pressed juice of the belladonna berry as a cosmetic (to dilate the pupil of the eye).

Description

Belladonna (deadly nightshade) can be found everywhere in Europe, especially in the southern regions, in the temperate regions of northern Asia and in North Africa. It also grows wild in North America as an adventitious plant. It is selectively cultivated in a number of countries.

Belladonna is a perennial, strongly poisonous plant belonging to the Solanaceae family. Its root is 0.4–0.6 m long, 40–60 mm thick at the root collar, yellowish-brown outside, and greyish-white inside. It has a taproot with a mild odour and a repellant taste. It is slightly fleshy when fresh, and hard after drying, and its broken surface is starchy. The older roots have multi-headed scale-like basal leaves, and buds from which shoots will emerge in the spring. The stem is 1–2 m high, depending on the growth location, erect, cylindrical, and branching at the top. The lower leaves are alternate, and the leaves above appear in pairs, the larger of which is located outside and stands horizontally, while the smaller one is inside and drooping. The leaves are petioled, ovate with an acute tip, and entire-edged.

The flowers originate in the axils of leaves, are solitary, and appear at the ends of branches in a loose, leafy, spike shape. The corolla is 2–3 cm long, cylindrically bell-shaped, and dull yellow or purple. Flowering takes place from June to the appearance of frost.

The fruit is a berry with a number of seeds. The berry has a diameter of 10–15 mm, and a nearly globular shape; it is shining black and its juice is bluish-purple. The seeds are disc-shaped, brown and have diameters of 1.5–2 mm. The mass of 1000 seeds is 0.01–0.05 g.

Active substances

The roots contain about 0.2–0.8 per cent alkaloids. The alkaloid content is 0.1–1.32 per cent in the leaves and 0.3–0.9 per cent in the seeds. According to Lewis and Elvin-Lewis (1977), roots and leaves contain l-hyoscyamine, (±)-hyoscyamine (atropine) and l-scopolamine (hyoscine), which in overdoses can be fatal in man. Roots also contain apoatropine, belladonnine and cuscohygrine. Tannin has also been reported. Pyridinc, suninic acid, leucatropic acid, asparagine, choline, phytosterol, scopoletin and scopolin have also been reported in the leaves (Duke, 1985).

Uses

The drug obtained from roots (Belladonnae radix) and leaves (Belladonnae folium) is considered to be an antiasthmatic, antispasmodic, anodyne, diuretic, febrifuge, mydriatic, narcotic, nervine and sedative. According to Leung (1980), it can be used to treat Parkinson's disease. Atropine and its optical isomer hyoscyamine are used as pre-anaesthetic agents to check secretion in the throat and respiratory passages. It is used as an antidote to depressant poisons such as chloral hydrate, muscarine and opium. Hyoscine (5–11 per cent of total alkaloids) is used as a truth drug.

The root, prepared in various ways, is used to treat inflammatory tumours, tumours of the rectum, and other tumours. The leaf is used for mammary and tongue cancers, uterine cancer, carcinoma, running cancers, and scirrhus of the breast (Duke, 1985).

Gathering

The leaves are gathered from the time when the plant starts to flower up to the maturing of fruits. Drying is done with artifical drying equipment, on drying frames in a layer of 3–4 cm thickness, placed in lofts. In the latter case the belladonna leaves should be turned over several times until they are completely dry. The drying ratio is 6 : 1. The colour of properly dried leaves is bright green; the leaves have no odour and their taste is sweetish initially, then bitterly spicy.

The roots are gathered in the autumn prior to the onset of frosts. The content of alkaloid reaches the maximum value at this time. In bad weather, belladonna roots can also be gathered in the early spring prior to sprouting. In this case, however, the active substance content is generally less than that of roots gathered in the autumn. After shovelling out, the roots are cleaned of earth particles from the aerial parts and from the adventitious roots. The fleshy main roots are cut into 15–20-cm-long pieces; the thick ones are sliced into 2–4 parts and are dried, preferably in an artificial drier at 50–70 °C, until they give a short snapping sound if broken in two. The drying ratio is 3 : 1.

Cultivation

Belladonna is cultivated in a number of countries, mainly in clearings of forests. For the cultivation of belladonna, deep-layered soils or lime soils are applicable; these can be ploughed in the autumn. Propagation can be done by division of root heads, or by seed sowing and seedling nursing. When the roots are gathered the top 12–15 cm are divided, so that every part will contain well-developed buds. Planting is done in October with a row distance of 0.7–0.8 m. The root heads are planted in 15–25-cm-deep furrows so that at least 5 cm of soil covers them.

For the purpose of seedling nursing, the seeds are sown in warm seed-beds in early March. The germination ability of the seeds is very poor, and for this reason the seeds should be soaked for 18–20 hours in water at 30 °C, and then kept at a temperature near 0 °C for 24–28 hours. The seeds thus treated can also be sown in open seed-beds in April. The seedlings can be transplanted to the field in the autumn. The seed requirement is 0.7–1.0 kg/ha. The roots can be gathered in the second or third year if propagation was done by division of root heads, and they can be gathered in the third or fourth year if propagation was done with seedlings. The gathering of leaves can be started from the first year, at the start of flowering, and then continuously. No more than one-third of the lower leaves may be gathered at one time. The active substances content of the leaves reaches its maximum in September. The expected yield is 0.7–1.5 t/ha of dry leaves and 1.5 t/ha of dry roots.

Buckthorn (*Frangula alnus* Mill. syn. *Rhamnus frangula* L.)

Description

Buckthorn (alder buckthorn) is a native of Europe and western Asia. It is also grown in the northern USA. It can be found frequently in gallery forests, in alder fenwoods and in oak forests full of lily-of-the-valley. Buckthorn is a shade-loving shrub belonging to the Rhamnaceae family. The bark of the stems is grey, and densely covered with lenticels. The leaves are alternate, entire-edged, and 30–60 mm long; the adaxial surfaces are shiny. The flowers are tiny, greenish-white, and found in clusters (2–6) at the axils of leaves. The fruit is globose, 4–8 mm in diameter, fleshy, green initially, then red and violet–black later; it is a drupe containing three seeds. The mass of 1000 seeds is 11–20 g. Flowers as well as fruits can be found on the shrubs in May to August (Fig. 105).

Active substances

The activity of the bark is due to the 3–9 per cent anthraquinone glycosides (frangulin A and B, glucofrangulin A and B, emodin-1-glucoside, emodin-8-glucoside, emodin-8-0-β-gentiobioside, etc.). An alkaloid armepavine is present in fresh but not in dried bark. In the fruits, one encounters glucofrangulin, frangulin, emodinanthranol, jes-

Fig. 105 Buckthorn

trin, chrysophanol, aloeemodin, etc. Dry seeds contain 12.9–15.7 per cent protein and 10.7–15.8 per cent fat (Duke, 1981). The seeds contain amygdalin, robinin, kaempferol, helicin and tannin (List and Horhammer, 1979).

Uses

The purgative action of the bark (Frangulae cortex) is said to be similar to that of rhubarb. It works without irritating the system and can be used for all conditions causing or associated with constipation, including liver and gallbladder problems. Bark, used medicinally, should be aged for at least one year before use. (The active substances mentioned above will form during long storage or during drying at 80–100 °C.) According to Duke (1985), bark and leaves can be used as a yellow dye, and mixed with iron salts, as a black dye. Unripe berries give a good green colour; ripe, blue or grey. The leaves are eaten avidly by goats, and the flowers are visited by bees. The wood makes a relatively inflammable charcoal, which was used by old gunpowder makers. The wood is also for boot lasts, wooden nails and veneers.

Gathering

The drug is made from the bark, which is stripped from the 10–20-mm-thick branches of the plant. The bark is gathered in the early spring at the start of fluid circulation from the 3–5-year-old branches in such a way that two circular splits are made on the bark, 10–20 cm apart, and then a longitudinal split is made between the two circular splits. In this way the bark can be easily stripped, and will curl during drying. About 3 kg of fresh bark will give 1 kg of dry bark. The drug is grey outside, lenticellate, reddish if rubbed slightly, and yellow inside.

Cultivation

According to the studies made so far it can be stated that buckthorn will grow well mainly on lowland, humid soils. In our experience, the most successful mode of propagation is seed sowing in nursery gardens in the autumn. The densely rooted two-year-old seedlings should be transplanted to their final location. It is absolutely necessary to deep plough prior to transplanting. Owing to the requirements of mechanical interrow cultivation, planting is done with 0.80–1.00 m row spacing and 0.40–0.50 m plant distance.

Cardamom (*Elettaria cardamomum* Maton)

Description

Cardamom is a tall, herbaceous perennial with branched subterranean rhizomes, belonging to the Zingiberaceae family. Cardamom occurs wild in the evergreen monsoon forests of the Western Ghats in southern India and Sri Lanka (Purseglove *et al.*, 1981). From its rhizomes arise several erect leafy shoots and erect or decumbent panicles. There is a tough, rather woody horizontal rhizome with numerous fibrous roots in the surface layer. The leafy shoots are 2–5.5 m tall, composed of leaf sheets, and borne in thick clumps. The leaves are distichous with lanceolate acuminate lamina, dark green galbrous above, and paler beneath. The lower surface may be either glabrous or pubescent. The inflorescences arise from the rootstock at the base of the leafy shoots and are 0.6–1.2 m long. They are erect, recumbent, slender panicles. The flowers are hermaphrodite, about 40 mm long and 17 mm across, and white with violet streaks radiating from the centre. The fruit is a trilocular capsule, fusiform globose, pale green to yellow in colour, and varying in size according to the variety. It contains 15–20 seeds, which are dark brown, angled, rugose, and aromatic. The mass of 1000 seeds is 22–23 g.

Two botanical varieties have been recognized, based on the size of the fruit. The var. *major* Thwaites is the wild cardamom of Sri Lanka. It is a robust plant, about 3 m tall. The fruit, which is larger than that of the other variety, is elongated, 2.5–5.0 cm long, yellowish-green when ripe and darkish-brown when dry. It is known in the trade as 'long wild native cardamom'. The var. *cardamomum* (syn. var. *minor* Watt) includes most of the cultivated races. The height varies from 2.5 to 5 m. The fruits, which are about 10–20 mm long, are smaller then those of the other variety above, and are yellowish when dried. It is known as 'true cardamom' (Fig. 106).

Fig. 106 Cardamom

Active substances

The dried fruit of true cardamom contains essential oil, fixed oil, pigments, proteins, cellulose, pentosans, sugars, starch, silica, calcium oxalate and minerals. The principal quality determinant for cardamom is the content and composition of the essential oil, which governs the odour and flavour of the spice. Whole cardamoms usually have essential oil contents in the range 3.5–7 per cent. The major components of cardamom oils are 1,8-cineole (20–60 per cent) and α-terpinyl acetate (20–53 per cent). Other principal components are as follows: linalyl acetate, linalool and borneol, α-terpineol, α-pinene, limonene and myrcene.

Uses

Cardamom is an expensive spice, the price being only exceeded among spices by saffron and vanilla. The seeds have a pleasant aroma and a characteristic warm, slightly pungent taste. It is used for flavouring curries, cakes and bread, and for other culinary purposes. In India, cardamom is used as a masticatory. It is thought to have aphrodisiac properties. Cardamom is official in the British and US Pharmacopoeias and is used as an aromatic stimulant, a carminative and a flavouring agent. The essential oil is used for flavouring processed foods, in perfumery and for flavouring liqueurs and other beverages (Purseglove *et al.*, 1981).

Cultivation

Principal cardamom production areas are located in India (60 per cent of world production), Guatemala, Tanzania and Sri Lanka. Cardamom is grown at altitudes between 760 m and 1400 m in areas with an annual rainfall of 1500–1700 mm and a temperature range of 10–35 °C. Cardamom can be cultivated on chocolate-coloured forest loam extending to a considerable depth below the humus layer, to white quartz gravel with only a shallow zone of humus accumulation. It requires good drainage and cannot tolerate waterlogging. The canopy of the evergreen forest provides the shade requirement for the crop. It is sensitive to high winds and exposed areas.

Cardamom can be propagated vegetatively by division of the rhizomes, or from seedlings, which are usually raised in nurseries. For vegetative propagation, rhizomes from large clumps of growing plants are taken out, separated into small clumps, each consisting of at least one old and one young shoot, and planted in prepared pits.

For large-scale production it is usual to sow selected seeds. For sowing, seeds are collected from high-yielding, healthy plants with compact panicles. Fully matured and ripened capsules are immersed in water for some time, the seeds are extracted by gentle pressing, and the mucilaginous coating is removed by washing in cold water. The seeds are sown immediately after extraction to ensure better germination. The sowing season is September–October or November–December. The seeds are sown in raised seed-beds on which a fine layer of humus-rich soil is spread. For the nursery site, gently sloping land with access to a water source is generally selected (Nair, 1982). The seeds are either sown in lines or broadcast in beds of 1 × 6 m size raised to 0.3 m; 10 g/m^2 of seeds are used. The seeds are covered first with a thin layer of sand and later on with paddy straw. The beds should be irrigated frequently and overhead shading should also be provided. The seeds germinate within 30 days and continue up to 90 days. Six-month-old seedlings are transplanted from the primary nursery to the secondary nursery, where they are allowed to remain for about one year. In the secondary beds there is a spacing of about 0.3 m between plants. Mulching and overhead shade should also be provided at this stage. The main field selected for cardamom plantation is cleared of undergrowth, and overhead shades are regulated by selective thinning. Wherever required, suitable quick-growing trees are also planted to provide adequate shade. Pits of about 0.6–0.6 × 0.3 m size are dug with a spacing of 1.5 × 1.5 to 3 × 3 m, depending on the variety. The pits are filled with a mixture of surface soil, compost and cattle manure, 1–2 months in advance of the commencement of planting. During the first year, frequent weeding is necessary. In older plantations, weeding is restricted to 2–3 per year. The old and dry shoots from the plantation are removed and the base of the clumps cleared once a year at the time of weeding, during May–June. According to CPCRI (1977), cardamom responds to fertilizer application; the recommended amounts are: 75 kg/ha of N, 75 kg/ha of P_2O_5 and 150 kg/ha of K_2O.

Flowering starts from the third year of planting. The normal flowering season is from April–May to August–September. The capsules are ready for harvesting by July–August or August–September. The harvesting season continues up to October–November.

The capsules are dried either in the sun or in specially built drying devices varying from sheltered platforms, heated by a slow fire from below, to a large drying house or a kiln heated by flue pipes. The fruits are spread thinly and stirred frequently to ensure uniform drying. Three to five days are required to dry the capsules properly in the sun; under artificial conditions drying may take only 48 hours. The dried capsules are rubbed by hand or with rough coir matting or wire mesh and winnowed to remove foreign matter. The dried and cleaned capsules are graded according to colour and size (Nambiar, 1978).

Sastri (1952) gives the average yield of dry capsules from a well-maintained cardamom estate as 112–168 kg/ha.

Catnip (Nepeta cataria L.)

Description

Catnip (catmint) is a perennial, herbaceous plant belonging to the Lamiaceae family. It is native to Eurasia and widely naturalized in North America. It is well known as an attractant and behaviour-modifying drug for both domestic and wild members of the cat family. The species name 'cataria' is from the Latin word for cat, indicating an early recognition of the special attraction of cats to catnip. Its erect, square, branching stem is hairy and grows 0.6–1.2 m high. The oblong or cordate pointed leaves have scalloped edges and grey or whitish hairs on the lower side. The bilabiate flowers are white with purple spots and grow in spikes from June to September. The seeds are very small, and brownish-black. The mass of 1000 seeds is 0.25–0.40 g (Fig. 107).

Active substances

The leaves and inflorescence of catnip (Catariae herba) contain essential oils, sterols, acids and tannins. The major constituent of essential oil is nepetalactone (70–99 per cent); other compounds are nepetalic acid, nepetalic anhydride, dispentine and β-caryophyllene. According to Duke (1985), var. *citriodora* Becker contains acetic acid, butyric acid, citral, citronellol, geraniol, limonene, nerol, tiglic acid and valeric acid.

Uses

Catnip has been used for ornamental and culinary purposes and as a domestic folk medicine remedy. The dried leaves and flowering tops are used as a spice to prepare a sedative tea and as a titbit for domestic cats. Leaves and shoots are used for flavouring sauces, soups and stews. It has been used as an antispasmodic, carminative,

Fig. 107 Catnip

diaphoretic, emmenagogue, nervine, stomachic, stimulant and mild sedative. The herb has also found use in the treatment of diarrhoea, colic, the common cold and cancer (Makovkina, 1972). According to Tarnopol and Ball (1972), extracts catnip exhibit juvenile hormone activity. Smoking catnip can produce euphoria and visual hallucinations, thought to be caused by the nepetalactone content.

Cultivation

The reported life zone of catnip is 7–19 °C with an annual precipitation of 0.4–1.3 m and a soil pH of 4.9–7.5. The plant thrives in well-drained soils. Propagation can be done by plant division or seed sowing. The seedlings prepared by plant division should be transplanted to the field in the autumn. Its seeds can be sown directly in the field or in open-air seed-beds. Direct sowing is done in April with a row distance of 0.5–0.7 m. The seed requirement is 1.5–2.0 kg/ha. For the purpose of seedling nursing, the seeds should be sown in open-air seed-beds with a row distance of 0.25–0.40 m in the spring. The depth of sowing is 15–20 mm. The seedling should be transplanted to the field in June with irrigation. It is planted with 0.5–0.7 m row distance and 0.3–0.4 plant spacing.

Catnip can be harvested once in the year following sowing or transplanting, and later on it can be cut twice a year. The herbs are harvested during full bloom and allowed to dry in the shade to preserve colour and fragrance. Harvesting should be done by hand or mower–loader machines. The expected yield is 1.5–2.0 t/ha of dry drug.

Cinnamon and Cassia (*Cinnamomum* species)

Cinnamon and cassia are two of the oldest spices known to man. Both condiments were probably known to the ancients at a very early period in history, for they are mentioned frequently in the Bible (Rosengarten, 1969).

Description

The genus *Cinnamomum* Schaeffer belongs to the family Lauraceae, which has some 32 genera and 2000–2500 species. They are mainly evergreen trees of the tropics and subtropics (Purseglove *et al.,* 1981).

Cinnamon or true cinnamon is *Cinnamomum verum* Presl. (syn. C. *zeylanicum* Bl.), which reaches a height of 8–17 m in the wild state. The bark and leaves are strongly aromatic. In an unharvested state, the trunk is stout, 0.3–0.6 m in diameter, with a thick, grey bark and branches set low down. Under conditions of cultivation, the continual removal of shoots by cropping almost to ground level results in the formation of a dense bush with a number of leafy coppiced shoots of height 2–2.5 m. The stiff exstipulate opposite leaves are somewhat variable in form and size. The petiole is 10–20 mm long, and grooved on the upper surface. The lamina is usually 50–180 mm long and 30–100 mm wide, and ovate or elliptic; the base is more or less rounded and the tip tends to be somewhat acuminate. The flowers are borne in lax axillary and terminal panicles on the ends of the twigs. The peduncles are creamy white in colour, softly hairy and 50–70 mm long. The individual flowers are very small, pale yellow in colour, with a foetid smell; each is subtended by a small, ovate, hairy bract. The trees flower in January and the fruits ripen six months later in Sri Lanka. The fruits are avidly eaten by birds, and so it is usual to net the trees if seeds are required for planting. The fruit is a fleshy, ovoid drupe, black, and 10–20 mm long when ripe.

True cinnamon is indigenous in Sri Lanka and southern India. Sri Lanka still produces the largest quantity and best quality; the Seychelles is the second largest producer (Fig. 108a).

Cassia, also known as Chinese cassia or Chinese cinnamon, is *Cinnamomum cassia* Presl. It is native to southern Vietnam and the eastern Himalayas. Cassia is also an evergreen tree, aromatic in all its parts, resembling true cinnamon, but larger in size. It reaches a height of 18 m with a girth of 1.5 m when left uncut. The bark is grey and smooth and is thick in the adult tree. The leaves are oblong–elliptic, dark shiny green, about 150 mm long and 75 mm wide. The flowers in a lax spreading panicle are small and are borne in a cymose arrangement of threes on short pedicels. The fruits are smaller than those of true cinnamon and are black, pulpy, aromatic, elliptic, drupes with a single seed. It is cultivated mainly in the provinces of Kwangsi and Kwangtung

Fig. 108 Cinnamon

in southern China. The spice named cassia should not be confused with the large genus *Cassia* in the Leguminosae or pea family from which senna leaves are obtained (Fig. 108b).

Active substances

Cinnamon bark contains up to 4 per cent essential oil, tannins, resins, mucilage, gum, sugars, two insecticidal compounds (cinnzelanin, cinnzelanol), and coumarin (Leung, 1980). Its leaf also contains essential oil (0.5–0.7 per cent). The main constituent of cinnamon bark oil is cinnamic aldehyde (60–75 per cent); it also contains eugenol (5–18 per cent), eugenol acetate, cinnamyl acetate, methyl eugenol, linalool, etc. The main component of cinnamon leaf oil, by contrast, is eugenol (60–80 per cent); other constituents are cinnamaldehyde, cinnamyl alcohol, cinnamyl acetate, etc. Cassia bark contains 1.0–2.5 per cent essential oil; the leaves and twigs also contain essential oil (0.2–0.7 per cent). The major constituent of the cassia oil (sometimes known as Chinese cinnamon oil) is cinnamaldehyde (70–95 per cent).

Uses

Both cinnamon and cassia are important baking spices today; in ground form they are used in cakes, breads, buns, cookies and pies. Stick cinnamon, an important ingredient in pickling, may be used to flavour stewed prunes, spiced peaches and beverages. The dried unripe fruits of cassia are called 'cassia buds'. Highly aromatic, they resemble little cloves, are popular in confections, and may be used to add cinnamon flavour to sweet pickles (Rosengarten, 1969). Cinnamon bark oil is used in small quantities in both the flavouring and perfumery industries. As a flavouring agent, its main uses are for meat seasonings, baked goods, apple pies, fast-food seasonings, sauces, pickles, confectionery, various beverages of the cola type, dental and pharmaceutical preparations and tobacco. It is also used in certain expensive perfumes, but its application in perfumes is limited by its skin-sensitizing properties (ITC/UNCTAD, 1986). Cinnamon leaf oil is mainly used for seasonings and savoury snacks, generally dispersed on a neutral base, and as a cheap fragrance in soaps and insecticides. Cassia oil is obtained by distillation of the leaves, twigs and the inferior bark; it is used in soft drinks of the cola type, although a little finds its way into liqueurs and sauces. The oil also has minor uses in perfumery. The essential oils are antiseptic, and ether-soluble fraction antioxidants. Cinnamaldehyde is an antipyretic, a hypothermic and a sedative. Cinnamon oil has antifungal, antiviral, bactericidal and larvicidal properties (Leung, 1980).

Cultivation

Cinnamon requires a warm and wet climate with average temperatures around 27 °C, and an annual rainfall between 2000 and 2540 mm. The best cinnamon grows in fine white quartz sand or in lateritic gravelly soil. Rocky and stony ground is unsuitable. The crop is usually grown without shade.

Cinnamon may be propagated by seeds or vegetatively. The seeds should be planted as soon as possible, as viability is quickly reduced. The highest rate of germination of 94 per cent was obtained by sowing on the third day after the seeds were harvested. On the fourteenth day after harvesting, germination was reduced to 52 per cent (Kannan and Balakrishnan, 1967). Seeds sown during the first week germinated within 20–25 days. The seeds may be sown in nurseries or in the field. The nursery beds should be of well-prepared rich, sandy soil free from roots and stones, with an adequate supply of fresh water nearby, and should be lightly shaded.

After about four months in the nursery, clumps of seedlings may be transplanted into baskets and transplanted into the field four or five months later. Cinnamon may be propagated vegetatively by cuttings, layering, or division of old rootstocks.

The usual spacing is approximately 2 m between stands. Further care consists mainly of weeding, which may be done three or four times per year during the first two years. Thereafter twice per year should be sufficient. The plants are-coppiced after two years, the stems being cut to within 0.10–0.15 m from the ground and covered with earth. This encourages the formation of shoots, of which 4–6 are allowed to grow for a further two years before harvesting. The stems, which are kept straight by pruning, are cut when they are 2–3 m high and 15–20 mm in diameter. After cutting, the plants are pruned of all unwanted or distorted shoots and stumps, and more soil is drawn up around them.

The stems are cut during the rains to facilitate peeling, at the time when the red flush of the young leaves is beginning to turn green and the sap is flowing freely. The cutter will test the stems to see if the bark lifts easily. In Sri Lanka, harvesting is done mainly in May–June and October–November, although harvesting on a limited scale continues throughout the year. The leaves and twigs are cut off and are used for mulcing, or the leaves may be retained for distillation. The cut shoots are bundled and taken to the plantation headquarters for peeling and preparation of the spice.

Samarawira (1964) gives the yield in Sri Lanka from the first crop after 3–4 years of 56–67 kg/ha of quills, subsequently increasing to 168–224 kg/ha of quills. Yields usually decline after ten years. About 63 kg/ha of chips and approximately 2.5 t/ha of undried leaves are also obtained.

Chinese cassia is grown on terraced hillsides at altitudes of 90–300 m. It is usually propagated by cuttings. Sometimes it is grown from seeds, the seedlings being raised in nursery beds and transplanted when 1–2 years old. The trees are first cut when they are 5–7 years old and 1.5–2.5 m tall, according to Brown (1956). The stems for bark production are 25–40 mm in diameter. They are cut a few centimetres above ground level and are stripped of their leaves and twigs, which may be used for distillation. The bitter-tasting epidermis is removed, and the bark is dried in the sun where it turns brown and curls into a hollow tube or quill. Harvesting normally occurs during April–May.

Common Juniper (*Juniperus communis* L.)

Description

Common juniper is found in dry, rocky soils in Europe and Asia, as well as in North America from the Arctic circle to Mexico. It is an evergreen, dioecious, 1–3-m-tall shrub belonging to the Cupressaceae family. On older branches the bark is brown and filamentous. The leaves are alternate, stand in trimerous whorls, are 10–20 mm long, acicular, and greyish or bright green. The flowers are small; the staminate flowers are catkin-like, and the pistillate flowers are ovate. It flowers in April–May. Its fruit is a galbulus, which is formed by the interlacing of three fleshy cone scales; for this reason it is a false berry. It remains green in the first year and ripens in the second year, in August to early September. The ripe fruits are almost globose, their diameter is 5–10 mm, their colour is brownish-black, and they are blue pruinose. The shell of the berry is thin and leathery; the inside of the berry is soft and porous, containing three ovate seeds, which are extremely hard and stone-like. Oil-holding cells are located on the surface of the seeds. The taste of the fruit is sweet, with a slightly bitter aftertaste. Its odour is similar to that of turpentine.

Active susbstances

The fruits (Juniperi fructus) contain 0.2–3.43 per cent essential oil, which is composed mainly of monoterpenes (about 58 per cent) which include α-pinene, myrcene and sabinene as the major components. It also contains limonene, *p*-cymene, γ-terpinene, camphene, α-thujone, etc. The ripe fruits also contain about 30 per cent invert sugar, 9 per cent resin, juniperine, juniperic acid and flavone glycosides. Fresh berries also contain vitamin C, which decomposes during drying and storage.

Uses

The berries are the principal flavouring agent in gin, and are also used in alcoholic bitters. The extracts and oils are used in most food categories, including alcoholic and non-alcoholic beverages, frozen dairy desserts, candy, baked goods, gelatins, puddings, meat and meat products. The berries, wood and oil are used in folk remedies for cancer, indurations, polyps, swellings, tumours and warts (Hartwell, 1970). In therapeutics, the berries are used as a diuretic, and as an ingredient of tea blends for stomachic, sudatory, and gallbladder stone- and kidney-stone dissolving effects. Marmalade is also prepared from juniper which is consumed for a similar purpose. Juniper berries are also used in households as spice when marinading meats.

The juniper essential oil (Aetheroleum juniperi) is made by the distillation of juniper berries. This essential oil is also used for other purposes. It is mainly used for scenting bath-salts, for inhaling in the case of bronchitis, and for relieving rheumatic pains of joints, as an embrocation. The juniper tar (Pix juniperi) is extracted by heating the woody parts of juniper. This is used in dermatology.

Gathering

The ripe fruits are gathered in the second year, from September to November. Canvas baskets are placed under the shrubs and the ripe fruits are knocked down carefully so as not to damage the shrubs. Unripe berries, leaf and branch parts are removed from the gathered fruits by winnower machines, and then the fruits are spread in 20–50 mm layers for drying. They become dry within 4–6 weeks. The drying ratio is 2:1. The dry drug is strongly hygroscopic. For this reason it should be stored only in a dry place in proper packaging.

Sometimes the wood shavings (Juniperi lignum) are also gathered; these have a spicy odour and mild spicy taste. Wood shavings are gathered in the early spring, and dried in a heated room; the drying ratio is 2:1.

Cultivation

Common juniper can be well cultivated; it especially lends itself to the exploitation of sandy soils. It is justifiable to establish juniper plantations. The seedlings of 2–3 years age, established from seeds in nursery beds, are transplanted in the autumn or in the spring with a row spacing of 1.0–1.5 m and plant distance of 0.5–0.6 m. The fruits can be gathered by the third to fourth year after planting.

Fig. 109 Rose hips

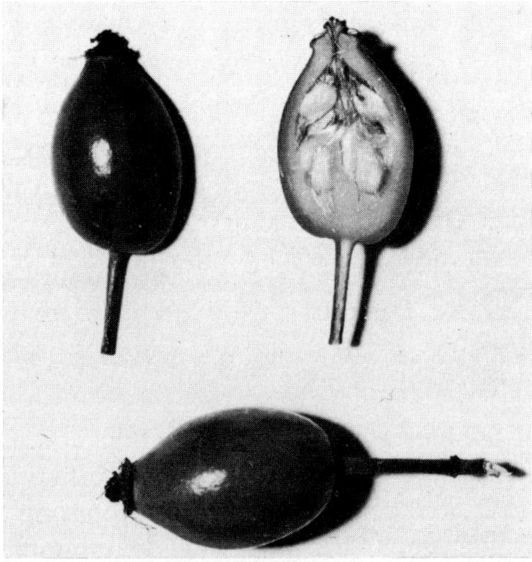

Dog Rose (*Rosa canina* L.)

Description

Dog rose (brier hip, hip rose) is a perennial shrub belonging to the Rosaceae family. It is very widely distributed, and can be found on dry, rocky, eroded hillsides, in dry oak forests and in shrubberies.

The stem is erect, up to 1–3 m high. The branchlets are mostly drooping and spiky. The odd-pinnate leaflets are 20–70 mm long, and are alternate. The flowers are solitary; they form a knot-like cyme. The petals are pink; flowers appear at the end of May and in June. Its well-known fruit is actually a spurious fruit; it is ovate or globose, and 10–20 mm long. Often a peduncle residue can be found at one end of the berry and a calyx residue at the other. In the initial period of ripening, the colour of the fleshy part of the berry is coral, and it is purple–red in the later period. It contains lots of stone-fruits embedded among hairs; these fruits are frequently mistaken for seeds (Fig. 109).

Active substances

The rose hips (Fig. 109) (Cynosbati fructus) contain 0.3–1.5 per cent vitamin C. As well as the vitamin C, they contain other water-soluble vitamins, mainly flavone derivatives, which complement the effect of vitamin C. Essential oil, pectin, sugar, fatty acids and other organic acids (malic acid and citric acid) can also be found in them. The seeds contain 7–8 per cent fatty acids.

Uses

Rose hip is an astringent, a carminative, a diuretic, and a tonic. It is particularly beneficial for the digestive system, and produces a diuretic effect without irritating the kidneys. It is also recommended for kidney or bladder inflammation. By eliminating uric acid accumulation, rose hip also helps in gouty and rheumatic complaints. A decoction of crushed achenes is also sometimes used for fever and as a beverage tea. Rose hips are a very important source of vitamin C, which can be preserved and stored for a long time.

Gathering

Rose hips are gathered when their colour is purple–red, but the flesh is still hard. The frost-bitten fruits are not worth gathering, because the flesh is made soft by the action cold, and cannot be used for preparing drug. Unripe and blackened berries have no value at all. After gathering, the hips are split into two and the seeds are removed, and cleaned of hairs, and then the remaining flesh is dried.

For the production of large amounts of fruit-flesh, machines are used nowadays, which perform the operation of removing seeds as well. After the fruits are dried without previously cutting them, they are broken up and the seeds and trichomes are shaken out in rotating cylinders which are provided with lattice walls. In order to prevent the loss of vitamin C, the most valuable active substance of dog rose, the hips should be dried quickly, as the vitamin C might decompose through the action of enzymes in the fruit. The fast decomposition of vitamin C starts when the cells die, and lasts until a sufficient amount of humidity is present to maintain the action of enzymes. The intrinsic vitamin C of the split hips can be retained only in driers, which operate with fast-flowing hot air at 80–90 °C. If the hip contacts the metal parts during processing or handling, the vitamin C might start decomposing. Heavy metals, especially iron and copper, act as catalysts for the enzymes which decompose vitamin C. Hips preserved by deep-freezing will retain their original vitamin C content. The drying ratio of the fruit is 3 : 1.

The drug of suitable quality has a vivid red colour, has no peduncle, has a shining surface, and its humidity content is 12 per cent at most. Even the high-quality drug will become brown and valueless if stored for a prolonged time. For this reason, storage for more than one year is not advisable.

Cultivation

The cultivation of dog rose is overdue, owing to the lack of a manual labour requirement, to the need to protect and exploit lands exposed to erosion, and to the regular need for it.

Dog rose can be propagated from the seeds of fruits gathered at the end of August or early in September. The seeds are sown within a short time after gathering, or are placed into wet sand in layers, or put into open ground where they are exposed to frost, and sown only in the spring. The seeds put aside without layering can be retained for years. The pretreated seeds are sown with a row distance of 0.4–0.6 m. From the seeds germinating in the spring, seedlings of pencil thickness will develop until the autumn, and can be transplanted to the field. In the case of closed planting, the advisable row distance is 5–6 m. Efforts should be made to have the plants closed together as soon as possible. In this way the producing surface of the plants will be towards the row spaces. The closing of the rows can be achieved most effectively if planting is done with a 0.5 m plant distance. Prior to planting, the area might be treated with 5–6 kg/ha of Aktikon herbicide. The herbicide treatment of older plants can be performed with 10–12 kg/ha of Aktikon. The herbicides can be spread to the row spacing and under the foliage in the rows in the spring when the vegetation starts.

The expected fruit yield is 8–10 t/ha of fresh hips.

European Elder (*Sambucus nigra* L.)

Description

European elder or black elder is shrub or small tree belonging to the Caprifoliaceae family, and can be found along the borders of forests in shady, humid places all over Europe.

One-year-old branches are green, and older branches are covered with grey or greyish-brown bark. On the surface of the branches, protruding lenticels can be seen. The branches are white and soft inside. The leaves are compound, consisting of 5–7 leaflets, odd-pinnate, opposite, large and up to 400 mm long. The leaflets are orbiculate, the abaxial surface being light green and shiny. The inflorescence is a dense, flat dichasium 100–250 mm in diameter. The flowers have a characteristic rank odour, and are yellowish-white. The flowers will become dark after drying and their odour will also change. Flowering takes place from mid-May to mid-June. The fruit is a glossy black, thin-shelled, sticky drupelet.

Active substances

The flower contains 0.1–0.2 per cent essential oil, glycosides (sambunigrin, rutin), other flavonids, tannins, sugars and vitamins (A–C). The fruit is an excellent source of vitamin C, and contains pigments. The strong sambunigrin glycoside can also be found in the fruits, leaves and bark. The leaves contain 3.5 per cent rutin, and the bark contains baldrianic acid. Young shoots contain iridoid morroniside (Leung, 1980).

Uses

The bark and root are diuretic, emetic and purgative. The leaves and shoots are diuretic. The flowers are diaphoretic. The fruit is aperient. A tea made from the leaves and young shoots increases the production of urine and helps to eliminate excess water from the body. The tea of the flowers promotes perspiration and is used particularly for colds and for rheumatic complaints. The berries can be made into a jam which is mildly laxative and suitable for irritated or inflamed intestines and for small children.

Gathering

The most frequently gathered part of the European elder is the inflorescence. The ripe berries and sometimes the bark and the leaves are also gathered.

The gathering of flowers can be started when the extreme flowers of the dichasium are open, but the central flowers are still in bud. The flowers may be gathered only in dry weather. Dewy flowers or flowers gathered just after rain will become brown during drying. Fresh flowers should not be stuffed into sacks or cases, because the ruptured flowers will go red or brown in spite of careful drying. The best practice is to gather the flowers loosely in baskets, then to place the flowers in one layer with their peduncles upwards on drying frames or on paper-covered boards. In the case of artificial drying, the temperature may be 35–40 °C maximum. The best appearence of the drug can be achieved with predrying by the sun or in an artificial drier. The dried flowers should be rubbed away from the peduncles by hand or with machines. Stem parts and dust can be removed from the leaves with the help of screening. Plastic bags or wooden cases lined with paper are the best means of packaging. To produce 1 kg of dry drug, 5–6 kg of fresh flowers are necessary.

Ripe fruits are gathered together with the peduncles, similarly to the gathering of flowers. Drying is also similar. Drying should be done with the peduncles not removed, because in this way the juice of the berries will not flow out, drying will be uniform, and berries will not stick together. After drying, the 'berries' should be removed by rubbing them from the peduncles, as with the flowers. The drying ratio is 3 : 1.

The dried flower (Sambuci flos) has a characteristic, strong, aromatic odour. The taste is sweety, slightly hot. The well-treated drug has a light butter colour. The dried fruit (Sambuci fructus) is black, withered to a soft consistency, and sweet–sour in taste.

Cultivation

Propagation is generally done by seed sowing. The seeds gathered in August–September are placed in layers, and then are sown in a nursery in the autumn or in the spring. The seeds, which have a mass of 2.5–3.5 g per 1000, will germinate relatively quickly, and the initial development of the plants is also fast. The one-year-old seedlings can be transplanted to the field either in the autumn or in the spring.

Propagation can be done vegetatively in July by layering the thin, green shoot, or in September by layering lignified, thin branches. Plantations are established by placing the seedlings or rooty cuttings with a 2.5–3 m row distance and 1.0–1.5 m plant distance.

Fenugreek (*Trigonella foenum-graecum* L.)

Description

Fenugreek is an erect, annual, herbaceous plant belonging to the Leguminosae family. It is native to southern Europe and Asia, and is undoubtedly one of the oldest cultivated medicinal plants. Fenugreek is widely grown today in the Mediterranean countries, Argentina, France, India, North Africa and the USA as a food, condiment, medicinal, dye and forage plant.

The plant reaches a height of 0.4–0.8 m. The leaves are pinnately trifoliate and light green, and the leaflets are obovate. The flowers are lemon-yellow or white, and small and papilionaceous, and the calyx tube is campanulate. The cylindrical peaked pod is 100–150 mm long and each pod contains about 10–20 small seeds. The seeds are 2.5–6 mm long, oblong or square in outline, hard, and yellow, olive green or yellowish-brown to dark brown in colour. The odour of the seed, especially if powdered, is spicy, strong and characteristic.

Active substances

The main active substance of the seeds is diosgenin, a steroid sapogenin, which is the starting compound for over 60 per cent of the total steroid production by the pharmaceutical industry. Other sapogenins found in fenugreek seed include yamogenin, gitogenin, tigogenin and neotigogens. According to Fazli and Hardman (1968), the sapogenin content calculated as diosgenin in fenugreek seed varies from 0.8 to 2.2 per cent, expressed on moisture-free basis. Other constituents of fenugreek include mucilage (50 per cent), bitter, fixed oil (7 per cent), alkaloids (choline and trigonelline), yellow colouring matter and essential oil, the amount of which is very small, less than 0.02 per cent.

Uses

The maple aroma and flavour of fenugreek has led to its use in many baked goods, chutneys, confections and imitation maple syrup. It is an important constituent of curry powder. Young seedlings and other parts of the fresh plant material are eaten as a vegetable. Fenugreek is also used as livestock feed.

As a medicinal plant, it has traditionally been considered to be a carminative, demulcent, expectorant, laxative and stomachic. It has also been employed against bronchitis, fevers, sore throats, wounds, diabetes and ulcers, and in the treatment of cancer (Hartwell, 1970). Fenugreek has been used to promote lactation and as an aphrodisiac.

Cultivation

The reported life zone of fenugreek is 8–27 °C with an annual precipitation of 0.4–1.5 m. It thrives in full sun and rich, well-drained soils. As a leguminous plant, it needs little if any nitrogen fertilizer, and the plant can enrich soils with nitrogen.

Fenugreek is cultivated commercially by seeds. The seeds are sown at the end of March or at the beginning of April. The seeds are sown in rows with a spacing of 0.3–0.5 m, and in a quantity of 20–25 kg/ha. Seed germination takes place 7–10 days after sowing. The plants flower about 30–37 days after sowing and the duration of the flowering phase is 7–18 days. The pods mature within 60–90 days after sowing. After the seeds mature, the entire aerial parts are uprooted and dried, so that the seeds can then be threshed, winnowed, dried once more and stored. One-run harvesting with a combined harvester is acceptable if most of the crops have already ripened, in the case of large-scale cultivation. The seed yields vary from 600 kg to 1 or 2 t/ha, depending on the growth conditions and methods of cultivation.

Ginger (*Zingiber officinale* Rosc.)

Ginger is one of the earliest oriental spices known to Europe and is still in large demand today.

Description

Ginger is a herbaceous perennial, usually grown as an annual plant, belonging to the Zingiberaceae family. It originated in South East Asia, but is nowhere known in a wild state. The principal countries producing ginger are India (the world's largest producer), China, Jamaica, Sierra Leone, Nigeria, Australia and Thailand. Ginger is a slender herb, 0.3–1.0 m high, with a robust branched rhizome borne horizontally near the surface of the soil, bearing leafy shoots close together. The fleshy sympodial rhizome is hard and thick, often palmately branched, and usually pale yellow within. The leafy shoots are annual, erect, and 5 mm in diameter. They are formed from the long leaf sheets and are glabrous except for short hairs near the base of each leaf blade. The aerial pseudostems usually bear 8–12 distichous leaves (Purseglove *et al.*, 1981). The inflorescence arises directly from the rootstock and is spicate, and 150–250 mm long. The bracts are appressed, ovate or elliptic, and are green with a pale submarginal band and incurved translucent margins. The fruit, which is very seldom produced, is a thin-walled, three-valved capsule, with small, black arillate seeds (Fig. 110).

Active substances

The rhizome contains pungent substances (gingerol, shogaol and zingerone), and 1.0–4.0 per cent of a pale yellow, spicy, essential oil. The constituents of the essential oil are sesquiterpene hydrocarbons (α-zingiberene, arcurcumene, etc.), sesquiterpene

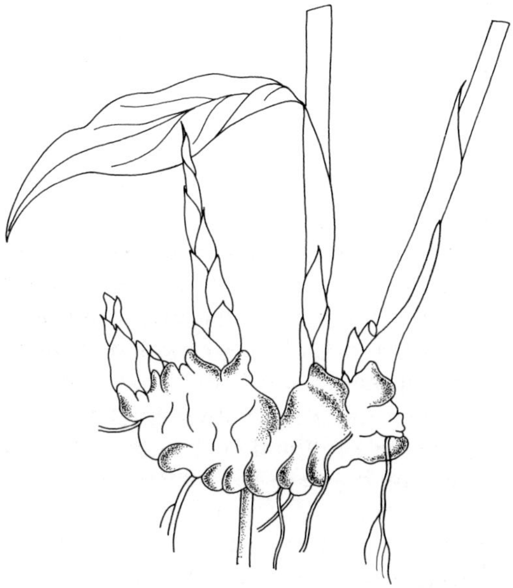

Fig. 110 Ginger

alcohols (zingiberol, zingiberenol, etc.), monoterpene hydorcarbons (*d*-camphene, *p*-cymene, myrcene, etc.), oxygenated monoterpenes (*d*-borneol, bornyl acetate, geraniol, etc.) and miscellaneous compounds (acetaldehyde, methyl acetate, etc.). Other components of the ginger rhizome are fixed oil, resin, proteins, cellulose, pentosan, starch, and mineral elements (Purseglove *et al.* 1981).

Uses

Ginger is widely used for culinary purposes in gingerbread, biscuits, cakes, puddings, soups and pickles. It is a frequent constituent of curry powder. It is one of the most widely used spices in Chinese cookery. It is used in the production of ginger beer, ginger ale and ginger wine. Ginger oil is used largely as a flavouring material in various alcoholic and non-alcoholic beverages. It is also used in confectionery and perfumes, and is occasionally admixed with ginger oleoresin to restore an ideal balance between aroma and pungency in the oleoresin. Ginger tea or tincture, taken hot, promotes cleansing of the system through sweating, and is also said to be useful for suppressed menstruation. It is used in medicine as a carminative and aromatic stimulant to the gastrointestinal tract and externally as a rubefacient and counter-irritant. It has a reputation as an aphrodisiac.

Cultivation

Ginger is mainly cultivated in the tropics from sea level to 1500 m. It requires a consistently warm and moist climate, brilliant sunshine and heavy rainfall. In Kerala the rainfall is about 3.0 m/year on average, distributed over 8–10 months during the ginger-growing season. It thrives best on medium loams with a good supply of humus. It is very sensitive to waterlogging. Ginger is a soil-exhausting crop, requiring heavy fertilization. Whenever possible, it is grown in rotation with other crops.

Fertile seed is rarely produced, and ginger is propagated commercially only by division of the rhizome. Small pieces of selected rhizomes, 25–50 mm long, are planted 40–50 mm below the soil surface on ridges at intervals of 0.2–0.3 m. The crop is very conducive to weed growth except when mulched, and adequate weed control is essential during the early stages. Weeds can be controlled chemically or mechanically. Kasasian (1971) states that no firm recommendations have been made for the use of herbicides in this crop, but ginger appears to be highly tolerant to simazine applied as a pre-emergent. Most of the non-chemical control of weeds is carried out manually, as tractor-mounted cultivating implements cause damage to the crop.

The first shoots appear above ground 10–15 days after planting and this continues for a period of about four weeks. In areas of low rainfall, irrigation is necessary. Harvesting begins 9–12 months after planting, when the stalks begin to wither. The rhizomes are twisted and dug up carefully with a hoe, spade or digging fork. These 'hands' are then cleaned, washed, scraped, boiled, peeled carefully with specially designed knives, and dried in the sun for about eight days.

A yield of 1.5–2.0 t/ha of dried ginger may be obtained under favourable conditions. The dried ginger constitutes about 25 per cent of the raw rhizome's weight.

Ginseng (*Panax* species)

Ginseng is the most well known of all Chinese medicinal products. Korean and Chinese ginseng *(Panax ginseng)*, American ginseng *(P. quinquefolium)* and Siberian ginseng *(Eleutherococus senticosus)* are the three species of major importance in the preparation of herbal products. The word 'ginseng' is derived from the Chinese words 'shen seng', meaning 'man-root' as its roots have a resemblance to the human body (Soon-Keun, 1982).

Description

The *Panax* species are perennial herbs belonging to the Araliaceae family; they bloom and bear berries three years after sowing. Yearlings are about 90–150 mm tall, and bear a palmately compound leaf with five leaflets. Two-year-old plants have two such leaves while three-year-old plants have three. One palmately compound leaf is added each year (Fig. 111).

Panax ginseng C. A. Meyer (syn. *P. schinseng* Nees; *Aralia ginseng* P. and P.) is indigenous to northern China and Korea and is cultivated in China, Japan, Korea and the USSR. Korean ginseng is 0.6–0.8 m high, with a simple, erect stem that is deep red and bears compound, whorled, digitate leaves and pink, polygamous flowes. The fruit is a small, red berry containing one or two seeds. The roots are bifurcate, and the roots and rhizomes vary from 80 to 200 mm in length and from 5 to 20 mm in

Fig. 111 Ginseng

diameter; they are greyish-white to amber-yellow with a wrinkled and furrowed surface, resembling a small parsnip but with more protruding roots. Ginseng has contractile roots that ensure the correct positioning of the regeneration bud of the plant. The roots are crowned with a vertical rhizome which increases in length at the upper end and bears the regeneration bud at its apex. The roots shrink as the buds rise above soil level.

Panax quinquefolium L. is indigenous to the woods of Quebec and Manitoba and extends southwards to the Gulf Coast. American ginseng grows to about 0.3 m in height. The flowers are greenish-yellow, and borne in clusters in midsummer, and the fruit is a red berry containing 1–3 wrinkled seeds. The mature rhizome varies from 50 to 100 mm in length, and reaches up to 25 mm in diameter.

Siberian ginseng, *Eleutherococcus senticosus* M. (syn. *Acanthopanax senticosus* Harms.) is a common Far Eastern forest shrub. It is found in eastern Siberia, northern China and Korea.

Active substances

The pharmacological properties of ginseng are believed to be due to its saponins (triterpene glycosides), although the other constituents, including the essential oils such as panacene, non-essential oils such as panaxynol, organic acids, ginseng pectin, choline, amino acids and peptides, vitamin B_{12}, and minerals and trace elements, may also be responsible. There are about twelve primary glycosides, known as ginsenosides or panaxosides, in Korean ginseng, and slightly fewer in American ginseng. Partial hydrolysis yields at least eighteen different glycosides. The major aglycones are dammarane-type triterpenes, and acid hydrolysis yields panaxadoil and panaxatriol. There are also traces of oleanic acid derivatives and ocotillone-type ginsenosides (ITC UNCTAD/GATT, 1982).

Uses

Ginseng is held in high esteem because of its use as a medicinal plant. The genus name *Panax* is derived from the Greek word meaning 'panacea' or 'all-healing', the species name *ginseng* is said to mean 'wonder of the world'. Both terms refer to the medicinal virtues of the plant. In Chinese medicine, ginseng is prescribed as a tonic, a stimulant and an aphrodisiac. It is used in cases of neurasthenia, dyspepsia, palpitations and asthma, and is incorporated into tonics for amnesia, headaches, convulsions, dysentery and cancer. It has been claimed that ginseng enhances the natural resistance and recuperative powers of the body, having both a stimulant and sedative activity. It is called an adaptogen, because it is claimed to have the ability to maintain the body's stamina at a regular level without detrimental side-effects. Ginseng is marketed as capsules, tablets, as an instant tea and in wines by the herbal and health food industries, and is used in lotions, creams and perfumes by the cosmetics industry.

Ginseng root is marketed in two forms in world trade: white ginseng (dried and raw), and red ginseng (boiled or steamed and then redried).

Cultivation

Ginseng is not an easy crop to cultivate. The main producing countries of ginseng are Korea, China, Japan, the USA, Canada, and the USSR. Korea is the largest producer in the world.

The reported life zone of *Panax* species is 9–15 °C with an annual precipitation of 0.7–1.3 m. The plant thrives in shady locations on rich, moist, well-drained forest soils. For cultivation, plants must be grown in open woods or under laths or other structures that provide 75 per cent shade, and where large quantities of forest litter, leaf mould or other compost materials have been incorporated into the soil.

Artificial fertilizer is not used, but ginseng leaves are used as a mulch.

Korean ginseng is generally grown from seed. The larger seeds are selected from mature plants and the seeds are planted immediately; they must not be allowed to dry out. The embryo is not formed at the time of ripening, and germination of the seeds under natural conditions takes 18—22 months. Seeds stratified immediately have sprouted after eight months. However, treatment with gibberellic acid (0.05 or 0.1 per cent) has reduced the time needed for germination and increased the viability to 90–100 per cent. Winter mulches are necessary to protect the seedling from cold weather. Korean ginseng is prone to fungal attack and the seeds are treated with fungicides. The roots of seedlings—1–2 years old—are also washed with appropriate fungicides when they are transplanted to their permanent location. The plants may be sprayed, e.g. with Bordeaux mixture, during each growing season.

The roots are generally harvested after 5–6 years' growth in September and October. The roots are cleaned and then dried in the sun. The yield is 2.0–5 t/ha of dried root (Soon-Keun, 1982).

American ginseng is also grown from seed. According to Hemmerly (1977), the seeds harvested at maturity in July and August may be stratified in damp sand (125 mm deep) for eighteen months in a cool basement and then planted out in the autumn or early winter into raised beds shaded by wooden laths. The seeds are sown 25 mm deep with a density of 150–160 kg/ha, and they germinate the following spring. The crop is weeded by hand and sprayed frequently with insecticides; the leaves are used as a mulch in the autumn. The roots may be harvested in the autumn after four years' growth, using a potato digger. Drying takes about six weeks in the shade, but the roots may also be dried in a specially designed oven at 32–35 °C for three weeks. Care must be taken not to mark the roots and to keep them intact. The dried roots should be stored in a dry, airy and rodent-proof place until sold. Cultivated American ginseng is reported to yield, on average, 1.7–2.5 t/ha of dried root.

Hawthorn *(Crataegus oxyacantha)*

Description

Hawthorn (English hawthorn) is a shrub or small tree belonging to the Rosaceae family. It grows mainly on hilly, stony areas of continental Europe and in England. Frequently it is widely grown as a hedge plant. Hawthorn is 2–4 m high. The young branches are reddish-brown, and shiny, and have 0.5–1.5-cm-long thorns. The leaves are alternate, petioled, and ovate or rhomboid shaped. The inflorescence is a dense, umbellate raceme. The petals are white or slightly pinkish; the flowers have two stigmas each. Flowering is in May–June. The fruit is a pome which is almost globose, red, and has thin flesh and a floury taste when ripe. It contains two drupes.

Active substances

The leaves, flowers and fruits all contain cardioactive compounds. The leaves contain flavonoids, triterpenoid acids and calcium oxalate. The flowers contain 0.15–0.17 per cent essential oil, quercitin, quercitrin and trimethylamine. The fruits contain citric acid, crataegus acid, and vitamin C. Hawthorn marmalade (made from berries) has up to 0.15 per cent vitamin C. Its main active substance is vitexin rhamnoside.

Uses

Hawthorn is reported to be astringent, cardiotonic, cyanogenetic, depurative, diuretic, fumitory, stomachic and tonic; it is a folk remedy for arteriosclerosis, dropsy, dyspnea, hypertension, nephrosis and sore throat (Grieve, 1974; Duke and Wain,

1981; Tyler, 1982). According to Tyler (1982), it dilates the blood vessels, especially the coronary vessels, reducing peripheral resistance and thus lowering blood pressure. It is thought to reduce the likelihood of angina attack. It apparently has a direct, favourable effect on the heart itself, which is especially noticeable in cases of heart damage.

Gathering

Generally, the flowering branch ends (Crataegi summitas) and mature fruits (Crataegi fructus) are gathered. The flowering ends of branches are cut with scissors, and dried in the shade. In order to avoid the flowers falling off, the flowering ends should not be turned over during drying. The drying ratio is 4–6 : 1. The mature fruits are generally gathered in July–August, and dried in the shade. The drying ratio of the fruits is 3 : 1.

Hollyhock (*Alcaea rosea* (L.) Cav.)

Description

Hollyhock is an ornamental and medicinal plant cultivated all over Europe since the sixteenth century. It is an annual or biennial herbaceous plant belonging to the Malvaceae family. It is indigenous to China and southern Europe.

Hollyhock has a deeply penetrating taproot with many adventitious roots. The stem is 1.5–1.8 m high, and branching. The basal leaves are orbiculate, and entire-edged; the stem leaves are long-petioled, and mostly lobed. Both sides of the leaf blades are covered with thick rough hairs. The flowers are large, full or half full, black and 70–80 mm in diameter. The fruit is flat, and splits into sections. The seed is reniform, and greyish-brown. The mass of 1000 seeds is 8–12 g (Fig. 112).

Active substances

The dried flowers (Malvae arboreae flos) contain pigments (althaein, delphinine, malvidin) belonging to the anthocyanins, as well as mucilage materials and tannins.

Uses

It is demulcent, diuretic, and emollient. It is used in therapeutics and a throat wash against diseases of the respiratory and digestive organs. The drug is used to colour foods, drinks and cakes. Hollyhock flowers are used as an emollient ingredient in various cosmetics.

Fig. 112 Hollyhock

Cultivation

Hollyhock likes sunny places which are protected from the wind. It requires no special conditions from the soil. The most appropriate soils are deep–layered, medium–firm, rich in nutrients, and have good water management capacity. Hollyhock can be propagated both by sowing in the field and by seedling growth in open seed-beds. The disadvantage of the reliable and inexpensive method of sowing in the field is that flowering starts later, only at the end of July. Sowing in the field shall be done as early as possible—at the beginning of March—or the plants will remain in the rosette state for the winter. Sowing should be done in rows 0.6–0.7 m apart, with seeds 8–10 m using 3 kg/ha of sowing seeds. The optimum depth of sowing is 30–35 mm. The too-dense sprouting may be thinned when the plants have two or three leaves with the help of a harrow or cultivator moved perpendicularly to the rows. In order to keep the field free of weeds, interrow cultivation and hoeing the rows two or three times is sufficient.

The flowers should be gathered in dry weather after the dew has evaporated. The flowers are mostly picked without sepals (sine calycibus) after flowering, when the petals close together. Sometimes the flowers are picked together with sepals (cum calycibus) at the time of full bloom. Gathering is done continuously, according to the rate of flowering. The flowers gathered in baskets should be immediately dried. Drying can be done with the flowers spread in a well-ventilated location or in an artificial drier at 40–50 °C.

The expected flower drug yield is 250–400 kg/ha.

Horehound (*Marrubium vulgare* L.)

Description

Horehound is a spreading perennial herb belonging to the Lamiaceae family. It is native to central and western Asia, southern Europe and northern Africa. Horehound is produced commercially in France and several other European countries.

It is also known as common horehound, white horehound and hoarhound. The plant reaches a height of almost 0.7 m. A fibrous, spindle-shaped rootstock sends up numerous bushy, square, downy stems. It is characterized by white, pubescent leaves, woolly stems, and continually blooming white flowers. The leaves are opposite, petioled, usually wrinkled, roundish–ovate, and rough on top; small, white, two-lipped flowers feature a spiny calyx and grow in axillary whorls from June to September. The seeds are greyish-brown. The mass of 1000 seeds is 0.8–1.0 g.

Active substances

The dried shoots (Marrubii herba) contain bitter principle, marrubiin, tannins, resins and waxes; an essential oil containing monoterpenes, and a sesquiterpene, have been isolated from the plant. The chief constituent of horehound is marrubiin. The name 'marrubium' refers to the bitter qualities of the herb, and 'hoar' refers to the white pubescence covering the plant.

Uses

The leaves and stems of horehound are often boiled and used in the preparation of candied products, cough drops and syrups. Extracts of horehound are used in bitters and liqueurs. It has traditionally been used against asthma, coughs, colds, bronchitis, sore throats and skin irritations. The plant has also been used as a diaphoretic, diuretic, expectorant, laxative, stimulant, stomachic, tonic and vermifuge. According to Hartwell (1969), horehound has been used in the treatment of tumours. The essential oil is a carminative and expectorant, while the marrubiin has gastric activity. Consumption of large quantities of horehound can induce diarrhoea and nausea (Spoerke, 1980).

Cultivation

Although still collected from the wild, horehound is primarily grown in cultivated plantations for commercial uses. The reported life zone of horehound is 7–24 °C with an annual precipitation of 0.3–1.3 m and a soil pH of 4.5–8.3. The hardy plant grows in full sun on poor, dry calcareous soils with good drainage. Propagation can be done by plant division, or by seed sowing. The seedlings prepared by plant division should be transplanted to the field in the autumn. For the purpose of seedling nursing, the seeds should be sown in open-air seed-beds with a row distance of 0.20–0.30 m in April. The depth of sowing is 20–25 mm. The seedlings should be transplanted to the field in August or September with irrigation. They are planted with a 0.4–0.6 m row distance and a 0.3–0.4 m plant spacing.

The crop is harvested and dried just prior to open bloom. Cultivated stands, which generally last for 4–5 years, can be cut two or three times each year. Harvesting should be done by hand or with mower–loader machines. The expected yield is 2–3 t/ha of dry drug.

Liquorice (*Glycyrrhiza glabra* L.)

Description

Liquorice or licorice, is a perennial plant belonging to the Leguminosae family. Liquorice consists of the dried, peeled and unpeeled roots and stolons of varieties of *Glycyrrhiza glabra* L. and other *Glycyrrhiza* species. Such varieties include *G. glabra* var. *typica* Regel et Herd. (Spanish liquorice), *G. glabra* var. *glandulifera* Wald et Kit. (Russian liquorice), *G. glabra* var. *beta-violacea* (Persian liquorice) and *G. uralensis* Fisch. (Chinese liquorice). *G. glabra* is indigenous to the Mediterranean region, and to central through southwestern Asia, and to parts of the USSR and China. Modern plantation growing and harvesting of liquorice has been successful in the USSR, in California (USA) and in Yorkshire (UK). Commercial production takes place in several European countries, the USA, the USSR and China, and the chief suppliers to world markets are China, the USSR, Iran and Iraq.

Liquorice is a shrub, 0.5–1.5 m in height, with compound leaves and bearing yellow, blue or violet flowers in ancillary clusters. The fruits (seedpods) are 75–100 mm long, and contain 3–5 seeds. The seeds are round, flat, and shining chestnut-brown in colour. The mass of 1000 seeds is 10 g.

The plant develops a taproot and an extensive root system; the stolons from one plant may extend as much as 7.0 m in all directions.

Active substances

Liquorice contains about 7 per cent of the triterpenoid glycoside, glycyrrhizin; this consists of the calcium and potassium salts of glycyrrhizinic acid, the diglucopyranosiduronic acid of the triterpenoid glycyrrhetic acid (synonymous with glycyrrhetinic acid). Glycyrrhizin is the principal flavouring constituent and its content varies according to the variety of liquorice, area of origin and season. Spanish liquorice contains 5.9–10.6 per cent glycyrrhizin; Russian contains 9.9 per cent; Persian contains 7.4–13.2 per cent and Chinese contains about 7 per cent. Other constituents include other triterpenoid acids and flavonoid glycosides, including isoliquiritin, which imparts a yellow colour to the peeled drug, glucose (up to 4 per cent), sucrose (2.4–6.5 per cent), bitters, resins, mannite, asparagine (2–4 per cent), fat and starch.

Uses

The extract is extensively used as a demulcent and expectorant in cough remedies. It is also used in tonics, laxatives, anti-smoking preparations, and various other proprietary medicines. Liquorice extract is used in the treatment of Addison's disease. It is used as a flavouring agent in galenicals and other drugs to mask the unpleasant taste of other ingredients (e.g. in cascara and quinine preparations), and it is used in medicinal teas, again chiefly as a taste improver. Powdered liquorice is used as an absorbent pill excipient and is also taken as snuff.

Carbenoxolone sodium, prepared from glycyrrhetic acid, is used as an anti-inflammatory agent in the treatment of gastric ulcers and to accelerate healing, and it is also claimed to help cases of duodenal ulcer (Larkworthy, 1977).

The chief use of liquorice as a flavouring agent is in the tobacco industry. Ammoniated glycyrrhizin is reported to be 50 times as sweet as sugar and is used as both a sweetener and a flavouring agent is low-calorie foods. It is also used with sucrose, with which it has a synergistic effect, thereby increasing the sweetness to 100 times that of sucrose alone. The liquorice flavour of glycyrrhizin tends to come through in more delicate flavours but is masked by chocolate, root beer, maple, caramel and other 'heavy' flavours which it tends to intensify. Other uses are in icings, baked goods, fountain syrups and dietetic foods, and as a stabilizer in beverages. When used in confectionery, it is often mixed with anise oil.

Cultivation

Liquorice is found in more temperate regions, and is an important crop in the steppe lands and around the shores of the Caspian Sea, where winters are severe. The reported life zone of liquorice is 6–25 °C, with an annual precipitation of 0.3–1.1 m and a soil pH of 5.5–8.2. The soil for liquorice should be deep, mellow, moist, rich and free from stones and generally heavily manured.

Although the plant can be grown from seed, propagation is usually effected from stolons, which are cut into short lengths, each piece with two or three buds. It has been suggested that when starting cultivation in a new area, a good plan is to work up a stock of plants from seed and to use these as a source of cuttings for the main plantings. Planting commonly takes place in the early spring. Division of the crown or rooted cuttings of the underground stems are planted 0.45–0.6 m apart in rows, with 0.9 m between the rows. The plants should be irrigated until they are established. The established plants need little care except for an occasional thinning. Weeds should be kept down during the growing period and the shoots should be cut back to the ground before each winter.

The rhizomes are ready for harvesting after 3–5 years and hence it is desirable to grow cash crops such as carrots or potatoes between the rows during the first 2–3 years. The roots and stolons are lifted in the autumn as soon as the rains have commenced. Lifting is a laborious operation, as it entails the digging of trenches about one metre deep along the sides of the rows so that the soil is loosened enough to release the whole root system. After the small branches are trimmed off, the long, straight roots are washed and cut to lengths of 0.9–1.5 m and then slowly dried under cover for some months. If digging is exhaustive, yields can be as high as 50 t/ha of fresh root.

The root system is so dense that even after careful harvesting sufficient root remains in the ground to regenerate the plant, regeneration taking 2–5 years to provide enough growth for economic harvesting.

If liquorice is cultivated commercially, it is important that the flowering shoots be removed as soon as they appear, as the roots will be of inferior quality if the seed is allowed to ripen.

Linden Tree (*Tilia* species)

Description

About thirty species of genus *Tilia,* belonging to the Tiliaceae family, live in the northern temperate regions.

Linden trees or lime trees are mostly deciduous large trees. Their shoots are frequently stellate and hairy. The leaves are simple, entire, long-petioled, mostly asymmetric, alternate, saw-toothed along the edges, and cordate or oblique at the base. The flowers are small, yellowish or greenish-white, mostly pendent, and borne as pseudo-umbels in the axils of leaves. The peduncle grows in the green, tongue like, tunicate bracts. The flower is bisexual, fragrant, five-parted, and has many stamens. The fruit is a small, globose or elongated capsule containing one or two seeds.

The small-leaved linden—*Tilia cordata* Mill. syn. *T. parvifolia* Ehrh.—grows as high as 20–25 m, and has elongated large foliage. The slightly asymmetric leaves are dark green above and bluish or greyish-green underneath. Both sides of the leaves are bare, except in the vein-axils on the underside, where tufts of rusty hairs can be found. The tertiary veins do not protrude, or only protrude very slightly. It is a native of Europe. It likes deep-layered soil which is not too dry. Linden has a significant ability for adaptation; it tolerates shade, dryness and cold. It produces abundant flowers in June–July, and is the best honey producer among the lindens.

The large-leaved linden—*Tilia platyphyllos* Scop. syn. *T. grandifolia* Ehrh.—is 30–40 m high, and has symmetrical cone-shaped or wide egg-shaped foliage. Both sides of the leaves are vivid green. The leaves are hairy; yellow hairs can be found in the vein-axils of the underside. Tertiary veins protrude on the underside. The flowers bloom early, at the beginning of June. It is a native of central and southern Europe and Asia Minor. It requires soils with a high nutrient content, and does not tolerate

Fig. 113 Linden flower

dry locations; however, it likes warmth and shade, and is sensitive to late frosts. It grows relatively quickly.

The silver linden—*Tilia argentea* Desf. syn. *T. tomentosa* Moench.—is a tree 20–30 m high, with regular cone-like foliage. The leaf blade is dark green above, and the underside is uniformly covered with dense, stellate, silver–grey hair. It is a fast-growing species. Among the three linden species, this is the one which flowers latest (Fig. 113).

Active substances

The dried flowers (Tiliae flos) contain mucilage, tannin, quercetin-3-gluco-7-rhamnoside, kaempferol-3-gluco-7-rhamnoside, quercetin rhamnoxyloside, isoquercitrin, quercitrin, afzelin, tiliroside, chlorogenic acid and caffeic acid. It contains 0.05–0.1 per cent essential oil consisting of eucosane, eugenol, farnesol, farnesyl acetete, geraniol, linalool, and other compounds. The flowers also contain the free amino acids cysteine, cystine, phenylalanine, etc.

Uses

The oldest and most widely used herbal tea base material is the flower of linden. Authorities generally agree that linden flower tea is both a pleasant-tasting beverage and a useful diaphoretic (Tyler, 1982). It is reported to be a diuretic, a haemostat, a nervine, a sedative, a spasmolytic, a stomachic and tranquilizer. It is also used in cosmetics and for the preparation of mouth washes and bath lotions. Honey from the flowers is one of the world's best flavoured, and is used almost exclusively in liqueurs and medicines.

Gathering

The flowers are gathered together with bracts at the time of full bloom, and prior to the blooming of the central flowers. Sometimes the flowers are also gathered without bracts. Care should be taken not to damage the tree during flower gathering. The best way is to cut the ends of branches with scissors, and gather the flowers manually from the branches. The gathered flowers are dried in well-ventilated shade or in an artificial drier at 40 °C. The drying ratio is 4 : 1.

Marshmallow (*Althaea officinalis* L.)

Description

This is a perennial plant belonging to the Malvaceae family. It grows wild everywhere in Europe in humid places and along rivers, lakes and ditches.

The root is simple or branching, greyish-brown outside, and white inside. The rhizome is generally multi-headed. The stem is 0.60–1.20 m high, simple or branching, and cylindrical. The leaves are ovate, lobed, toothed, greyish-green, and velvety hairy. Flowers are borne in the axils of upper leaves and bracts; their colour is white or pale pink, sometimes purplish, with darker veins. Flowering is from June to September. The fruit is disc-like, and is composed of sections. The mass of 1000 seeds is 3–5 g (Fig. 114).

Active substances

The dried roots (Althaeae radix) and leaves (Althaeae folium) contain 20–30 per cent and 5–10 per cent mucilage, respectively, as well as starch and pectin.

Uses

The drug is used as a remedy for tracheal catarrh and coughing in therapeutics. It is also an important ingredient of various tea blends. The brew of the flower drug is also used as a soothing agent for inflamed eyes. The extract is used to prepare cough-soothing syrup and sweets.

Fig. 114 Marshmallow

Gathering

Marshmallow can be regarded as the only plant which is exploited entirely, i.e. every part of it can be used for some purpose. Leaves, flowers, shoots and roots are gathered. Only the well-developed, sound and clean leaves are gathered, without petioles. The leaves can be gathered all through the summer. To obtain 1 kg of dry drug, 4–6 kg of fresh leaves are needed.

The shoots are cut at the time of flowering in July, 10–20 cm from the ground when the plant is 70–80 cm high. The cut plants will shoot again, and they can be cut for the second time in September. The leaves or the leafy stems should be dried after the cutting, spread on a drying frame in a shady location, in lofts, or in an artificial drier at 40–50 °C. The completely opened flowers are gathered together with sepals when the weather is dry, after the dew has evaporated. To obtain 1 kg of drug, 7 kg of freshly cut flowers are needed.

The roots are gathered in the autumn, at the start of the dormant period, or early in the spring. The gathered roots are cleaned of soil, and then the root head and the rotten parts are removed. This is preferably followed by washing in running water for a short period, and then the roots are left to dry for 2–3 hours. The roots are then peeled, sliced into two if their thickness is over 1.5 cm, cut into 15–20 cm pieces, and dried. The best way of drying roots is artificial drying at 35–40 °C. To obtain 1 kg of dry root, 4–5 kg of fresh roots are needed.

Cultivation

The cultivated plant has thicker, softer, less fibrous roots than the wild-growing one, and also contains more mucilage substance; the quality of the drug is also better. It can be successfully cultivated in soils which are rich in nutrients, thick-layered, and loose, and which contain humus and high humidity. It can be used for the exploitation of floodlands.

Propagation can be done by sowing to the final location, by seedling nursing, or by root heads. Sowing to the final location is done in the autumn or in the spring, with 0.5–0.6 m row distance, and using 4–5 kg/ha of sowing seeds. It is preferable to sow in the autumn before the frosts. Spring sowing should be started early in the spring. The optimum sowing depth is 10–20 mm.

Propagation by root heads is done at the time of terminating the older plantations, or in the case of root gathering. Root heads the size of a thumb planted into 0.6–0.8-mm-deep furrows over a breeding area of 0.6×0.4 m.

The nursing of plants sown in the final location consists of regular weeding and sometimes watering, as well as thinning.

The aerial parts of the plant can be cut with windrowing machines, but the flowers can be gathered only by hand. The roots can be gathered in the autumn after turning out by a plough not provided with a steering plate. The expected yield is 1.5–2 t/ha of dry roots or 0.8–1.0 t/ha of dry leaves, and 80–100 kg/ha of dry flowers. An amount of 5 kg of fresh roots is necessary to prepare 1 kg of drug cut to cubes.

Mullein (*Verbascum phlomoides* L.)

Description

Mullein (orange mullein) is an annual or biennial herbaceous plant belonging to the Scrophulariaceae family. It can be found all over Europe in sunny, shrubby areas, along roads in meadows, and mainly on sandy soils.

The root is a spindle root, rarely branching. The stem is 0.60–2.0 m high, erect, five-edged, and lignified at flowering. The biennial variety develops a large rosette in the first year; the basal leaves are as long as 0.40 m, elongated or ovate, and deeply lobed. The stem leaves are alternate. The flowers are petiolate, and 0.25–0.35 m long. Every leaf is velutinous owing to the stratified stellate hairs. The inflorescence is 0.30–0.80 m long, with a simple or branching terminal spike. The flowers are solitary or appear in clusters of 3–5. The fruit is an opening capsule. The seed is very small, and reddish-brown. The mass of 1000 seeds is 0.08–0.18 g.

Active substances

The flowers contain about 3 per cent mucilage, 10–11 per cent sugar, saponins, carotenoids (β-carotene, crocin) and flavonoids (hesperidin, verbascoside).

Uses

The dried flowers (Verbasci flos) are a very important ingredient of expectorant teas and chest-rubs, and additionally have sudatory and diuretic effects. Mullein tea is a good remedy for coughs, hoarseness, bronchitis, bronchial catarrh and whooping cough. According to Lewis and Elvin-Lewis (1977), a nineteenth century home remedy included dried flowers or roots of mullein used as cigarettes for asthmatics. For external use on inflammations or painful skin conditions, the tea or a fomentation of the leaves boiled or steeped in hot vinegar and water is used.

Gathering

The flowers generally open during the night and wither until noon. For this reason they are gathered in the morning between 9 a.m. and 11 a.m. in sunny weather, when the flowers are in full bloom. Owing to the large number of flowers, flowering lasts for several weeks, even on one plant. Gathering is done by hand. Gathering can be done either with or without sepals. The flowers gathered without sepals are more valuable. The corolla is pulled out of the calyx of every flower and gathered in small baskets or boxes. The picked flowers should not be pressed together, because they would break, heat and go brown very soon. The flowers should be dried without delay after gathering, on the same day. Drying can be done in sunshine or in an artificial drier at 45–50 °C. The drug should not be turned over during drying. The slowly dried or humid drug goes brown, thus becoming unusable. For this reason, transportation and storage should be such that hygroscopic material is always present, or should be in airtight plastic bags.

Cultivation

Mullein is cultivated in several European countries. In Hungary, cultivation was started in recent years, mainly for the exploitation of sandy soils. At the same time a project was started to select the annual plants with large and vivid yellow flowers.

According to Hungarian and other experience, it can be reliably cultivated by sowing directly in the field. It can be sown either in the autumn or in the spring. The autumn (end of September to early in October) sowing is always more reliable, as the seeds sown very shallowly start to germinate early in the spring (as long as a sufficient amount of water is available in the top 10–20 mm layer of the soil).

The optimum row distance is 0.50–0.60 m (Csáki, 1982). The too densely sprouted plants should be thinned after they have strengthened sufficiently by harrowing perpendicularly to the rows. The requirement for sowing seed is 4–5 kg/ha.

According to reports, the yield can be significantly increased by applying 30–50 kg/ha of N, 40–50 kg/ha of P, and 70–80 kg/ha of K as basic fertilization.

Flower gathering can be started in mid-June, and can be done continuously up to the end of August. The expected yield is 5.0–6.0 t/ha of fresh flowers, which gives 0.5–0.6 t/ha of dry drug.

Oregano (*Origanum* species)

Description

Oregano is the common name for a general aroma and flavour primarily derived from several species of *Origanum* belonging to the Lamiaceae family. European oregano (frequently called wild or winter marjoram) is derived principally from *Origanum vulgare* L. Greek oregano (called winter sweet or pot marjoram) is derived from *O. heracleoticum* L. or *O. creticum* L. They are perennial herbs native to the dry, rocky calcareous soils in the mountainous areas of southern Europe and southwestern Asia, and the Mediterranean countries. The erect plants reach a height of 0.8–1.4 m. The creeping rootstock produces a square, downy, purplish stem with opposite, ovate leaves that are dotted with small depressions. The stems and leaves are pubescent. White or purple, two-lipped flowers grow in terminal clusters from July to September. The seeds are very small and dark brown; the mass of 1000 seeds 0.08–0.1 g.

Active substances

The dried herbs of European oregano *(Origani herba)* contain 0.2–0.8 per cent essential oil. The essential oil content of Greek oregano is 0.5–1.5 per cent. The essential oil of European oregano is composed mainly of carvacrol and thymol. Greek oregano oil contains carvacrol, thymol, α-terpineol, β-caryophyllene, *d*-linalool, and other compounds.

Uses

Oregano species are used as flavourings in meat and sausage products, salads, stews, sauces and soups. The essential oil and oleoresin, used extensively in place of the plant material, are found in food products, cosmetics and alcoholic liqueurs. They have traditionally been used as a carminative, diaphoretic, expectorant, emmenagogue, stimulant, stomachic, and tonic. *Origanum* oil is a powerful disinfectant, and carvacrol and thymol are considered to be anthelmintic and antifungal agents (Windholz, 1976).

Cultivation

Although much of the commercial material is collected from wild plants, fields can be cultivated on light, dry, well-drained soil that are somewhat alkaline. Propagation can be done by plant division, or by seed sowing. The seedlings prepared with plant division should be transplanted to the field in September–October. The seeds should be sown in open-air nursing-beds with a row distance of 0.25 m in April. The seed requirement is 1 g/m^2. The depth of sowing is 5–10 mm. The seedlings can be transplanted to the field in June or in July with irrigation. They are planted with 0.6–0.8 m row distance and 0.4–0.5 m plant spacing. The herbs are harvested during the start of blooming and allowed to dry in the shade for the preservation of colour and fragrance.

Cultivated stands, which generally last 5–6 years, can be cut 2–4 times each year. The expected yield is 2.5–3.5 t/ha of dry drug.

Plantain (*Plantago* species)

Description

Plantain, belonging to the Plantaginaceae family, is the general name for several small herbs used medicinally because of their mucilaginous properties. Common plantain is derived from *P. major* L., and lance-leaved plantain is *P. lanceolata L.* Indian or blood plantago (Isabgol) comes from *P. ovata* Forsk., and Spanish psyllium or fleawort is from *P. psyllium* L.

Common plantain is a perennial plant that is, appropriately enough, common in waste places, lawns, and roadsides all over Europe and North America. Its leaves are broadly ovate, entire or toothed, and characterized by a thick, channelled footstalk. The flower stalks grow 0.2–0.5 m high and are tipped with long, slender spikes of greenish-white flowers whose colour is overshadowed by brownish sepals and bracts. Flowering time is from May to October. Seeds are very small, and brown. The mass of 1000 seeds is 0.24–0.32 g.

Lance-leaved plantain is a perennial plant that inhabits meadows, roadsides, agricultural lands and yards in Europe, Canada and the USA. The erect, hairy, lanceolate leaves grow from the rootstock on marginated petioles in a basal rosette. Several grooved flower stalks may grow 0.2–0.8 m high, and are tipped by a short spike of tiny, white flowers whose brownish sepals and bracts give the spike its predominantly dark colour. Flowering time is from April to November. Seeds are black; the mass of 1000 seeds is 1.0–1.5 g.

Indian plantain is an annual herb which attains a height of 0.3–0.5 m under cultivation. It is native to the eastern Mediterranean region and to India. It is a nearly stemless, softly hairy plant, with large, narrowly linear leaves, 75–200 mm long and about 60 mm broad, appearing whorled due to the short terete stem. The spikes are 12–40 mm long and about 50 mm broad, cylindrical to ovoid in shape, and bear 45–70 flowers. The fruit is an ellipsoid capsule, boat-shaped, with smooth, rosy-white seeds. The mass of 1000 seeds is 1.3–1.8 g.

Spanish psyllium is an annual herb, native to the eastern Mediterranean region and naturalized in the eastern USA. It is about 0.6 m tall with hairy leaves, a dense spike of flowers, and a dehiscent seed capsule. The seeds are brown; the mass of 1000 seeds is 0.5–1.5 g.

Active substances

The dried leaves of common and lance-leaf plantain *(Plantaginis folium)* contain aucubin, gum, mucilage, resin, and tannin. The seed of Indian and Spanish plantain is particularly valued for the thin white seed hask, which contains about 30 per cent mucilage, mainly xylose, arabinose and galacturonic acid. In addition to the mucilage, the seed contains small amounts of aucubin and tannin. The mucilage, or seed gum, may be extracted from the husk with boiling water followed by mechanical separation from cellulose and other insoluble materials.

Uses

Leaves of common and lance-leaved plantain are astringent, demulcent, expectorant, and haemostatic. They are a useful remedy for cough irritations and hoarseness and for gastritis and enteritis. They are good for all respiratory problems, especially those involving mucous congestion. A decoction of the dried leaves promotes the coagulation of blood. The major use of Indian and Spanish plantain seeds and husks is as a bulk-forming laxative. They may be used in the treatment of chronic constipation and they may also be used for their emollient and demulcent properties to assist in the production of a smooth, solid faecal mass after a colotomy and to treat irritation of the colon caused by amoebic and bacillary dysentery and diarrhoea. The action of the drug is purely mechanical.

Cultivation

Common and lance-leaved plantain are cultivated in European countries. Plantation can be established by seed sowing. The seed can be sown directly in the field in the spring to 0.4–0.5 m row distance. The sowing seed requirement is 6–8 kg/ha. The leaves can be cut prior to the initiation of flowering, two to four times each year. The expected yield is 4–6 t/ha dry drug. The life of the plantation is 3–4 years. The Indian and Spanish plantain are grown in India, the United States, France and Spain. They are grown from seed, and sown to field directly. They grow well in a variety of soils but thrive in rich well-drained loamy soil. Some 7 kg/ha of seeds has been found satisfactory. Line sowing at 0.15 m spacing has given good results. Irrigation is carried out immediately after sowing and the young seedlings are irrigated approximately every ten days. The crop is ready for harvesting three and a half to four months after sowing. Seed maturity is determined by pressing an ear between two fingers when ripe full grown seed should come out. The crop is cut 0.2 m above ground. It is bound in heaps for a few days, after which it is threshed and winnowed. On average, 1 t/ha of seeds are collected.

Rosemary (*Rosmarinus officinalis* L.)

Description

Rosemary is an evergreen, perennial shrub belonging to the Lamiaceae family. It is native to the chalky, calcareous hills in the Mediterranean region. Reaching a height of up to 1.5 m, the plant is characterized by linear, narrow leaves whose undersides are matted with thick hair. The flowers are small, blue, and borne in whorls produced in the axils of the leaves. The flowers can develop throughout the vegetation season, although profuse blooming occurs during the early spring or later in cooler climates.

Active substances

The flowering tops contain 0.3–2.0 per cent essential oil. The essential oil content of the dried leaves is higher, about 10–15 per cent. The main constituents of the essential oil are cineole, α-and β-pinene, camphor, bornyl acetate, camphene, linalool, *d*-limonene, borneol, myrcene, α-terpineol and β-caryophyllene (Lawrence, 1977).

Uses

Dried leaves, whole or ground, are used as seasonings for soups, stews, sausages, meat fish and poultry. The essential oil is used in food products, perfumes and cosmetics, such as soaps, creams, deodorants, hair tonics and shampoos. Rosemary is also used in non-alcoholic beverages. The plant and extracts have antibacterial and antioxidant activity and can be used to extend the storage time of fats and meats. The drug (Rosmarini folium) has been used as an external stimulant and as a relaxant for nervousness, muscle spasms and headaches. It was used in wines as a carminative, and it is thought to act as a stimulant to the kidneys. It has also been used against rheumatism.

Rosemary is often used as a ground cover along roads and on embankments because of its beauty and deep root system, which helps stabilize the soil and allows the plant to withstand hot, dry periods. It is considered a good source of nectar for bees.

Cultivation

Leading areas of rosemary production are the Mediterranean countries, the USA and England. The reported life zone for rosemary is 9–28 °C, with an annual precipitation of 0.3–2.7 m and a soil pH of 4.5–8.7. The drought-tolerant plant grows in rocky to sandy soils, as long as there is adequate drainage and a minimum soil depth of about 0.2 m. The plant is not cold-hardy. Commercially it is propagated by cuttings. The ripened shoots, which can be used for developing rooted cuttings, are cut with 150–160 mm length throughout the vegetation period. The cut and properly prepared shoots are planted into open-air nursery beds to a 60 × 40 mm growing area. The cuttings are regularly irrigated. Two to three months are required for the rooting of cuttings. The rooted cuttings are transplanted to the field in rows 1.0–1.5 m apart, with about 0.5 m between the plants in the rows.

Plantations of rosemary are usually harvested once or twice each year, depending upon the geographical area and whether the harvest is for dry leaves or essential oil. A first harvest can be obtained in the transplanting year, but harvesting is usually delayed until six months after transplanting. Leaves are dried in the shade directly after harvesting to maximize retention of colour and aroma.

The essential oil is extracted from flowering tops by steam distillation or the use of organic solvents. The oleoresin is also commercially available.

The expected yield is 1.5–2.0 t/ha of dry leaves or 10–15 kg/ha of essential oil.

Wormwood (*Artemisia absinthium* L.)

Description

Wormwood was known as a medicinal plant even in ancient times. It can be found in temperate areas of Europe and Asia as a wild-growing plant in relatively dry climates. It was introduced to America and New Zealand together with the seeds of other cultivated plants, and it became naturalized in these places.

Wormwood is an erect-growing, perennial herb belonging to the Asteraceae family. It can be found on dry, stony wastelands, sometimes in great numbers. The root and rhizome is 15–20 cm long, abundantly branching, and multi-headed. The stem is up to 1.0–1.5 m high, lignified at its base, and covered with silver-grey, silky hairs. The basal leaves are long-petioled and thrice-divided pinnate. The stem leaves are smaller; the lower ones are petioled and twice-divided, while the upper ones are sessile and once divided. All leaves are covered with silvery, greyish-white, silky hairs. The inflorescence is a dense, spiky, leafy panicle composed of capitula. The capitulum inflorescences are globose, and 3–5 mm in diameter. The flowers are yellow; the petals are cylindrical and slightly bell-shaped. Flowering is in July and August. The fruit is 1 mm long, 0.3 mm wide, obovate, pale brown, and slightly shiny. The mass of 1000 seeds is 0.07–0.09 g.

Active substances

Wormwood contains 0.2–0.9 per cent essential oil (Aetheroleum absinthii) which has the following main ingredients: thujone, thujyl alcohol, phellandrene, cadinene, etc. The blue or bluish colour of the essential oil is given by camazulene, which forms from the sesquiterpene lactones (artabazine, etc.) during distillation. Chemical taxons involving no azulogene are also known within the species, the essential oils of these are colourless. The herb also contains the bitter glucosides absinthin, absinthic acid, anabsinthin, astabsin, artametin, etc. The bitter taste is due to absinthin and anabsinthin.

Uses

Wormwood has been used as a flavouring agent in alcoholic beverages, such as vermouth, bitters and liqueurs. The dried leaves, flowering tops and essential oil of wormwood have traditionally been used as anthelmintics, antiseptics, antispasmodics, carminatives, sedatives, stimulants, stomachics and tonics. It has also been used to improve blood circulation, as a cardiac stimulant, as a pain reliever for women during labour, and as an agent against tumours and cancers (Hartwell, 1968). The plant is known to be poisonous, and the essential oil is a central nervous system depressant. The plant is recognized as a moth and insect repellant.

Gathering

Two kinds of drugs are prepared from the wormwood; wormwood leaves of shelled wormwood (Absinthii folium) and flowery wormwood (Absinthii herba). The leaf drug is composed of basal leaves and leaves removed from the stems, which are gathered from the end of May to August. In order to prepare flowery wormwood drug, the plant is cut at the time of full flowering. The top of the stem is cut in this case; a maximum of 50 cm should be cut. Wormwood is dried in the shade or in an artificial drier at relatively low temperatures. The drying ratio is 3:5.

For the production of essential oil, fresh and dried wormwood are also used. According to experiments, the ester content of essential oil prepared from dried wormwood is higher than that of essential oil distilled from fresh wormwood.

Cultivation

Wormwood is cultivated in large fields in the USA, the USSR, France and other countries. Wormwood has no special soil requirements; it is a relatively undemanding plant, and grows on poor, sandy soils as well as on more firm soils. The seeds retain their germination capacity for 1–2 years. For the purpose of seedling nursing, the seeds should be sown in open-air seed-beds in the spring. Both the seedlings prepared by plant division and the seedlings grown from seeds should be transplanted to their final location in the autumn, in October. The transplanting should be done 0.3–0.4 m apart in rows 0.5–0.6 m apart. Wormwood can be cut once in the year following transplanting, and later on it can be cut twice a year. Harvesting should be done when the plants are in full flower, by hand or with machines (mower–loader or windrower machines). The expected yield is 1.0–1.5 t/ha of dry drug. The life of the plantation is 6–8 years.

Yarrow (*Achillea millefolium* L.)

Description

Yarrow (milfoil, thousand-leaf) is native to Europe and Asia and is naturalized in North America. Sometimes it can be found in large numbers along roads, in meadows of humid or dry character, in hayfields and on fallow lands.

Yarrow is a perennial plant belonging to the Asteraceae family. The roots of yarrow are fibrous, creep like stolons, and develop a number of stems. The stem is 15–70 cm high, cylindrical, and simple or branching. The leaves are alternate, oblong, lanceolate, 2–4 cm wide on average, and 2–4 times divided pinnate. The leaf parts are pointed, and the width is under 1 mm. The inflorescence is a flat-topped panicle composed of heads of 3–5 mm diameter. There is no pappus. The corolla is white, sometimes pink. Flowering lasts from June to late August. The fruit is an achene; its shape is flattened, oblong, obovate, and it has no pappus. The mass of 1000 seeds is 0.13 g.

Active substances

The inflorescence of yarrow contains 0.2–0.5 per cent essential oil, while the leaves and stem parts contain only 0.02–0.07 per cent essential oil. The essential oil distilled from the inflorescence is dark blue owing to the 20–30 per cent (or sometimes more) azulene content. Additionally, the essential oil contains *d*-α-pinene, cineole, β-pinene, l-limonene, l-borneol, bornyl acetate, l-camphor and various aldehydes and acids. The plant also contains terpenoid alkaloids, sesquiterpene lactones, flavonoids, tannins, etc., besides essential oil.

Uses

The dried flowers can be utilized in flower arrangements. The leaves and flowers have a bitter, stringent taste when used in culinary applications.

Owing to its stomachic, antispasmodic, carminative, diaphoretic, stimulant, digestive and hypotensive effects, it is an ingredient of many tea blends. It is also used for controlling bleeding. Its brew is used mainly externally to remedy inflammations of the gum, eye and other organs as a washing agent and lotion. The essential oil of yarrow is mainly used by the cosmetics industry to produce creams and hair shampoos. Extract of yarrow exhibits antibiotic activity and may also act as an antineoplastic drug.

Gathering

Three types of drug are produced from yarrow. The inflorescence (Millefolii flos) should be gathered at the time of full bloom with short (3–4 cm) stem parts. The flowering tops should also be gathered at the time of full bloom; this is a drug known as Millefolii herba. The leaves (Millefolii folium) are gathered when the plants are in the rosette stage.

The fresh plant parts are dried in the shade or in an artificial drier. In order to retain the original colour of the plant parts, fast drying is very important. The essential oil (Aetheroleum millefolii) is produced from the inflorescence and from the flowery tops.

Cultivation

Yarrow grows on many soil types if there is adequate drainage of the temperate zone. Yarrow plantation can be established by plant division: the older (4–5 year old) plants are gathered and divided into parts, then planted into soil deep-ploughed in the autumn. The spring planting is preferred. Yarrow can be sown directly in the field at the end of summer or early autumn to rows 40–50 cm apart similarly to camomile. The sowing seed requirement is 2–3 kg/ha. The plants sown directly in the field give a complete yield in the third year (Ruminska, 1973).

For the purpose of seedling nursing the seeds are sown in spring (April) into open-air seed-beds, in rows 0.15–0.20 m apart.

The seedlings are nursed through the vegetation period, and are then transplanted to the field in October. The optimum breeding area is established to be 60 cm row distance and 25–30 cm plant distance. The transplanted plants flower in the year of transplantation, but complete yield may only be expected from the second year on.

The expected yield is 1.0–4.0 t/ha dry herb.

References

Brown, E. G. (1956). 'Cinnamon and cassia: Sources, production and trade', *Collection of Plant Animal Production,* **6,** 96–116.

CPCRI (1977). 'Towards higher yield in cardamom' *Proceedings of the National Seminar on Cardamom,* CPCRI, Kasaragod.

Csáki, Gy. (1982). 'Studies on the ontogenesis and morphology of *Verbascum phlomoides* L' (in Hungarian), *Herba Hungarica,* **21,** (2–3), 67–72.

Duke, J. A. (1981). *Nutrition File.* Computer index to nutritional data in 21 a to e.

Duke, J. A. (1985). *Handbook of Medicinal Herbs,* CRC Press, Boca Raton.

Duke, J. A. and Wain, K. K. (1981). *Medicinal Plants of the World.* Computer index with more than 85 000 entries. 3 Vols, P. 1654

Fazli, F. R. Y., and Hardman, R. (1968). *Tropical Science,* **10,** (2), 66–78.

Grieve, M. A. (1974). *Herbal Reprint,* Hatner Press, New York 1931, 916.

Hartwell, J. L. (1968). 'Plants used against cancer. A survey', *Lloydia,* **31,** (2) 71–170.

Hartwell, J. L. (1969). 'Plants used against cancer. A survey', *Lloydia,* **32,** (3), 247–96.

Hartwell. J. L. (1970): 'Plants used againts cancer. A 'survey', *Lloydia,* **33,** (2), 97–194.

Hemmerly, T. E. (1977). 'A ginseng farm in Lawrence County, Tennessee', *Economic Botany,* **31,** 160–2.

ITC UNCTAD/GATT (1982). *Markets for Selected Medicinal Plants and their Derivatives,* ITC UNCTAD/GATT, Geneva.

ITC UNCTAD/GATT (1986). *Essential Oils and Oleoresins. A Study of Selected Producers and Major Markets,* ITC UNCTAD/GATT, Geneva.

Kannan, K., and Balakrishnan, S. (1967). 'A note on the viability of cinnamon seeds (*Cinnamonum zeylanicum* Nees), Madras', *Agricultural Journal,* **54,** 78–9.

Kasasian, L. (1971). *Weed Control in the Tropics,* Leonard Hill, London.

Larkworthy, W. (1977). 'Deglycyrrhiziised liquorice in duodenal ulcer', *Brit. Med. J.,* 29 October, 1123.

Lawrence, B. M. (1977). 'Recent progress in essential oils', *Perfumer and Flavorist,* **2,** (4), 53–6.

Leung, A. Y. (1980). *Encyclopedia of Common Natural Ingredients Used in Food, Drugs, and Cosmetics,* John Wiley and Sons, New York.

Lewis, W. H., and Elvin-Lewis, M. P. F. (1977). *Medical Botany,* John Wiley and Sons, New York.

List, P. H., and Horhammer, L. (1979). *Hager's Handbuch der Pharmazeutischen Praxis, 6,* Springer Verlag, Berlin.

Makovkina, A. I. (1972). *'Nepeta transcaucasica,* a source of geraniol, citronnellol and citral' (in Russian). In Naumenko, M. A. (ed.) *Mezhdunarodnyi kongress po efirnym maslam,* 4th, Tiflis, 1968, Pishcheraia Promishlenostü, Moscow, pp. 97–8.

Nair, M. K. (1982). 'Cultivation of spices'. In C. K. Atal, and B. M. Kapur, (eds) *Cultivation and Utilization of Aromatic Plants,* Regional Research Laboratory, Jammu-Tawi.

Nambiar, M. C. (1978). *Report on the Spice and Essential Crops of India,* FAO Rome.

Purseglove, J. W., Brown, E. G., Green, C. L. and Robbins, S. R. J. (1981). *Spices,* Vol. 1 and 2, Longman, London and New York.

Rosengarten, F. (1969). *The Book of Spices,* Livingston Publishing Company, Wynnewood, Pennsylvania.

Ruminska, A. (1973). *Rosliny Lecznicze,* Panstwowe Wydawnictwo Rolnicze i Les'ne, Warsaw.

Samarawira, I. St E. (1964). 'Cinnamon', *World Crops,* **16,** 45–8.

Sastri, B. N. (ed.) (1952). *The Wealth of India: Raw Materials,* Vol. III, New Delhi, Council of Scientific and Industrial Research.

Soon-Keun, H. (1982). 'Ginseng Cultivation.' In C. K. Atal and B. M. Kapur (eds) *Cultivation and Utilization of Medicinal Plants,* Regional Research Laboratory, Jammu–Tawi, 419–35.

Spoerke, D. G. (1980). *Herbal Medications,* Woodbridge Press Publishing Company, California.

Tarnopol, J. H. and Ball, H. J. (1972). 'A survey of some common Midwestern plants for juvenile hormone activity', *J. of Econ. Ent.,* **65,** 980–2.

Tyler, V. E. (1982). *The Honest Herbal–A Sensible Guide to the Use of Herbs and Related Remedies,* George F. Stickley, Philadelphia.

Windholz, M. (ed.) (1976). *The Merck Index. An Encyclopedia of Chemicals and Drugs,* Merck and Co., inc., Rahway, N. J.

Table 17 Main data of the essential oil plants

Plant	Essential oil content of dry substance (%)	Average yield of essential oil from: Fresh (kg/t)	Dried (kg/t)	Specific gravity of essential oil
Angelica				
seed	0.6–1.5	6.0–12	7–12	0.85–0.89
root	0.5–1.0	1.0–3	3–10	0.85–0.91
Anise seed	1.5–3.5	—	10–25	0.98–0.99
Caraway seed	3.0–7.0	—	20–35	0.90–0.93
Cardamom fruit	3.5–7.0	—	20–35	0.91–0.94
Carot seed	0.8–1.6	—	5–12	0.89–0.94
Cassia bark	1.0–2.5	7–10	15–20	1.05–1.07
shoot	0.2–0.7	1.5–3.0	—	1.04–1.06
Catnip	0.1–0.2	1.0–1.5	3–5	0.89–0.97
Celery plant	2.5–3.0	0.8–2.0	—	0.84–0.90
Cinnamon				
bark	3.0–4.0	15–20	30–40	0.88–0.95
leaf	0.5–0.7	1.0–2.0	—	0.91–0.93
Clary sage	0.04–0.2	0.4–1.8	—	0.89–0.92
Coriander seed	0.6–1.5	—	4–12	0.86–0.88
Cumin seed	2.5–5.0	—	25–40	0.90–0.93
Dill				
seed	2.0–3.0	15–28	20–35	0.89–0.92
herb	0.8–1.6	2–5	—	0.87–0.91
Fennel seed	2.0–6.0	—	30–40	0.96–0.97
Fir shoots	0.15–0.55	2.2–2.6	—	0.86–0.88
Garden thyme	1.0–2.5	1–3	—	0.90–0.93
Garlic	0.25–0.29	1.0–2.5	—	1.02–1.04
Ginger rhizome	1.0–4.0	2.5–3.5	—	0.88–0.95
Hop cone	0.35–1.0	1–2	5–10	0.82–0.87
Hyssop	0.3–1.0	0.8–2.5	—	0.92–0.96
Juniper berry	0.2–3.4	6–12	9–13	0.86–0.87
Lavandin	0.9–3.0	8–14	—	0.87–0.89
Lavender (spike)	0.5–1.5	4–8	—	0.88–0.90
Lovage				
root	0.5–1.0	0.5–2.0	—	1.00–1.05
herb	0.5–1.7	4–10	—	0.98–1.04
Marjoram	0.5–1.3	2–7	6–12	0.88–0.91
Mugwort	0.02–0.2	0.4–0.8	—	0.90–0.93
Onion	0.01–0.04	0.05–0.2	—	1.01–1.03
Parsley				
herb	0.01–0.05	0.3–3.0	—	0.89–1.00
seed	2–7	—	20–50	1.03–1.10
Pastinaca seed	0.4–0.6	4–6	6–8	0.86–0.83
Peppermint	1.0–1.5	1.5–4.0	5–8	0.92–0.93
Radish root	0.14–0.21	1–2	—	1.01
Roman camomile	0.6–1.0	1.0–2.5	3–6	0.90–0.91
German camomile	0.2–1.5	0.3–1.0	2–6	0.91–0.94
Rosemary tops	0.3–2.0	2–6	—	0.89–0.92
Sage	1.0–2.5	3–6	8–14	0.91–0.92
Spearmint	0.5–1.0	1–3	—	0.92–0.94
Summer savory	1–2	—	9–13	0.89–0.91
Sweet basil	0.5–1.1	0.2–1.3	4–6	0.95–0.96
Sweet-flag	0.6–2.5	3–7	10–40	0.93–0.96
Tarragon	0.5–2.8	1–3	—	0.90–0.94
Valerian	0.5–1.7	1–2	5–9	0.92–0.96
Wormwood	0.2–0.9	0.5–2.0	—	0.87–0.93
Yarrow	0.2–0.5	2–5	3–4	0.89–0.92

Table 18 Main growing data of medicinal plants

Name of plant species	Life time	Propagation		Row space (m)	Plant distance (m)	Requirement for 1 ha		Time of harvest	Used part of plant	Yield	
		Time	Method			Seed (kg)	Seedling (pieces)			Drug (t/h)	Essential oil (kg/ha)
Angelica (*Angelica archangelica*)	Biennial	Nov.	Seed sowing to final location	0.60	—	8–10	—	Autumn	Root	1.5–1.7	5–8
Anise (*Pimpinella anisum*)	Annual	March	Seed sowing to final location	0.25–0.30	—	13–15	—	July	Seed	0.4–0.6	
Blessed thistle (*Cnicus benedictus*)	Annual	March–April	Seed sowing to final location	0.40–0.50	—	17–20	—	June and Aug.	Herb	1.8–2.2	
Caraway (*Carum carvi*)	Biennial	March–April	Seed sowing to final location	0.30–0.40	—	6–10	—	June–July	Fruit	0.4–1.5	
Clary sage (*Salvia sclarea*)	Biennial	April	Seed sowing to final location	0.40–0.50	—	5–7	—	July	Inflorescence	—	4–10
Common foxglove (*Digitalis purpurea*)	Biennial	March	Seedling growing under plastic cover	0.15	—	0.2					
		April–May	Planting to final location	0.60	0.30–0.40	—	45 000–60 000	Sept.	Leaf	1.5–2.0	
Coriander (*Coriandrum sativum*)	Annual	March	Seed sowing to final location	0.25–0.30	—	16–20	—	July	Fruit	0.6–0.8	7–9
Dill (*Anethum graveolens*)	Annual	March–April	Seed sowing to final location	0.12–0.24	—	12–14	—	July–Aug.	Herb, seed	0.4–0.5 0.7–1.0	30–50 15–20
Fennel *Foeniculum vulgare*)	Perennial	March	Seed sowing to final location	0.36–0.48	—	8–10	—	Sept.	Fruit	0.4–1.2	

cont. of Table 18

Name of plant species	Life time	Propagation		Row space (m)	Plant distance (m)	Requirement for 1 ha		Time of harvest	Used part of plant	Yield	
		Time	Method			Seed (kg)	Seedling (pieces)			Drug (t/h)	Essential oil (kg/ha)
Garden thyme (*Thymus vulgaris*)	Perennial	March May	Seedling growing in free bed	0.30	—	—	—				
		May	Planting to final location	0.50	0.25	—	80 000	May–June and Sept.	Herb	1.5–2.5	
German camomile (*Matricaria recutita*)	Wintering annual	Aug.–Sept.	Seed sowing to final location	0.12	—	3.0–4.6	—	April–May	Inflorescence, herb	0.3–0.6	1.5–4.0
Grecian foxglove (*Digitalis lanata*)	Biennial	Nov.–Dec. or Feb.–March	Seed sowing to final location	0.45–0.50	—	5	—	Sept.	Leaf	1.5–2.5	
Henbane (*Hyoscyamus niger*)	Biennial	March	Seed sowing to final location	0.50–0.60	—	4–6	—	July–Aug.	Leaf	0.8–1.0	
Hollyhock (*Alcaea rosea*)	Biennial	Feb.–March	Seed sowing to final location	0.75	0.25	3	—	July–Sept.	Flower petal	0.25–0.4	
Hyssop (*Hyssopus officinalis*)	Perennial subshrub	March–April	Seed sowing to final location	0.50–0.70	—	3–5	—	July and Sept.	Herb	0.8–2.0	8–15
Lavender: true (*Lavandula angustifolia*)	Perennial	Nov. or March–April Sept.–Oct.	Seedling growing in free bed	0.35–0.40	—	1–1.5	—				
hybrid (*Lavandula intermedia*)	Perennial	Autumn	Planting to final location	1.80 2.00	0.60 0.50	— —	9200–10 000	Aug.	Inflorescence	0.5–0.7	50–70

cont. of Table 18

Name of plant species	Life time	Propagation		Row space (m)	Plant distance (m)	Requirement for 1 ha		Time of harvest	Used part of plant	Yield	
		Time	Method			Seed (kg)	Seedling (pieces)			Drug (t/h)	Essential oil (kg/ha)
Lemon balm (*Melissa officinalis*)	Perennial	March–April	Seedling growing in free bed	0.25–0.30	—	0.5	—	June and Sept.	Herb	2.0–4.0	
		Aug.–Sept.	Planting to final location	0.60	0.40	—	42 000				
Lovage (*Levisticum officinale*)	Perennial	March–April or Oct.–Nov.	Seed sowing to final location	0.50–0.60	—	6–8	—	Autumn	Root	1.5–2.5	5–6
Marjoram (*Majorana hortensis*)	Annual	April	Seed sowing to final location	0.40–0.50	—	6–8	—	July and Sept.–Oct.	Herb	1.8–2.0	
Mustard: white (*Sinapis alba*)	Annual	March	Seed sowing to final location	0.25–0.30	—	15–17	—	June	Seed	0.8–1.0	
brown (*Brassica juncea*)		March	Seed sowing to final location	0.25–0.30	—	12–14	—	April	Seed	1.1–1.5	
Nightshade (*Solanum laciniatum*)	Annual	March	Seedling growing under plastic cover	0.20	—	—	—				
		April–May	Planting to final location	0.60–0.70	0.30–0.40	— —	40 000 50 000	June and Aug.–Sept.	Herb	3.0–5.0	
Peppermint (*Mentha piperita*)	Perennial	Sept.–Nov.	Planting to furrows	0.50–0.70	—	—	1.4–1.7 t	July and Sept.	Herb	2.5–3.0	20–40
		May–June	Planting to final location			—	80 000–90 000				
Periwinkle (*Vinca minor*)	Perennial	Aug.–Sept.	Planting to final location	0.20–0.25	0.05–0.10	—	40 000–600 000	Aug.–Sept.	Herb	0.7–1.5	

cont. of Tabl 18

Name of plant species	Life time	Propagation		Row space (m)	Plant distance (m)	Requirement for 1 ha		Time of harvest	Used part of plant	Yield	
		Time	Method			Seed (kg)	Seedling (pieces)			Drug (t/h)	Essential oil (kg/ha)
Poppy (*Papaver somniferum*)	Annual	Feb.–March	Seed sowing to final location	0.35–0.40	—	3.5	—	July–Aug.	Seed, head	4–10 4–8	
Pot Marigold (*Calendula officinalis*)	Annual	March	Seed sowing to final location	0.40–0.50	—	5–6	—	May–Aug.	Flower inflorescence	0.2–0.3 0.5–0.8	
Pyrethrum (*Chrysanthemum cinerariaefolium*)	Perennial	March–April	Seedling growing in free bed	0.30–0.40	—	0.5					
		Sept. and April–May	Planting to final location	0.60	0.20–0 30	—	60 000–70 000	May–June	Inflorescence	0.7–1.2	
Red periwinkle (*Catharanthus roseus*)	Annual	Jan.–Feb.	Seedling growing in plastic cover	0.05–0.1	—	—	—				
		May	Planting to final location	0.60	0.25	—	70 000–80 000	Sept.	Herb	0.9–1.5	
Roman camomile (*Anthemis nobilis*)	Perennial	April–May	Planting to final location (plant division)	0.60	0.20–0.25	—	60 000–65 000	July	Herb		6–10
Rue (*Ruta graveolens*)	Perennial	April–May or June–Sept.	Seedling growing in free bed	0.35–0.30	—	—	—				
		Autumn or spring	Planting to final location	0.60 0.50 or	0.20 0.30	—	70 000–90 000	May–June and Oct.	Herb	3.5–4.0	
Safflower (*Carthamus tinctorius*)	Annual	March–April	Seed sowing to final location	0.40–0.50	—	18–20	—	June–July	Flower petal	0.12–0.20	

cont. Table 18

Name of plant species	Life time	Propagation		Row space (m)	Plant distance (m)	Requirement for 1 ha		Time of harvest	Used part of plant	Yield	
		Time	Method			Seed (kg)	Seedling (pieces)			Drug (t/h)	Essential oil (kg/ha)
Sage (*Salvia officinalis*)	Perennial subshrub	April–May	Seedling growing in free bed	0.15–0.20	—	—	—				
		Oct.	Planting to final location	0.70	0.40		30 000–36 000	May–June and Aug.–Sept.	Herb leaf	1.0–1.5 0.5–0.8	8–10
Summer savory (*Satureja hortensis*)	Annual	March–April	Seed sowing to final location	0.50	—	4–8	—	July and Aug.	Herb	1.0–1.5	
Sweet basil (*Ocimum basilicum*)	Annual	March–April	Seed sowing to final location	0.40	—	4–6	—	June and Sept.	Herb	1.8–2.0	
Tarragon (*Artemisia dracunculus*)	Perennial	Sept.–Oct. or May	Planting to final location (plant division)	0.60	0.40	—	40 000–50 000		Herb	2.0–2.5	
Thorn apple (*Datura innoxia*)	Annual	March	Seed sowing to final location	0.50–0.60	—	12–13	—	June–July and Aug.–Sept.	Herb	1.5–2.0	
Valerian (*Valeriana officinalis*)	Perennial	June–July	Seedling growing in free bed	0.15	—	0.5	—				
		Sept.–Oct. or April	Planting to final location	0.50	0.20–0.25	—	80 000–100 000	Oct.–Nov.	Root	1.0–2.5	
Variegated thistle (*Silybum marianum*)	Annual	March–April	Seed sowing to final location	0.60–0.70	—	10–12	—	July–Aug.	Seed	1.0–1.6	

Table 19 Main data of wild-growing plant gathering in temperate zones

Name of plant	Name of drug	Life time	Gathering period	Drying ratio	Main active ingredients	Usage
Achillea	Millefolii	Perennial	June–Sept.	5–6 : 1	Proazulene,	Medicine, essential
millefolium	flos				essential oil	oil, herb tea
	herba		June–Sept.		Essential oil	Spice, medicine
Acorus	Calami	Perennial	Autumn	4 : 1	Glycosides	Liqueur industry
calamus	rhizoma					
Adonis	Adonidis	Perennial	April–May.	5 : 1	Glycosides	Medicine,
vernalis	herba				saponins	herb tea
					flavonoids	
Aesculus	Hypocastani	Tree				
hippocastanum	semen		Sept.	2 : 1	Saponins	Medicine
	folium		June–July	4–5 : 1	Favonoids	
Agrimonia	Agrimoniae	Perennial	July–Aug.	4 : 1	Flavone derivatives,	Herb tea,
eupatoria	herba				tanning materials	extracts
Agropyron	Graminis	Perennial				
repens	rhizoma		Autumn or	2–3 : 1	Polysaccharides,	Tea blends
			spring		mucilage materials,	
					saponins	
Althaea	Althaeae	Perennial				
officinalis	folium		June–Aug.	6 : 1	Mucilage, starch	Tea blends,
					pectin	medicines
	radix		Autumn or	3–4 : 1	Mucilage, starch	Tea blends
			spring		pectin	medicines
Anthriscus	Cerefolii	Annual			Essential oil,	Spice plant,
cerefolium	herba		May–April	5 : 1	fatty oil, bitter	components of
	semen		July–Aug.	1.2 : 1	materials	tea blends
Artemisia	Absinthii	Perennial			Essential oil,	Herb tea,
absinthium	folium		May–July	5–6 : 1	bitter material	viticulture,
	herba		July–Aug.	2–3 : 1	proazulene	liqueur industry
Atropa	Belladonnae	Perennial			L-Hyoscyamine,	Herb tea,
belladonna	folium		June–Aug.	3–4 : 1	atropine bella-	extracts
	radix		autumn or		donine	
			spring			

cont. of Table 19

Name of plant	Name of drug	Life time	Gathering period	Drying ratio	Main active ingredients	Usage
Betula pendula	Betulae	Tree			Essential oil, saponins, resin	Tea blends, cosmetics and household industry
	folium		June–Aug.	4–5 : 1		
Centaurium umbellatum	Centaurii	Biannual			Bitter materials, gentiopikrin	Extracts, liqueur industry, tinctures, tea blends
	herba		July–Aug.	4–5 : 1		
Chelodonium majus	Chelodonii	Perennial			Alkaloids, acids	Medicine, extracts, tincture
	herba		April–July	5–6 : 1		
	radix		Autumn or spring	3–4 : 1		
Cichorium intybus	Cichorii	Perennial			Triterpenes, inulin, acid	Food, herb tea medicine
	herba		July–Aug.	5–6 : 1		
	radix		Autumn or spring			
Colchicum autumnale	Colchici	Perennial			Colchinine, alkaloid	Plant-improving, medicine
	semen		June–July	1–2 : 1		
	tuber		Autumn	3–4 : 1		
Convallaria majalis	Convallariae	Perennial	May–June		Glycosides, saponins	Medicine
	folium			5–6 : 1		
	flos		April–June	7 : 1		
	rhizoma		Autumn or spring	3 : 1		
Datura stramonium	Stramonii	Annual			L-Hyoscyamine, atropine alkaloid	Medicine, asthmatic cigarette
	folium		June–Sept.	6 : 1		
Equisetum arvense	Equiseti	Perennial			Flavonoids silicic acid	Tea blends, household, chemical industry
	herba		June–Aug.	3–4 : 1		
Filipendula ulmaria	Ulmariae	Perennial			Phenolic glycoside, flavone derivative	Tea blends
	herba		July–Aug.	4–5 : 1		
Frangula alnus	Frangulae	Shrub			Anthraquinone	Medicine, herb tea, extracts,
	cortex		March–April	3 : 1		

cont. of Table 19

Name of plant	Name of drug	Life lime	Gathering period	Drying ratio	Main active ingredients	Usage
Galium odoratum	Asperulae odoratae herba	Perennial	April–June	6 : 1	Coumarin, tanning material, bitter material	Viticulture, tobacco industry, spice, tea
Glechoma hederacea	Hederae terrestris herba	Perennial	April–May	4–5 : 1	Essential oil, bitter material	Medicine
Glycyrrhiza glabra	Liquiritiae radix	Perennial	March–April or Oct.–Nov.	3–4 : 1	Saponins, coumarins, flavonoids	Extracts, medicine, food
Gypsophila paniculata	Saponariae albae radix	Perennial	Autumn or spring	3–4 : 1	Saponins	Tea blends, medicine
Hypericum perforatum	Hyperici herba	Perennial	June–July	4–5 : 1	Essential oil, hypericin, red pigments	Herb teas, spice, medicine
Inula helenium	Inulae radix	Perennial	Autumn or spring	3–4 : 1	Inulin essential oil	Medicine, herb tea
Juniperus communis	Juniperi fructus	Evergreen shrub	Sept.–Nov.	2 : 1	Essential oil, sugar, resin	Medicine, herb tea liqueur industry
Leonurus lanatus	Leonuri lanati herba	Annual	July–Aug.	4–5 : 1	Alkaloids, bitter materials	Medicine, herb tea, extract
Malva neglecta	Malvae folium	Annual or perennial	June–Sept.	5 : 1	Mucilage, tanning materials	Herb teas, medicine
Marrubium vulgare	Marrubii herba	Perennial	May–June	4–5 : 1	Bitter material tanning material, essential oil	Medicine, extracts

cont. of. Table 19

Name of plant	Name of drug	Life time	Gathering period	Drying ratio	Main active ingredients	Usage
Matricaria recutita	Chamomillae flos	Wintering annual	April–June	5 : 1	Essential oil, bitter material, glycosides	Herb tea, medicine, cosmetics industry
Menyanthes trifoliata	Trifolii fibrini folium	Perennial water plant	May–June	6 : 1	Bitter material meliantin vitamin A, phytosterol	Herb tea, tea blends
Plantago lanceolata	Plantaginis folium	Perennial	April–June	4–5 : 1	Glycosides, mucilage materilas	Medicine, herb tea, extract
Populus nigra	Populi gemma	Tree	Feb.–March	2 : 1	Essential oil, phenolic glycoside flavone derivative	Herb tea, cosmetics and household industry
Potentilla anserina	Potentillae anserinae herba	Perennial	June–Sept.	5 : 1	Flavone glycoside, tanning materials	Tea blends
Primula veris	Primulae	Perennial			Saponins	Tea blends, medicine
	flos		April–May	5–6 : 1		
	radix		Autumn or spring	3–4 : 1		
Prunus spinosa	Pruni spinosae flos	Shrub	March–April	5 : 1	Flavone derivative	Herb teas
Pulmonaria officinalis	Pulmonariae folium	Perennial	May–June	5 : 1	Mucilage materials silicic acid	Tea blends, medicine
Quercus robur and Q. petraea	Quercus cortex	Tree	Spring	3 : 1	Tanning materials, resin	Medicine, leather industry, paint industry

cont. of Table 19

Name of plant	Name of drug	Life time	Gathering period	Drying ratio	Main active ingredients	Usage
Robinia pseudoacacia	Robiniae acaciae flos	Tree	May–June	7 : 1	Loxalbumin glycoside	Herb tea, spice, perfume industry
Rosa canina	Cynosbati fructus	Shrub	Sept.–Nov.	3 : 1	Vitamin C, flavonoids	Tea and herb tea
Rubus caesius	Rubi fruticosi folium	Shrub	May–Aug.	4–5 : 1	Tanning materials, organic acids	Tea blends, tea
Sambucus nigra	Sambuci	Shrub			Organic acids,	Herb tea,
	flos		May–June	6 : 1	flavone, glycoside	food industry
	fructus		Aug.–Sept.	4–5 : 1	Essential oil	
Solidago virgaurea	Virgaureae herba	Perennial	June–Aug.	4–5 : 1	Flavone derivatives, essential oil	Medicine preparations
Symphytum officinale	Symphyti	Perennial			Alkaloids,	Medicine,
	folium		June–Aug.	5 : 1	allantoin	cosmetics
	radix		Autumn or spring	3–4 : 1		
Tanacetum vulgare	Tanaceti flos	Perennial	June–July	5 : 1	Thujone, essential oil, bitter material	Herb teas, extracts
Taraxacum officinale	Taraxaci	Perennial			Terpene, triterpene,	Extract tea blends,
	herba		Sept.–June	6 : 1	bitter material	bitter tincture
	radix		Autumn or spring			
Tilia cordata and T. platyphyllos	Tiliae flos	Tree	May–June	4 : 1	Essential oil, flavone glycoside, saponin	Herb teas, tea blends, sweets industry
Tussilago farfara	Farfarae	Perennial			Mucilage material,	Herb tea,
	flos		March–April	6 : 1	flavone derivatives	medicine
	folium		May–July	5–6 : 1		

cont. of Table 19

Name of plant	Name of drug	Life time	Gathering period	Drying ratio	Main active ingredients	Usage
Urtica dioica	Urticae	Perennial			Vitamin B_2 and B_5 chlorophyll	Sweets industry, cosmetics, medicine
	folium		June–Sept.	5 : 1		
	radix		Autumn or spring	3–4 : 1		
Verbascum phlomoides	Verbasci	Biennial			Saponins, mucilage materials	Tea blends
	flos		June–Sept.	7 : 1		
Verbena officinalis	Verbenae	Perennial			Glycosides, tanning materials, bitter materials	Medicine, spice, canning industry
	herba		July	5 : 1		
Viola tricolor	Violae tricoloris	Annual or biennial			Glycoside, saponins	Tea blends
	herba		May–Sept.	5 : 1		
Viscum album	Visci	Hemiparasite perennial			Choline	Extracts, tea blends, medicine
	stipes		Nov.–April	3 : 1		

Table 20 Trade names and their active substances of the pesticides mentioned in the book

Trade name	Active substances name	Chemical name of active substances according to the chemical Abstracts Index Guide 1987
(1)	(2)	(3)
Afalon	linuron	Urea, N'-(3,4-dichlorophenyl)-N-methoxy-N-methyl-
Aktion	aktinit PK	1,3,5-Triazine-2,4-diamine, 6-chloro-N-ethyl-n'-(1-methylethyl)-
Aresin	monolinuron	Urea, N'-(4-chlorophenyl)-n-methoxy-N-methyl-
Asolox	asulam	Carbamic acid, [(4-aminophenyl) sulfonyl]-methyl ester
Bi 58 EC	dimethoate	Phosphorodithioic acid, esters 0,0-dimethyl S-[2-(methylamino)-2-oxoethyl] ester
Buvilan EC	ethalfluralin	Benzenamine, N-ethyl-N-(2-methyl-2-propenyl)-2,6-dinitro-4-(trifluoromethyl)-
Chinoin fundazol	benomyl	Carbamic acid, [1-[(butylamino)-carbonyl]-1H-benzimidazol-2-yl]-methyl ester
Chinetrin 25 EC	23% permethrin	Cyclopropanecarboxylic acid, 3-(2,2-dichlorethenyl)-2,2-dimethyl-, (3-phenoxyphenyl) methyl ester
Cineb 80	2% tetramethrin	Cyclopropanecarboxylic acid, 2,2 dimethyl-3-(2-methyl-1-propenyl)-, (1,3,4,5,6,7-hexahydro-1,3-dioxo-2H-isoindol-2-yl) methyl ester
	4% PBO	1,3-Benzodioxide, 5-[2-(2-butoxyethoxy)-ethoxy methyl]-6-propyl-
	zineb	Zinc, [[1,2-ethanediylbis [carbamodithioato]] (2-)]-
Dachtal W–75	chlorthal-methyl	1,4-benzenedicarboxylic acid, 2,3,5,6-tetrachloro-, dimethyl ester
Decis 2,5 EC	deltamethirin	Cyclopropanecarboxylic acid, 3-(2,2-dibromoethenyl)-2,2-dimethyl-, cyano (3-phenoxyphenyl) methyl ester
Dicuran	chlortoluron	Urea, N'-(3-chloro-4 methylphenyl)-N,N-dimethyl-
Dithane M 45	mankozeb	Manganese, [[1,2-ethanediylbis-[carbamodithioato]] (2-)]-, mixt. with [[1,2-ethanediylbis [carbomodithioato]] (-2)] zinc
Dual 720 EC	metolachlor	Acetamide, 2-chloro-N-(2-ethyl-6-methylphenyl)-N-(2-methoxy-1-methyl-ethyl)-
Ethrel	ethephon	Phosphonic acid, (2-chloroethyl)-
Flubalex	benefin	benzenamine, N-butyl-N-ethyl-2,6-dinitro-4-(trifluoromethyl)-
Fusilade	fluazifop–butyl	Propanoic acid, 2-[4-[[5-(trifluoromethyl)-2-pyridinyl]oxy]-phenoxy]-, butyl ester
Geonter	terbacil	2,4(1H, 3H)-Pyrimidinedione, 5-chloro-3-(1,1-dimethylethyl)-6-methyl-
Gramoxone A	paraquat–dichlorid	4,4'-Bipyridinium, 1,1'-dimethyl-, dichlorid
Hungazin DT	simazine	1,3,5-Triazine-2,4-diamine, 6-chloro-N,N'-diethyl-
Karathane LC	dinocap	2-Butenoic acid, 2(or 4)-isooctyl-4,6 (or 2,6)- dinitrophenyl ester
Kerb 50 WP	pronamide	Benzamide, 3,5-dichloro-N-(1,1-dimethyl-2-propynyl)-
Maloran 50 WP	chlorbromuron	Urea, N'(4-bromo-3-chlorophenyl)-N-methoxy-N-methyl-
Merkazin	prometryn	1,3,5-Triazine, 2,-diamine, N,N'-bis(1-methylethyl)-6-(methylthio)-
Mezotox 25 EC	nitrophen	Benzene, 2,4-dichloro-1-)4-nitrophenoxy)-
Olitref	trifluralin	Benzeneamine, 2,6-dinitro-N,N-dioropyl-4-(trifluoromethyl)-

cont. of Table 20

Trade name	Active substances name	Chemical name of active substances according to the chemical Abstracts Index Guide 1987
(1)	(2)	(3)
Patoran 50 WP	metobromuron	urea, N'-(4-bromophenyl)-N-methozy-N-methyl-
Phosdrin	meviphos	2-Butenoic acid, 3-[(dimethoxyphosphinyl)oxy]-, methyl ester
Pirimor	pirimicarb	Carbamic acid, dimethyl-, 2-(dimethylamino)-5,6-dimethyl -4-pyrimidinyl ester
Plakin	asulam	Carbamic acid, [(4-aminophenyl)sulfonyl]-, methyl ester
Pyrotox ULV	tetramethrin	Cyclopropanecarboxylic acid, 2,2-dimethyl-3-(2-methyl-1-propenyl)-(1,3,4,5,6,7-hexahydro-1,3-dioxo-2H-isoindol-2-yl) methyl ester
Reglone	diquat–dibromid	Dipyrido [1,2-a: 2', 1'-c] pyrazinediium, 6,7-dihydro-, dibromide
Ronstar	oxadiazon	1,3,4-Oxadiazol-2(3H)-one, 3- [2,4-dichloro-5-(1-methylethoxy)phenyl]-5-(1,1-dimethylethyl)-
Rovlinka 30	dioxacarb	Phenol, 2-(1,3-dioxolan-2-yl)-methylcarbamate
Sencar 70 WP	metribuzin	1,2,4-Triazin-5 (4H)-one, 4-amino-6-(1,1-dimethylethyl)-2-(methylthio)
Sys 67 Prop	2,4–DP	Propanoic acid, 2-(2,4-dichlorophenoxy)-
Thiodan 50 WP	endosulfan	6,9-Methano-2,4,5-benzodioxoathiepin, 6,7,8,9,10-11-hexachloro-1,5,5a,9,9a-hexahydro-, 3-oxide
Tilt 250 EC	propiconazole	1H-1,2,4-Triazole, 1-[[2-(2,4-dichlorophenyl)-4-propyl-1,3 dioxolan-2-yl] methyl]-
Vigil	diclobutrazol	1H-1,2,4-Triazole-1-ethanol, β-[(2,4-dichlorophenyl) methyl]-α-(1-1-dimethylethyl)-
Wofatox	methyl parathion	Phosphorothioic acid, esters, 0,0-dimethyl 0-(4-nitrophenyl) ester
Zolone	phosalone	Phosphorodithioic acid, esters, S-[(6-chloro-2-oxo-3 (2H)-benzoxazolyl) methyl] 0,0-diethyl ester

Subject index

Absinthii folium, herba 312, 322
Absolutes 103
Acanthopanax senticosus 293
Accelerated distillation 102
Achillea millefolium 53, 313, 322
Acorus calamus 55, 322
Active substances 19
Adonidis herba 322
Adonis vernalis 45, 322
Aesculus hippocastanum 322
Aetheroleum absinthii 311
  anethi 137
  angelicae radix 147
  anisi vulgaris 143
  basilici 221
  carvi 154
  chamomillae 246
  chamomillae romanae 262
  coriandri 132
  cumini 160
  dracunculi 254
  foeniculi 163
  hyssopi 196
  juniperi 45, 284
  lavandulae 176
  levistici radicis 151
  majoranae 213
  melissae 184
  menthae crispae 195
  menthae piperitae 187
  millefolii 314
  salviae 210
  saturejae 218
  sclareae 206
  thymi 200
  valerianae 170
Aglycones 20
Agrimonia eupatoria 46, 322
Agrimoniae herba 46, 322
Agropyron repens 322
Alcaea rosea 296, 318
Alder buckthorn 273
Alkaloids 20, 40
Allelopathy 66
Allium sativum 54
Alternaria burnsii 162
Althaea officinalis 52, 303, 322
Althaea rosea 52
Althaeae folium, radix 303, 322
American ginseng 292
  valerian 172
Amaryllidaceae 20
Ammi visnaga 16
Anacardiaceae 48
Anethi fructus, herba 137
Anethum chryseum 137
  graveolens 49, 77, 137, 317
  involucratum 137
  sowa 137
Angelica 147, 316, 317
Angelica archangelica 49, 147, 317
  officinalis 148
Angelicae folium, fructus, herba 147
  radix et rhizoma 147
Anise 143, 316, 317
Anisi stellati fructus 143
Anisi vulgaris fructus 143
Annual caraway 154
Anthemis nobilis 53, 262, 320
Anthriscus cerefolium 322
Apiaceae 48
Apocynaceae 49, 72
Araceae 54
Araliaceae 49
Aralia ginseng 292
Artemisia absinthium 54, 311, 322
  balchanorum 77
  dracunculoides 255
  dracunculus 54, 254, 321
Artificial drying 89

Asperulae odoratae herba 324
Asteraceae 53
Athalia rosae 235
Atropa belladonna 51, 72, 271, 322
  caucasica 59
Azotoids 36

Basilici herba 220
Belladonna 271
Belladonnae folium 106, 272, 322
  radix 272, 322
Betula pendula 323
Betulae folium 323
Biennial caraway 154
Biogenetic system 22
Biosynthetic pathway 23
Bitter fennel 163
  substances 21
Black caraway 161
  cumin 161
  elder 287
  mustard 233
  peppermint 187
Blessed thistle 238, 317
Blood plantago 308
Brassicaceae 53
Brassica juncea 17, 233, 319
  napus 234
  nigra 53, 233
Breeding 69
Brevicoryne brassicae 235
Brier hip 285
Bromus tectorum 253
Brown mustard 233
Buckthorn 273
Bunium persicum 161

Calami rhizoma 322
Calendula officinalis 54, 241, 320
Calendulae flos 241
Canadian mint 196
Cannabidaceae 54
Cape periwinkle 166
Caprifoliaceae 49
Capsicum annuum 76
Caraway 154, 316, 317
Cardamom 275, 316
Cardui benedicti herba 238
Cardui mariani folium, fructus 243
Carot 316
Carthami flos 235
Carthamus tinctorius 77, 235, 320
Carum carvi 48, 154, 317
Carvi fructus 154
Cassia 280, 316
Catharanthus roseus 49, 72, 166, 320
Catmint 278
Catnip 278, 316
Celery 316
Centaurii herba 323
Centaurium minus 50
  pulchellum 50
  umbellatum 323
  vulgare 50
Cerefolii herba, semen 322
Certifying drugs 107
Ceutorrhynchus macula-alba 127
Chamaemelum nobile 262
Chamomillae flos 246, 325
  romanae flos 262
  vulgaris flos 107
Chamomilla recutita 246
Chelodonii herba, radix 323
Chelodonium majus 46, 323
Chemicals 82
Chemosyndrome 12
Chemotaxa 16
Chinese cassia (cinnamon) 280
  ginseng 292
  liquorice 299
Chrysanthemi flos 258
Chrysanthemum cinerariaefolium 54, 77, 258, 320
Cichorii herba, radix 323
Cichorium intybus 323
Cinnamoids 27
Cinnamomum cassia 280
  verum 280
  zeylanicum 280
Cinnamon 280, 326
Citronella 184
Clary sage 206, 316, 317
Classes of substances 23
Classification of active substances 20
Claviceps purpurea 111
Clavicipitaceae 45
Clear eye 206
Cnicus benedictus 54, 238, 317
Cocculus laurifolius 15
CO_2 extraction 104
Cohobated water 101
Colchici semen, tuber 323
Colchicum autumnale 54, 323
Cold air drying 89
Common basil 220
  fennel 166
  foxglove 231, 317
  juniper 283

plantain 308
thyme 200
Compositae 53
Concrete 103
Conium maculatum 48
Containerized distillation 100
Convallaria majalis 54, 323
Convallariae flos, folium, rhizoma 323
Conveyor driers 91
Cordyceps purpurea 111
Coriander 131, 316, 317
Coriandri fructus 131
Coriandrum melphitense 132
sativum 48, 77, 131, 317
Corydalis 46
Cotinus coggygria 48
Crataegi fructus, summitas 296
Crataegus oxycantha 295
Croci stigma 236
Crocus sativus 236
Cruciferae 53
Cucurbitaceae 52
Cucurbita pepo 52
Cumin 160, 316
Cumini fructus 160
Cuminum cyminum 49, 160
odorum 161
Cupressaceae 45
Cyclamen purpurascens 53
Cymbopogon jawarancusa 77
Cynosbati fructus 285, 326
Cypripedium calceolus 172

Dalmatian pyrethrum 258
sage 210
Datura godronii 59
innoxia 51, 63, 66, 321
innoxiae herba 106
stramonium 76, 323
tatula 61
Deadly nightsade 271
Depressaria nervosa 159
Digitalis lanata 51, 61, 65, 73, 228, 318
purpurea 51, 228, 231, 317
Dill 137, 316, 317
Dill herb oil 138
Dill seed oil 138
Dioscorea deltoidea 61
floribunda 77
Diplodina dracunculi 258
Distillation 93
Diterpenoids 33
Dog rose 285
Dracunculi herba 254

Driers 90
Drying 86
Drying methods 89
Drying ratio 92
Drug 4
Duration of light 61

Eastern mustard 233
Ecotype 14
Effects of drying 92
Elder 287
Eleagnaceae 47
Elettaria cardamomum 275
Eleutherococus senticosus 292
Elixirum thymi 200
English hawthorn 295
mustard 233
Environmental factors 56
Ergot 111
Equiseti herba 323
Equisetum arvense 323
Erigeron aurantiacus 14
Erwinia carotovora 136
Eryophyes pencendani f. *carvi* 159
Erysiphe polygoni 162, 175
umbelliferarum 159
Escherichia coli 154
Essential oils 20
Essential oil plants 4
Eucalyptus citriodora 16
European elder 287
oregano 307
valerian 171
Extracting the essential oil 93

Fabaceae 47
Families of substances 23
Farfarae flos, folium 326
Fennel 162, 316, 317
Fenugreek 289
Filipendula ulmaria 323
Finocchio 163
Fir 316
Fleawort 308
Florence fennel 163
Florentine flask 100
Foeniculi fructus 163
Foeniculum vulgare 49, 77, 162, 317
Foxglove 228
Frangula alnus 48, 273, 323
Frangulae cortex 274, 323
French mustard 233
tarragon 255
Fusarium oxysporum f. *cumini* 162

Galega officinalis 47
Galium odoratum 324
Garden thyme 200, 316, 318
Garlic 316
Genera of substances 23
Gentianaceae 50
German camomile 246, 316, 318
German tarragon 255
Ginger 290, 316
Ginseng 292
Glaucium flavum 46, 77
Glechoma hederacea 324
Glycosides 20
Glycyrrhiza glabra 47, 299, 324
  uralensis 299
Graminis rhizoma 322
Grecian foxglove 228, 318
Greek oregano 307
Green peppermint 187
Growing location 65
Gypsophila paniculata 324

Harmful residue 84
Hawthorn 295
Hederae terrestris herba 324
Hemlock 48
Henbane (Henblai) 224, 318
Herbicides 83
Hibiscus sabdariffa 52
Hippophae rhamnoides 47, 77
Hip rose 285
Hollyhock 296, 318
Hop 316
Hot-air drying 92
Horehound 298
Humulus lupulus 77
Hungarian camomile 246
Hybrid lavender 177
Hydrodiffusion 101
Hydrodistillation 93
Hyosciami folium, semen 224
Hyoscyamus agrestis 224
  muticus 15, 75
  niger 51, 64, 77, 224, 318
Hyperici herba 324
Hypericum perforatum 324
Hyppocastani semen, folium 322
Hyssopi herba 196
Hyssopus officinalis 50, 196, 318
Hyssop 196, 316, 318

Illicium verum 143
Indian fennel 163
Indian mustard 233
  plantago 308
  valerian 171
Intensity of light 60
Intermittently operated vessel 97
Inula helenium 324
Inulae radix 324
Isabgol 308

Japanese mint 196
  valerian 171
Juniper 283, 316
Juniperi fructus 283
  lignum 284
Juniperus communis 45, 283
  sabina 45
Jusquaime 224

Kesso root 171
Korean ginseng 292

Labiatae 50
Ladyslipper 172
Lamiaceae 50
Lance-leaved plantain 308
Large-leaved linden 301
Large-leaved sage 210
Lavandin 176, 316
Lavandula angustifolia 50, 77, 176, 318
  burnati 177
  intermedia 176, 318
  latifolia 176
  officinalis 177
  spica 177
  stoechas 176
  vera 177
Lavandulae flos 176
Lavender 176, 316, 318
Lemon 184
Lemon balm 183, 319
Lemongrass 184
Lemon thyme 205
Leonurus cardiaca 50
  lanatus 50, 324
Leonuri lanati herba 324
Lepidium sativum 53
Leptinotarsa decemlineata 227
Leucojum aestivum 74
Levistici folium, fructus, herba, radix 151
Levisticum officinale 49, 151, 319
Licorice 299
Light 59
Liliaceae 54, 73

Linden tree 301
Lime tree 301
Liquiritiae radix 47, 324
Liquorice 299
Lisianthus 12
Lochnera rosea 167
Lovage 151, 316, 319

Madagascar periwinkle 166
Majoranae herba 213
Majorana hortensis 50, 213, 319
Malvaceae 52
Malvae arboreae flos 296
  folium 324
Malva neglecta 324
Malva silvestris 52
M. s. subsp. *mauritiana* 52
Marjoram 213, 316, 319
Marrubii herba 298, 324
Marrubium vulgare 50, 298, 324
Marshmallow 303
Matricaria chamomilla 53, 62, 64, 67, 246
  recutica 54, 246, 318, 325
Mechanization 80
Medicinal plants 4
Megacladosporium depressum 141
Meligethes aeneus 235
Melissae folium, herba 183
Melissa officinalis 50, 183, 319
Mentha aquatica 187
  arvensis 77, 196
  gentilis 77
  longifolia 187
  piperita 50, 77, 187, 319
  pulegium 196
  rotundifolia 187
  smithiana 192
  spicata 77, 187, 195
  viridis 187
Menthae crispae folium, herba 195
  piperitae folium, herba 107, 187
Menyanthes trifoliata 325
Milfoil 313
Millefolii flos, herba, folium 314, 322
Mitcham peppermint 187
Monosaccharides 24
Monoterpenoids 33
Morphogenetical variability 72
Mugwort 255, 316
Mullein 305
Mustard 232, 319

Natural drying 89
Natural substances 19
Nicotiana tabacum 59, 72
Nigella sativa 161
Nightshade 319
Nepeta cataria 278
  transcaucasica 75
Non-organic factors 66

Ocimum basilicum 50, 75, 220, 321
  canum 75
  gratissimum 77, 221
  suave 65
Oenothera erythrosepala 47
  lamarckiana 47
Oleum sabinae 45
Onagraceae 47
Onion 316
Operations after drying 93
Opium 120
Opium alkaloids 121
Orange mullein 305
Orders of substances 23
Oregano 307
Organic factors 66
Oriental mustard 233
Origani herba 307
Origanum creticum 307
  heracleoticum 307
  vulgare 77, 307

Packing drugs 106
Panax ginseng 292
  pseudoginseng 56
  quinquefolium 292
Papaveraceae 46, 73
Papaver hungaricum 46
Papaveris caput maturis 120
Papaver pseudo-orientale 76
  setigerum 46
  somniferum 46, 67, 73, 77, 119, 320
Parsley 316
Pastinaca 316
Pelargonium roseum 77
Pennyroyal 196
Peppermint 187, 316, 319
Periwinkle 166, 319
Peronospora arborescens 127
Peronospora valerianae 175
Persian liquorice 299
Phenoloids 25
Pimpinella anisum 48, 143, 317
Piptadenia falcata 12
Pix juniperi 284
Plantaginaceae 51
Plantaginis folium 309, 325

Plantago altissima 51
lanceolata 51, 308, 325
major 51, 308
media 51
ovata 308
psyllum 51, 308
Plantain 308
Plant substances 19
Plasmopara nivea 159
Plate-chamber driers 90
Poppy 119, 320
Polychemism 16
Polyketides 29
Populi gemma 325
Populus nigra 325
Pot marigold 241, 320
Pot marjoram 307
Potentilla anserina 325
erecta 46
Potentillae anserinae herba 325
Preparing operations 88
Primary processing 86
Primary substances 16
Primulaceae 53
Primulae flos, radix 325
Primula elatior 53
veris 53, 325
Pruni spinosae flos 325
Prunus spinosa 325
Puccinia dracunculina 258
menthae 199
petroselini 136
Pulegii herba 196
Pulmonariae folium 325
Pulmonaria officinalis 325
Pulvis insectorum 258
Purple foxglove 231
Pycanthemum lanceolatum 15
Pyretri flos 258
Pyrethrum 258, 320
Pyretrum carneum 15, 258
roseum 258

Quality of light 59
Quercus cortex 325
petrea 325
robur 325

Radish 316
Ramularia coriandri 136
Rancunculaceae 45
Racunculus arvensis 45
Rauwolfia canescens 59
Red peppermint 187
Red periwinkle 166, 320
Resinoid 103
Rhamnaceae 48
Rhamni cortex 48
Rhamnus catharticus 48
fallax 48
frangula 273
purshiana 48
Rhodiola linearifolia 15
Robinia pseudoacacia 326
Robiniae acaciae flos 326
Roman camomile 262, 316, 320
Roman fennel 163
Rosa canina 47, 285, 326
Rosaceae 46
Rosemary 310, 316
Rosmarini folium 310
Rosmarinus officinalis 310
Rue 128, 320
Rubi frucitosi folium 46, 326
Rubus caesius 326
fruticosus 46
Russian liquorice 299
tarragon 255
Rutaceae 47
Rutae folium, herba 128
Ruta graveolens 48, 128, 320
Rye-smut 111

Sabinae herba 45
Saccharides 23
Safflower 235, 320
Sage 210, 316, 321
St Benedict thistle 238
Sakhalin mint 126
Salviae folium 210
Salvia officinalis 50, 210, 321
sclarea 50, 77, 206, 317
Sambuci fructus, flos 288, 326
Sambucus nigra 49, 287, 326
Sanquisorba officinalis 77
Saponariae albae radix 324
Saturejae herba 218
Satureja hortensis 50, 77, 218, 321
montana 219
Sclerotium 113
Scopolia carniolica 51
Scrophulariceae 51, 73
Secale cornutum 111
Secalis mater 111
Secondary substances 19
Septoria digitalis 231
melissae 186
Sesquiterpenoids 33

Siberian ginseng 292
Silybum marianum 54, 243, 321
Silver linden 302
Sinapis alba 53, 232, 319
albae semen 232
nigrae semen 232
Small-leaved linden 301
Soil 63
Solanaceae 51
Solanum dulcamara 51, 60, 62, 65, 66
laciniatum 51, 59, 62, 66, 71, 319
Solidago virgaurea 326
Solvent extraction 103
Spanish liquorice 299
psyllium 308
thyme 205
Spearmint 195, 316
Special substances 19
Species aromaticae 147
Species of substances 23
Sphacelia segetum 111
Spice plants 4
Spike lavander 176
Spikinard 177
Spiritus balsamicus 147
Spurred rye 111
Star anise 143
Storing drugs 105
Steam distillation 93
Stephania glabra 59
Stramonii folium 323
Streptococcus aureus 154
fecalis 154
Substance taxonomy 23
Summer savory 218, 316, 321
Symphyti folium, radix 326
Symphytum officinale 77, 326
Systole coriandri 136
Sweet basil 220, 316, 321
fennel 163
flag 316

Tanaceti flos 326
Tanacetum cinerariaefolium 258
Tanacetum vulgare 326
Taraxaci herba, radix 326
Taraxacum officinale 326
Tarragon 254, 316, 321
Temperature 62
Terpenoids 31
Thorn apple 321
Thousand-leaf 313
Thyme 200
Thymi vulgaris herba 200
Thymus capitatus 205
citriodorus 205
Thymus serpyllum 50, 200
Thymus vulgaris 50, 73, 76, 200, 318
Thymus zygis 205
Tiliaceae 52
Tiliae flos 302, 326
Tilia argentea 52, 302
cordata 52, 301, 326
grandifolia 301
parvifolia 301
platyphyllos 52, 301, 326
tomentosa 52, 302
Tinctura thymi 200
Trifolii fibrini folium 325
Trifolium repens 11, 16
Trychophyton rubrum 31
True lavender 176
Tussilago farfara 326

Ulmariae herba 323
Umbelliferae 48
Universal substances 19
Urginea indica 73
Urtica dioica 327
Urticae folium, radix 327

Vacuum distillation 104
Valerian 170, 316, 321
Valerianaceae 49
Valeriana colline 49, 171
dioica 49
exaltata 171
jatamansi 171
officinalis 49, 64, 77, 170, 321
procurrens 171
sambucifolia 49
wallichi 171
Valerianae rhizoma et radix 106, 170
Variegated thistle 243, 321
Variety control 69
Veratrum album 54
Verbasci flos 306, 327
Verbascum densiflorum 51
phlomoides 51, 305, 327
thapsus 51
Verbenae herba 327
Verbena officinalis 327
Vinca minor 49, 67, 319
rosea 167
Vinum diureticum 147
Viola tricoloris herba 327
Viola tricolor 327
Virgaureae herba 326

Visci stipes 327
Viscum album 327

Warm-air drying 90
Water 63
Weed control 83
White mustard 233
  peppermint 187
Wild camomile 246
  fennel 163
  marjoram 307
  savory 219
  thyme 200
Woolly foxglove 228
Wormwood 311, 316

Xanthomonas heterocea 136

Yarrow 313, 316

Zingiber officinale 77, 290